D1641385

PREPARING SCHOOL LEADERS FOR THE 21ST CENTURY

An International Comparison of Development Programs in 15 Countries

STEPHAN GERHARD HUBER
Research Centre for School Development and Management, University of Bamberg, Bamberg, Germany

with contributions from:

Michael Chirichello, Janet Chrispeels, Peter Cuttance, S. Gopinathan, Jeroen Imants, Olof Johansson, Kathy Kimball, Kenneth Leithwood, Denis Meuret, Leif Moos, Jan Robertson, Heinz Rosenbusch, Michael Schratz, Anton Strittmatter, Mel West, and Huen Yu

LONDON AND NEW YORK

Library of Congress Cataloging-in-Publication Data

Huber, Stephan Gerhard, 1971-
Preparing school leaders for the 21st century : an international comparison of development programs in 15 countries / Stephan Gerhard Huber with contributions from Michael Chirichello ... [et al.].
p. cm. -- (Context of learning)
Includes bibliographical references and index.
ISBN 90-265-1968-0
1. School administrators--In-service training--Cross-cultural studies. 2. Educational leadership--Cross-cultural studies. I. Title. II. Series.

LB1738.5.H83 2004
371.2'011--dc22

2004041640

Printed in the Netherlands by Gorter, Steenwijk

Published by: Taylor & Francis The Netherlands, Lisse
http://www.tandf.co.uk/books

ISBN 90 265 1968 0
ISSN 1384-1181

To those who are planning and carrying out school leadership development in different countries and those conducting research in this field

Table of Contents

Abstract

This book is based on an international comparative research project about school leadership development in 15 countries. The project was conducted at the Research Centre for School Development and Management, University of Bamberg (Germany), in the years 1998 – 2001, with an update on recent developments in 2002. The methods used comprised two surveys, extensive documentation analysis, and additional country-specific investigations.

This volume looks at the growing importance placed on activities to prepare school leaders that has led to the design and implementation of school leader development programs in many countries in recent years. Its core purpose is to provide a systematic overview of the current practices of school leadership development in 15 countries world wide from an international perspective. Moreover, it will also provide a comparative discussion of the various leadership development models, identifying similarities and differences, based on a set of criteria selected. Additionally, central issues will be identified, international trends will be explored, and recommendations for future programs will be outlined.

The volume is organised into four parts, as follows:

Context of Research

Part I first reflects on roles, tasks, competences, and conceptions of school leadership. It draws on the two paradigms of school effectiveness research and school improvement approaches and points at school leadership as a key factor for the quality and effectiveness of the school and at the school leaders as important 'change agents' for school improvement. It then outlines the changing context in which school leaders find themselves as more and more countries devolve significant management decisions to the school level. It focuses on the new demands this brings to the school leader's role and on the complex range of school leaders' tasks. Efforts to systematise the complex scope of school leadership tasks will be briefly presented. To successfully fulfil these expectations, a broad range of knowledge, understanding, skills, and abilities, but also appropriate attributes and attitudes are necessary, constituting a complex amalgam of school leadership competence. This amalgam will be explored and various research studies are briefly highlighted. Then, contemporary thinking about how the leadership role can be most effectively exercised is explored in terms of leadership theories such as 'transactional leadership', 'transformational leadership', 'integral leadership', 'instructional leadership', and 'post-transformational leadership'. Taking these a step further in the light of 'organisational education', conceptual ideas of 'organisational-educational management and leadership', and 'cooperative and democratic leadership' are described.

In the second chapter, the aim and scope underpinning this international comparative study are stated. It is the central concern of this study to illustrate how school leader development is achieved internationally in a number of countries.

From this, the main research question of this work can be deduced and put as follows: 'how are (aspiring) school leaders qualified in different countries of the world, what kind of training and development opportunities for school leaders are offered?'

In the third chapter, the methodology of comparative research, and the methods used in the study are briefly described. Both are further elaborated in the appendix. The approach of the study and the implications of comparison and comparability are discussed, and, subsequently, methodical implications are outlined. These are the criteria of the selection of the sample countries and their training and development programs, the methods of data collection (international literature research, expert meetings, systematic collection of documents, a systematic questionnaire developed for this study, open and non-directive questioning, and additional research via telephone and email), and a depiction of the individual steps of the evaluation and analysis of the data. The structure of how the results are presented in the country reports is explained at the beginning of Part IV.

Comparison and Discussion of Findings

Part II comprises the comparative discussion, based on the set of criteria selected. This goes beyond a mere juxtaposition (for this see Part IV). Among those criteria are the respective programs' provider(s), target group(s), the timing, the nature of participation, and the programs' professional validity, aims, and contents, the teaching strategies and learning methods, and the pattern in terms of time structure. It examines similarities and differences in approach, and identifies common issues confronting those planning and carrying out school leadership development in different countries. Central aspects will be explored, investigated in greater detail, and commented on.

These include, as far as the 'provider' is concerned, the context of educational policy, interconnected responsibilities for the offerings, the interplay of a central institution and regional providers, as well as the issue of who is the staff (the trainers or teams of trainers).

Furthermore, target group and timing, the nature of participation and the professional validity of the programs are compared. The discussion focuses on issues such as the best timing, the selection of the participants, various forms of grouping them and the alternative of imparting competences to individual school leaders versus strengthening leadership competences of leadership teams and promoting school development.

Concerning the 'aims', the discussion focuses on the existence of explicitly and extensively formulated aims, the difference in orientation either towards the individual school leader or towards school leadership teams and school development, grounding the aims on a more broadly defined understanding of leadership and new paradigms for leadership, and the adjusting of aims according to 'organisational-educational' principles.

For the 'contents', the discussion dwells on the topics of legal and administrative issues, aspects of school-based leadership, communication and cooperation, the role and self-concept of the school leader, as well as the relevance of theoretical knowledge and a certain extent of academic orientation. Altogether, a multiple change of paradigms can be observed here.

Concerning the 'methods', often used types of program events, ways of learning, and study material provided are compared. The attention is then drawn to adult learning principles applied and participant, experience, and needs orientation realised in the programs. Frequently used teaching strategies and methods of learning, such as lectures and plenary sessions, reflective writing, group work, role playing, and simulation exercises, are reported on. New ways of learning such as collegial learning and learning communities, and innovative methods like problem-based learning (as an example for learning in the workshop), and internships as well as mentoring (exemplifying learning in the workplace) are discussed.

In the context of the 'pattern' of the programs, the number of training and development days, their scheduling, and the overall time span are compared. An example for a useful timetabling of the individual components of a program is presented, and the tendency towards establishing multi-phase designs and a modularisation of programs (which can be observed internationally) is elaborated.

Conclusions and Outlook

Part III outlines a number of current trends from a global perspective, 16 of them have been identified and are depicted here. Although some are mere changes in emphasis for the time being in terms of tendencies, others already have the character of paradigm shifts.

Even though judgments concerning the effectiveness of the programs cannot be made, preliminary assessments can be accomplished against the background of the aspects discussed in the previous chapters. For this, the insights gained from the data analysis, the comparison and the discussion are referred to. Additionally, the ideas from the theoretical background of Part I are used as a base as well. Examples for best and promising practice are cited and the necessity of a multi-stage adjustment of aims in school leader development is emphasised.

From the considerations resulting from this international comparative study, 19 recommendations for designing and conducting training and development programs for school leaders are deduced. These recommendations can be considered as stimuli for school leader development programs, but could also be used as standards for qualification programs, as a basis for accrediting providers and programs, or as specific certification criteria in the case of a certificate for quality assurance.

Additionally, research desiderata are formulated as a general outlook.

Country Reports

Part IV then offers detailed descriptions of current practices to develop school leadership competences. This overview is international in scope, comprising four different continents, that is Europe, Asia, Australia/New Zealand, and North America, and including 15 countries (see below). The resulting country reports have been validated each by an internal expert from the respective country, who has become the respective co-author. Hence, the reports are grounded on an external view and validated by an internal view. This systematic and consistent approach is seen as a strength of the country reports.

Each country report is organised in five sections: firstly, there is a brief description of the background, including information about the respective school system, recent changes within the last years affecting school leadership, and the specific role of school leaders in the respective country. Secondly (in the main section), there is some information about the general approach to school leader development and, in some cases, about the history of development activities or that of the current model, followed by the description of the current development model, either the state model, if centrally organised or provided, or one or two examples, if there are a variety of models. Tables summarising the key features will help to get a quick overview of each program Thirdly, there is a brief summary, which highlights the distinctive features of the particular country's approach and the program(s). Fourthly, recent developments that have occurred in the country or in the programs since the time the data was collected are described. Fifthly, a personal statement by the internal expert is given.

From Europe, the following countries and programs are included:

Sweden
  The State Basic Training Program

Denmark
  Example: The Diploma in Educational Leadership of Schools and other Educational Institutions at the Danish University of Education
  Example: Leadership in Development by 'Praxis'

England
  The National Professional Qualification for Headship

The Netherlands
  Example in the field of primary education: Meesters in Leidinggeven of the GCO Fryslân in Cooperation with the Katholieke Universiteit Nijmegen
  Example in the field of secondary education: Management- en Organisatie-opleidingen of the Nederlandse School voor Onderwijsmanagement

France
  National Program for School Leaders of Secondary Schools

Germany
  Example: Basic Courses and Additional Courses in Bavaria
  Example: In-Service Program in North Rhine-Westphalia
  Example: Qualification for Newly Appointed Members of the School Leadership Teams in Rhineland-Palatinate

Switzerland
  Example: Training Program of the Educational Region Central Switzerland

Austria
  Example: Part-time Complementary School Management Course of Salzburg

South Tyrol, Italy
  Executive Leader Training for Heads of School

From Asia, the following countries and programs are included:

Singapore
  Diploma in Educational Administration

Hong Kong
  Induction Course
From Australia and New Zealand, the following programs are included:
New South Wales, Australia
  Example: School Leadership Preparation Program of the Department of Education and Training
New Zealand
  Master of Educational Leadership of the University of Waikato

From North America, the following countries and programs are included:
Ontario, Canada
  Example: Principal's Qualification Program of the Ontario Institute for Studies in Education/University of Toronto
  Example: Principal's Qualification Program of York University
United States of America
  Example for the State of Washington: The Danforth Educational Leadership Program of the University of Washington
  Example for the State of New Jersey: the Educational Leadership Program of William Paterson University
  Example for the State of California: Foundation II of the California School Leadership Academy

The key features of the countries' approaches and the programs depicted in this study are summarised through a tabular juxtaposition at the end of Part IV.

The final chapter is a summary of the whole book.

Preface

In view of the ever-increasing responsibilities of school leaders for ensuring the quality of schools, school leadership has recently become one of the central concerns of educational policy makers in many countries. The pivotal role of the school leader as a factor in effective schools has been corroborated by findings of school effectiveness research over the last two decades. In most of the lists of key factors (or correlates) that school effectiveness research has compiled, 'leadership' plays an important part. Indeed the effectiveness lobby's original message that 'schools matter, schools do make a difference' has continued almost seamlessly into a sub-text that 'school leaders matter, school leaders also make a difference'. School improvement researchers have also demonstrated increasing recognition of the importance of school leaders for all stages of the school improvement process. The school leader is most often cited as the key figure in the individual school's development, either blocking or promoting change, acting as the internal change agent, overseeing the processes of growth and renewal. Moreover, the school leader's role has to be seen in relationship to the broad cultural and educational contexts in which the school is operating. Since schools are embedded in their communities and in the particular national educational system, and these in turn are embedded in the particular society, schools and their leaders have to cope with, to support or otherwise react to the social, economic and cultural changes and developments taking place. Schools, and consequently the expectations on school leaders, also change as a result of more subtle and indirect forces in society – social, political and economic changes – that are gathering pace across the world as the speed of international development increasingly reflects global factors. Moreover, direct changes in the educational system have a particularly strong impact on the school leader's role. These new conditions and demands certainly place new pressures on the leader, but the new tasks and challenges can also be viewed positively as bringing new opportunities.

It is perhaps not surprising that more and more attention is being given to the role of school leaders in creating the conditions for an effective school. At the same time, there has been a parallel growth in the attention given to how school leaders are prepared for this role. The training of future and current school leaders thus is high on the agenda of politicians of different political wings. At the turn of the century, there is broad international agreement about the need for school leaders to have the capacities needed to improve teaching, learning, and pupils' development and achievement. To establish and modify appropriate training and development opportunities has become a major focus of professional development programs in many countries.

At first sight, there may appear to be an international consensus about the important role of school leaders and their development. On looking more carefully, however, it is apparent that a number of countries have engaged in this issue more rigorously than others. While in some countries discussions of school leader development are mainly rhetoric (or have only just started), elsewhere concrete steps have been taken to provide significant development opportunities for school leaders.

Hence, at the beginning of the new century, a comparison of school leadership training and development opportunities in different countries is an instructive enterprise. Such an international comparative educational research project was conducted at the

Research Centre for School Development and Management, University of Bamberg (Germany), in the years 1998 – 2001, with an update on recent developments in 2002. The methods used comprised two surveys, extensive documentation analysis, and additional country-specific investigations. In this project it seemed to be particularly important to include those countries which feature a well-developed and differentiated school leadership research as well as long experience of training and developing school leaders.

Acquiring current and significant data (even if it is not easily accessible) is of central importance for a comparative study like this. The large international conferences, most of all the International Congresses for School Effectiveness and Improvement (ICSEI), offered great opportunities. During ICSEI 1999 in San Antonio, Texas, USA, for example, an intensive discussion developed after an interactive discussion session. This rather small event with around 20 persons resulted in establishing an informal international network. The exchange within this network has been maintained across the world through regular email contacts. During the following ICSEI 2000 in Hong Kong the discussion could be continued and intensified. A group of ten educationalists from ten countries was willing to design a symposium together (see Huber et al., 2000). The exchange of experiences and the professional discourse continued in the following year by means of a second symposium during ICSEI 2001 in Toronto (see Huber et al., 2001), a third during ICSEI 2002 in Copenhagen, and a presentation at AERA 2002 in New Orleans (see Huber et al., 2002). It will be resumed by a symposium, a workshop, and a network meeting during ICSEI 2003 in Sydney, Australia, and a group presentation at AERA 2003 in Chicago, USA, where further research projects will also be considered.

Therefore, I want to express my gratitude to all the colleagues and friends within the ICSEI context and especially the 'ICSEI Network for School Leadership and Leadership Development'. Particularly I want to thank this 'gang' of core people who participated in all of the ICSEI symposia. Of course, beyond this ICSEI group, further professional links and relationships to those planning and carrying out school leadership training and development have been established. They all helped to collect and validate the data and enriched the project with their knowledge and expertise. Special thanks to those of them who contributed as co-authors to this volume. For their support at the University of Bamberg, I want to express my gratitude to Prof. Dr. Heinz S. Rosenbusch, Prof. Dr. Detlef Sembill. Additionally, I want to thank the student assistants, interns and good friends at the Research Centre in Bamberg, who have time and again supported me in this and various other projects which had to be managed besides this one, particularly Susanne Niederhuber. Many thanks to Sigrid Hader-Popp, who is a great team-worker and, being a teacher and student counsellor herself, has always shared her experience of teaching and her view of school and the school system from the perspective of the 'workplace' with me in now more than 12 years of friendship. For all their financial support I want to thank the scholarship foundation Cusanuswerk, the German Research Association, the Federal Foreign Office of Germany, and Siemens TS.

Bamberg, 11th August 2002 Stephan Huber

Terms and Abbreviations

Use of terms in foreign languages
To ensure readability, terms from foreign languages were renounced to a considerable extent; when used, they are either in brackets without inverted commas or within the text in inverted commas.

Use of the term 'school leader'
The term 'school leader' is usually used for principal, headteacher, administrator, rektor or other terms describing the person who is in charge of an individual school.

Use of the term 'school leader development programs' and similar terms
Terms such as 'school leader development programs' or 'school leader training and development opportunities' are used in a general sense, regardless at which time in their careers the (aspiring) school leaders take part in the programs. The terms 'preparation', 'preparatory' or 'pre-service' are used exclusively for offerings before taking over a leadership position, 'induction' for those shortly after taking over, and 'in-service' or 'continuous professional development' for opportunities for established and experienced school leadership personnel. The term 'school leader qualification program' is used to describe programs which have a clear implication for application, selection, accreditation or licensure. However, sometimes it may also be used as a synonym for 'development program'.

Use of 'example' in country or program headlines
The term 'example' in headlines indicates that in the respective country there are either several programs available besides the ones depicted here or that in a federally organised country some states, provinces etc. have been selected.

Indication of the costs for the training and development programs
All costs are indicated in the country's currency as well as in euro for the sake of comparability. This calculation in euro was made on the basis of the exchange rate of March 2, 2001.

Abbreviations

AEB	Akademie für Erwachsenenbildung
AMA	American Management Association
ASD	Arbeitsgemeinschaft der Schulleiterverbände Deutschlands e.V., since 2000: Allgemeiner Schulleitungsverband Deutschlands e.V.
BDG	Bundeslehrer-Dienstrechtsgesetz
BDP	Biblioteca di Documentazione Pedagogica
BKK	Bundeskoordinationskommission für Schulmanagement und schulartübergreifende Angelegenheiten der Schulaufsicht
BKZ	Bildungsdirektorenkonferenz Zentralschweiz

BMBWK	Bundesministerium für Bildung, Wissenschaft und Kultur
BMUK	Bundesministerium für Unterricht und kulturelle Angelegenheiten
CASA	Committee for the Advancement of School Administration
CCSSO	Council of Chief State School Officers
CEDE	Centro Europeo dell' Educazione
CEO	Chief Executive Officer
COSMOS	Committee on the Organization, Staffing and Management of Schools
CPEA	Cooperative Project in Educational Administration
CSLA	California School Leadership Academy
CSLM	Certificate of School Leadership and Management
CSU	California State University
CPD	Continuous Professional Development
CTL	Certificate of Teaching and Learning
CVU	Centre for Videregående Uddannelse
DEA	Diploma in Educational Administration
DEETYA	Commonwealth Department of Employment, Education, Training and Youth Affairs
DES	Department for Education and Science
DF	Danforth Foundation
DfEE	Department for Education and Employment
DPPSP	The Danforth Foundation Program for the Preparation of School Principals
DPU	Danmarks Paedagogiske Universitet
EC	Education Commission
ECA	Extra-Curricular Activity
ECER	European Conference on Educational Research
ECR	Education Commission Reports
ED	Education Department
EdD	Doctor of Education
EDK	Schweizerische Konferenz der kantonalen Erziehungsdirektoren
ELC	Educational Leadership Center
EURAC	Verwaltungsakademie der Europäischen Akademie Bozen
GCO fryslân	Gemeenschappelijk Centrum voor Onderwijsbegeleiding yn Fryslân
GM-Schule	'grant-maintained' Schule
GradDipEL	Graduate Diploma in Educational Leadership
HEAD LAMP	Headteachers' Leadership and Management Programme
HKIEd	Hong Kong Institute of Education
ICSEI	International Congress for School Effectiveness and Improvement
ILS	Instituut voor Leraar en School
IQEA	Improving the Quality of Education for All
IRRE	Istituto Regionale di Ricerca Educativa
ISIP	International School Improvement Project
ISLG	Interdistrict School Leadership Group
ISLLC	Interstate School Leaders Licensure Consortium
KMK	Kultusministerkonferenz (Ständige Konferenz der Kultusminister der Länder der Bundesrepublik Deutschland)
LaSIS	Landesschulinformationssystem
LDG	Landeslehrer-Dienstrechtsgesetz
LEA	Local Education Authority
LeaD	Leadership in Development
LEP	Leaders in Education Program

LMS	Local Management of Schools
LPSH	Leadership Programme for Serving Headteachers
MEd Leadership	Master of Educational Leadership
MEN	Ministère de l'Education Nationale
MLT	Modular Leadership Teams
MOE	Ministry of Education
NAE	National Agency for Education (or: The Swedish National Agency for Education)
NASSP	National Association of Secondary School Principals Assessment Center
NCPEA	National Conference of Professional Educational Administration
NCSL	National College for School Leadership
NDC	National Development Centre for School Management Training
NFER	National Foundation for Educational Research
NIE	National Institute of Education
NPBEA	National Policy Board for Educational Administration
NPQH	National Professional Qualification for Headship
NSO	Nederlandse School voor Onderwijsmanagement
NSW	New South Wales
NTU	Nanyang Technological University
OCT	Ontario College of Teachers
OECD	Organisation for Economic Cooperation and Development
OFSTED	Office for Standards in Education
OISE/UT	Ontario Institute for Studies in Education, University of Toronto
PAL	Peer Assisted Leadership Program
PA-LCH	Pädagogische Arbeitsstelle des Schweizerischen Lehrerinnen- und Lehrervereins
PBL	Problem-Based Learning
PGDiplEd Leadership	Post Graduate Diploma in Educational Leadership
PIRLS	Progress in International Reading Literacy Study
PISA	Programme for International Student Assessment
PQP	Principal's Qualification Program
PSDP	Principal and School Development Program
QTLM	Quality Teaching and Learning Materials
RDSES	Royal Danish School of Educational Studies
SAR	Special Administrative Region
SD	Staff Development
SDP	School Development Plan
SIA	Commission on the Internal Work of Schools
SLC	School Leadership Center
SLPP	School Leadership Preparation Program
SLS	School Leadership Strategy
SLT	School Leadership Teams
SMI	School Management Initiative
SMP	School Management for Principals
SMTF	School Management Task Force
SVP	Südtiroler Volkspartei
TIMSS	Third International Maths and Science Study
TLC	Team Leadership Course
TTA	Teacher Training Agency
UCEA	University Council for Educational Administration

UCSB	University of California Santa Barbara
UE	Unterrichtseinheiten
VBE	Bundesverband für Bildung und Erziehung
VEW	Vergleichende Erziehungswissenschaft

List of Figures and Tables

I. Context of Research

Stephan Huber

1. School Leadership: Roles, Tasks, Competences, and Conceptions

> The headteacher plays a significant role in school management, being both focus and pivot at the centre of decision-making. Preparing, inducting and developing headteachers is a major responsibility of the education service. (DES, 1990, p. 3)

In the last few years, statements similar to this pronouncement from the Ministry of Education in the UK can often be encountered in many different countries. It appears in various versions, in official bulletins issued by the various ministries for education; it can even be heard in political speeches. It is not surprising that the press releases of school leaders' associations echo this sentiment. Moreover, there also seems to be international consensus about the importance of the role of school leaders and, therefore, their development among educationalists, teachers' associations and school boards or education authorities in the various countries. On looking more carefully, however, it is apparent that a number of countries have acted more positively than others regarding the consistent implementation of what is demanded in the statement quoted above. While in some countries discussions of school leader development are mainly rhetoric, elsewhere concrete steps have been taken to provide significant opportunities for school leader development.

School Leadership as a Key Factor for the Quality and Effectiveness of Schools

The pivotal role of the school leader as a factor in effective schools has been corroborated by findings of school effectiveness research for the last decades. Extensive empirical efforts of the quantitatively oriented school effectiveness research – mostly in North America, Great Britain, Australia and New Zealand, but also in the Netherlands and in the Scandinavian Countries – have shown that the leadership is a central factor for the quality of a school (see for example in Great Britain: Reynolds, 1976; Rutter et al., 1979, 1980; Mortimore et al., 1988; Sammons et al., 1995; in the USA: Brookover et al., 1979; Edmonds, 1979; Levine & Lezotte,

1990; Teddlie & Stringfield, 1993; in the Netherlands: Creemers, 1994; Scheerens & Bosker, 1997; Huber, 1999a, offers a critical overview).

The research results show that schools classified as successful possess a competent and sound school leadership (this correlates highly significantly). Empirical evidence is missing for unsuccessful schools, although Rosenbusch and Schlemmer (1997) indicate that failure often correlates with inadequate school leadership personnel. Obviously, an effective school is backed by a correspondingly effective school leadership. Although one cannot assume a direct correlation between school leadership actions and pupils' success, an indirect correlation through the effects of school leadership actions on school culture and the ways the teachers see themselves, on their attitudes and behaviours as well as their motivation, can be confirmed. This in turn affects instructional practices and therefore the quality of teaching as well as the quality of learning (see Leithwood & Montgomery, 1986; van de Grift, 1990; Sammons et al., 1995). The central importance of educational leadership is therefore one of the clearest messages of school effectiveness research (Gray, 1990). In most of the lists of key factors (or correlates) that school effectiveness research has compiled, 'leadership' plays such an important part, so much so that the line of argument starting with the message 'schools matter, schools do make a difference' may legitimately be continued: 'school leaders matter, they are educationally significant, school leaders do make a difference' (Huber, 1997b).

As an example for such a compilation of key factors for school effectiveness, the meta-study of the University of London can be referred to (Sammons et al., 1995). Eleven central features of effective schools were identified:

- professional leadership,
- shared vision and goals,
- a learning environment,
- concentration on teaching and learning,
- purposeful teaching,
- high expectations,
- positive reinforcement,
- monitoring progress,
- pupil rights and responsibilities,
- home-school partnership,
- a learning organisation.

Sammons and her colleagues describe 'professional school leadership' as firm and purposeful, sharing leadership responsibilities, involvement in and knowledge about what goes on in the classroom. That means that it is important to have decisive and goal-oriented participation of others in leadership tasks, that there is a real empowerment in terms of true delegation of leadership power (distributed leadership), and that there is a dedicated interest for and knowledge about what happens during lessons (effective and professional school leadership action focuses on teaching and learning and uses the school's goals as a starting point).

The other factors for school effectiveness (other than 'professional school leadership action') mostly fall within the influence of the school leader, too, such as a shared vision and goals (unity of purpose, consistency of practice, collegiality and collaboration), a learning environment (an orderly atmosphere, an attractive working environment), concentration on teaching and learning (maximisation of learning time, academic emphasis, focus on achievement), purposeful teaching (efficient organisation, clarity of purpose, structured lessons, adaptive practice), high expectations (high expectations all round, communicating expectations, providing intellectual challenge), positive reinforcement (clear and fair discipline, feedback), monitoring progress (monitoring pupil performance, evaluating school performance), pupil rights and responsibilities (raising pupil self-esteem, positions of responsibility, control of work), home-school partnership (parental involvement in their children's learning), and a learning organisation (school-based staff development etc.). Obviously, the school leaders have to be aware of how influential they could be to use the opportunities given.

School Leaders as Important 'Change Agents' for School Improvement

Studies on school development and improvement also emphasise the importance of school leaders, especially in the view of the continuous improvement process targeted at an individual school (see van Velzen, 1979; van Velzen et al., 1985; Stegö et al., 1987; Dalin & Rolff, 1990; Joyce, 1991; Caldwell & Spinks, 1992; Huberman, 1992; Leithwood, 1992a; Bolam, 1993; Bolam et al., 1993; Fullan, 1991, 1992a, 1993; Hopkins et al., 1994, 1996; Reynolds et al., 1996; Altrichter et al., 1998; Huber, 1999b offers a critical overview).

In many countries, the efforts made to improve schools have illustrated that neither top-down measures alone nor the exclusive use of bottom-up approaches have the effects desired. Instead, a combination and systematic synchronisation of both has proved most effective. This suggests that central reforms and decentral initiatives need to be combined. Approaching schools with external education policy measures is not automatically an efficient school improvement strategy. If changes are meant to have any significant influence on the pupils' success, strategies need to be developed from within the schools, that is through the design of school-based programs, offering internal training for teachers, and encouraging self-evaluation. Moreover, improvement is viewed as a continuous process with different phases, which follow their individual rules. Innovations also need to be institutionalised after their initiation and implementation at the individual school level, so that they will become a permanent part of the school's culture, that is the structures, atmosphere and daily routines. Ready-made solutions for this process as a whole, in terms of a cure-all, do not exist, however. Aid and supporting efforts need to be specifically tailored to suit the individual school. Hence, the goal is to develop problem-solving, creative, self-renewing schools that have sometimes been described as learning organisations. Therefore, the emphasis is placed on the priorities to be chosen by each school individually (see Hopkins et al., 1994) since it is the school that is the centre of the change process. Thereby, the core purpose of school, that is education and instruction, are at the centre of attention, since the teaching and learning processes play a decisive role for the pupils' success. Hence, both the individual teacher and the school leadership provided are of great importance. They are the essential change agents who will have significant influence on whether a school will develop into a 'learning organisation' or not.

For all phases of the school development process, school leadership is considered vital and is held responsible for keeping the school as a whole in mind, and for adequately coordinating the individual activities during the improvement processes (for the decisive role of school leadership for the development of the individual school see, for example, studies conducted as early as in the 1980s by Leithwood & Montgomery, 1986; Hall & Hord, 1987; Trider & Leithwood, 1988). Furthermore, it is required to create the internal conditions necessary for the continuous development and increasing professionalisation of the teachers. It holds the responsibility for developing a cooperative school culture, where the structurally conditioned segmentation of knowledge and competences of the isolated 'experts on individual subjects' is abolished in favour of an effective and efficient allocation of competences, forming and using teams within the staff. Regarding this, Barth (1990) and Hargreaves, D.H. (1994), among others, emphasise the 'modelling' function of the school leader concerning the creation of cooperative relationships within staff. School leaders are, then, considered key figures in their schools – they have the ability to foster or to block school development processes, they are best situated to be central 'change agents' and have the responsibility for the individual school's change process (see, for example, Fullan, 1988, 1991, 1992b; Schratz, 1998b).

New Demands on Schools and School Leadership

Of course, the school leader's role has also to be seen in relationship to the broad context in which the school is operating. As schools are embedded in their communities, and the country's educational system, and this again is embedded in society, schools and their leaders have to react to, to cope with and to support economic and cultural changes and developments. Sometimes they even have to anticipate them, and sometimes to counteract the problems arising from some of these developments. Altered social environments at work and at home as well as a growing multi-cultural world based on the versatility of a pluralistic, post-modern and globalised society, result in an increase in complexity in many areas of daily life. The accumulation of knowledge (which is developing exponentially), an information market which is not easily manageable and which features an ever-increasing supply of extracurricular information opportunities (via radio, TV, print media and, particularly recently, the internet) and a growing diversity and specialisation of the working environment are further aspects of this radical change (see Naisbitt, 1982; Coleman, 1986; Beck, 1986; Naisbitt & Aburdene, 1990; Krüger, 1996). Hence, the school as an institution cannot any longer be regarded as simply imparting traditional knowledge within a fixed frame. Rather, it is becoming an organisation which needs to renew itself continuously in order to take present and future needs into account (see Dalin & Rolff, 1990). This imposes the necessity on school leadership to consider itself as a professional driving force and mediator for the development of the school towards a learning organisation, an organisation which develops its own reforming and changing powers and re-invents itself (see, among others, Caldwell & Spinks, 1988, 1992; Fullan, 1993, 1995).

Additional stress within the range of tasks for school leadership is brought about by the changed structures of the education system, which inevitably strongly affect the individual school and therefore the role of school leadership as well. Tendencies towards decentralisation, transferring more decision power from the system level to the school level, result in an extended independence of school (see Bullock & Thomas, 1997). This may be interpreted as a burden for school leaders, but may also be considered positively, as a new chance and challenge.

Besides decentralisation, there are increasingly corresponding efforts to centralise. There is a legislative movement towards stronger central influence and control by means of intensified accountability, quality control through school inspections or external evaluation, a set national curriculum with national standardised tests, which allow for a direct comparison of pupil and school performances, and so on.

Hence, the roles and functions of school leaders have changed in many countries of the world. In addition to the traditional and already diversified range of tasks, completely new ones have been added, and the character of accustomed duties has changed as well. As a result, school leaders are confronted with an altogether new range of demands and challenges.

A Complex Range of School Leadership Tasks

The managing and leading tasks of school leadership are both complex and interrelated, so that there is no clearly defined, specific 'role' of school leadership, but at best a coloured patchwork of many different aspects. Some areas or role segments relate to working with and for people, others to managing resources like the budget. All are part of the complex range of tasks the school leader faces in the 21st century (see e.g. Huber, 1997b, 1999e):

In 'working with and for the people within their schools', school leaders, being 'change agents', are developers of organisations. They are responsible for school development processes, that is for their initiation, implementation, institutionalisation, and evaluation. Developing a model together with the staff, as well as the design and implementation of the school program, are part of this. School leaders are supposed to create favourable structures and conditions for this, including staff development opportunities. School leaders from independent schools have always been actual employers, with the corresponding rights and duties for the educational and non-educational staff. Now these duties are spreading increasingly to all headteachers, as more and more systems devolve power to the school level. They are responsible for the training and professional development of the staff, the support of the teachers and the development and maintenance of a cooperative and professional school culture. Additionally, school leaders are 'people persons', that is persons who want to have a very good relationship with teachers, pupils and parents. They encourage, counsel and convey appreciation. Their duties as 'teachers' are still regarded as an essential element of their work by many school leaders and this fulfils many important functions for them, for example, staying in touch with the core purpose of school, that is instruction and education, by classroom practice, which as well increases their credibility with staff.

Concerning the 'work with people outside of school', in many countries, through changed conditions, the tasks and responsibilities have extended to include, for example, intensive contact with the community, local employers and other persons in public life, the relationships with parents have changed as well. School leaders have also to view themselves as 'homo politicus', able to act diplomatically and 'committee-politically', possessing and applying political intuition. Furthermore, school leaders are 'representatives of their schools'. This aspect has gained momentum in a time where 'market conditions' are applied to schools in many countries, since the prestige of a school has a decisive influence on recruitment of both staff and pupils. School leaders are 'mediators', not only in internal relationships but also in contacts between the school and its environment, that is between internal and external interests.

In times of growing decentralisation, within many countries school leaders are increasingly confronted with tasks, which up to now have been performed by the next superior level, that is the administration of resources. This makes them 'administrators and organisers' of the school. These administrative tasks can consume a lot of time. Moreover, school leaders need to 'govern' or 'steer' the staff as 'people managers' and are responsible for the effective and efficient deployment of all teaching and non-teaching staff. In countries with a more decentralised system, they are responsible and accountable for quite an extensive school budget. Hence, school leaders are also 'financiers and entrepreneurs' since they decide (together with the respective committees or bodies) on the effective and efficient use of the resources within the assigned budget, and in compliance with central objectives of the system. Recruiting pupils, searching for sponsors and partnerships may also be part of their entrepreneurial activities.

International school leadership research already features a number of different alternatives for classifying school leadership tasks. Various approaches allocate school leadership action within various ranges of duties and assign responsibilities and activities to these (see the analysis of Katz, 1974, as an important 'precursor' for classifications of management tasks, but also classifications of school leadership tasks, for example, by Morgan et al., 1983; Jones, 1987; Leithwood & Montgomery, 1986; Glatter, 1987; Caldwell & Spinks, 1992; Esp, 1993; Jirasinghe & Lyons, 1996).

The Amalgam of 'School Leadership Competence'

The highly diverse and extensive activity areas and role segments of school leaders presuppose substantial competences. A competence approach to management was first applied in the American economy and industry in the 1970s and was borrowed from there for the analysis of school leadership and its requirements at the same time in the USA, and, later, in Great Britain, Australia, New Zealand and Canada as well.

Competence can be seen and defined in the context of the position to be filled, as the ability to effectively execute the activities and functions which are part of that position and its tasks at a level of performing that meets the standards postulated for the position.

Competence can be regarded as a fundamental characteristic of a person, which results in an effective and/or above average achievement. A holistic competence approach takes into account values and expectations, attitudes and attributes, motivation, knowledge and understanding, abilities and skills, aspects of how one sees oneself, and of one's social role (see, for example, Whitty & Willmott, 1991). In consideration of the complexity of school leadership tasks, it is reasonable to assume a complex competence structure (i.e. a complex and multi-faceted amalgam of school leadership competence). Moreover, the emergence and application of competences depend on the context that is they vary depending on the situation. A further differentiation is hinted at in the considerations concerning a dynamic 'life cycle' of competences. The idea here is that, over a period of time, different competences are required within an organisation, some are newly required (emerging), while the importance of others decreases (maturing) and others remain relatively constant (core), still other competences are only relevant for a certain amount of time (transitional). For school leaders to adequately cope with the requirements of the continuously developing school, varying competences are needed at different points in time.

Although the amalgam of school leadership competence cannot in fact be atomised into isolated single competences, it is nevertheless useful to illustrate individual areas of competence for a better understanding of the complexity: first of all, school leaders need to possess 'social competences', since being skilled in interpersonal affairs is an indispensable basis for successful professional interaction, the importance of which has increased due to the modified framing conditions in which schools are operating. Additionally, 'personal competences' are needed. They include personal abilities and attitudes, like being open to innovation and initiatives, flexibility in thinking and acting, the ability to live with changes and endure uncertainties, analytical skills, and also a wide range of concrete knowledge in different areas. 'Administrative competences' are required for performing the tasks related to being managers of the school. This also incorporates the appropriate knowledge concerning education, school law and regulations, but also efficient administration and organisational psychology etc.

As to the range of management and leadership tasks, one can find detailed lists of competences. They are mostly lists incorporating concrete compilations of the competences required for occupying a position, as in the compilations of McBer and Boyatzis. In the 1970s, McBer studied which characteristics distinguish the 'superior manager' from the 'average' manager on behalf of the American Management Association (AMA) (see e.g. Boyatzis, 1982; Boak, 1991). The British National Educational Assessment Centre (NEAC), for example, provides a compilation of competences for school leaders: NEAC assembled competences necessary for being a successful school leader in cooperation with the government as well as representatives from the economy/industry in 1995. They are derived from an analysis of leadership behaviour. They are divided into four groups: administrative, interpersonal, communicative and personal breadth competences. Jirasinghe and Lyons (1996), too, compiled a number of school leader competences on the basis of their analysis of activities and requirements. Five areas of competences were deduced from their work: the planning and the administrative process, dealing with people, managing the political environment, professional and technical knowledge, personal skills.

The compilation of competences necessary for a successful occupation of a position can reasonably be used for different purposes. Esp (1993) describes various projects, which make use of competence listings in various ways. Firstly, from competence standards clearly defined criteria for selecting staff can be deduced. Secondly, these compilations are employed in the area of human resource development. By means of competence listings, one can attempt to systemise the successes and goals of continuous professional development in the context of school development processes. Thereby, evaluations of the development of an individual teacher, a subject department or the school as a whole can be carried out. In this context, they are also used for self-assessment. The experiences of using competence listings by schools, as indicated by Esp (1993), were all very positive. Teachers and school leaders felt compelled to reflect on competences, which in turn helped them to understand their own roles and those of others more clearly and to develop plans for their personal professional development. Therefore, the compilations of competences stimulated the participants to discuss and reflect on their own needs, as well as the school's needs.

Leadership Theories

When considering the manifold tasks and responsibilities of school leadership, as well as the necessary competences, being of central importance to the quality and development of the single school, there is a danger that the individual school leader be propagated as a kind of 'multifunctional miracle being'. But, nobody can safely assume that school leaders are or will or should be the 'superheroes of school'. What may be deduced, however, is that their role can hardly be filled by persons with 'traditional' leadership concepts. The idea of the school leader as a 'monarchic', 'autocratic' or 'paternal' executive of school has increasingly been seen as inappropriate, but viewing a school leader as a mere 'manager' or 'administrative executive' is inadequate as well, despite the managerial pressures of the present situation.

As long as it is about seeing the school as a stable system where the existing structures need to be administered as well as possible to effectively and efficiently achieve fixed results, a static concept of leadership may work very well, with the school leader first and foremost ensuring that the school as an organisation functions well and smoothly. The term 'transactional leadership' has been applied to this concept of steady state leadership: The school leader is the manager of the transactions, which are fundamental for an effective and also efficient work flow within the organisation. The daily organisational office proceedings and the administration of buildings, financial and personal resources, the time resources of staff, as well as communication processes within and outside of school are all included in this definition of 'transactions' or 'interactions'. All this constitutes the daily routines of school leadership and should not be underestimated, since it represents parts of the workload required to create the appropriate conditions for teaching and learning processes to take place.

But, once rapid and extensive processes of change demand viewing and performing 'change and improvement' as a continuing process, different conceptions of leadership are required. Here, 'transformational leadership' (see, for example, Burns, 1978; Leithwood, 1992b; Caldwell & Spinks, 1992) is considered to point the way. 'Transformational leaders' do not simply administer structures and tasks, but concentrate on the people carrying these out, that is on their relationships and on making deliberate efforts to win their cooperation and commitment. They try to actively influence the 'culture' of the school so that it allows and stimulates more cooperation, coherence and more independent learning and working. Here, 'leadership' is emphasised over 'management'. School leadership, as it is understood here, is reputed to be particularly successful in school development processes and thereby opposes the concept of 'transactional leadership'. While administrators following the latter invest their energies into so-called 'changes of first order', that is attempt at improving administrative conditions, the 'transformational leader' focuses on 'changes of second order', that is the fundamental, essential leadership responsibilities, which include developing shared goals and improving internal communication, as well as developing efficient, cooperative decision-making and problem-solving strategies. That means that educational leaders should possess a high degree of motivating and integrative competences, which unleash and integrate the potential of all staff and make them more productive. This 'catalysing' function represents an important aspect of 'transformational leadership'. According to this conception, school leadership is not only concerned with processes, but also with the outcome. In addition to the smooth operation of the processes at school, leadership concentrates on the results, the success of the teaching and learning processes, on the outcomes and on the relation between the outcomes and the development processes themselves, which are supposed to result in improved outcomes.

Louis and Miles (1990) also distinguish between 'management', referring to activities in the administrative and organisational areas, and 'leadership', referring to educational goals and to inspiring and motivating others. For them, 'educational leadership' includes administrative tasks like, for example, managing and distributing resources or planning and coordinating activities as well as tasks concerning the quality of leadership, such as promoting a cooperative school culture in combination with a high degree of collegiality, developing perspectives and promoting a shared school vision, and stimulating creativity and initiatives from others.

In contrast, Imants and de Jong (1999) try to comprehend 'management' on the one hand and 'leadership' on the other not as contrary poles, but as complementary ones. They regard their leadership concept, 'integral school leadership', as an integration of management and leadership tasks. This means that steering educational processes and performing management tasks coincide and overlap. The underlying understanding of 'leadership' defines it as the deliberate 'control' of other people's behaviour. Therefore, educational leadership then means controlling the teachers' educational actions and the pupils' learning processes. Consequently, the central issue for a school leader is how to positively influence the teachers' educational actions and the 'learning activities' of the pupils. Thereby, the combination of educational leadership and administrative management, which is often perceived as contrary by school leaders, loses its contradictory character.

Studies conducted in North America, especially in the field of school effectiveness, have emphasised the relevance of 'instructional leadership' since the 1980s (see, for example, De Bevoise, 1984; Hallinger & Murphy, 1985). This leadership concept focuses most on those aspects of school leadership actions that concern the learning progress of the pupils. They include management-oriented as well as leadership-oriented activities like a suitable application of resources for teaching, agreeing upon goals, promoting cooperative relationships between staff (e.g. preparing lessons cooperatively), but, especially, the evaluation and counselling of teachers during lessons through classroom observation, structured feedback, and coaching.

In the German-speaking context, the notion of 'organisational education' (see Rosenbusch, 1997b) refers to the mutual influence of the school as an organisation on the one hand and the educational processes on the other hand. The core question of organisational education raises a two-fold issue: which educational effects do the nature and conditions of school as an organisation have on individuals or groups within the organisation – and vice versa, which effects do the conditions in and the nature of individuals or groups within the school have on the school as an organisation. Concretely speaking: how does school need to be designed in order to guarantee favourable prerequisites for education and support educational work? Hence, the influence of the school administration leadership and organisation on the teaching and learning process needs to be acknowledged. Administrative and organisational structures have to be brought in line with educational goals. This does not only concern the structure of the school system or the management of the individual school, but also the leadership style with aspects of the distribution of tasks and responsibilities among the staff. Hence, empowerment and accountability issues seem to be important and have to be considered seriously in the light of educational aims and goals. In the context of organisational education, school leadership action becomes educational-organisational action, and educational goals become superordinate premises of this action. This means that school leadership action itself must adhere to the four main principles of education in schools – that school leaders themselves assume or encourage maturity when dealing with pupils, teachers and parents, that they practise acceptance of themselves and of others, that they support autonomy, and that they realise cooperation. This adjustment of educational perspectives affects the school culture, the teachers' behaviour, and the individual pupils, particularly through the teaching and learning process on classroom level. Administrative and structural conditions have to be modified accordingly, and be in compliance with educational principles. Thereby, the unbalanced relationship (which is historically conditioned in many countries) between education on the one hand and organisation and administration on the other hand can be clarified.

This implies that school leadership needs to be based upon certain principles, which are oriented towards the constitutive aspects of a fundamental educational understanding (see Rosenbusch, 1997b):

- School leaders should adjust their educational perspective: educational goals dominate over administrative requirements, administration only serves an instrumental function.

- They should take the two levels of their educational work into consideration: first school leaders have to work with children and promote their learning, and second, as they also have to work with adults, they should promote their learning as well. Hence, conditions of adult education and adult learning have to be taken into account. This has to have an impact on their leadership and management style, particularly in professional dialogues, when knowledge is shared, expanded, and created. Therefore, school leaders have to integrate the two levels of child education and adult education in their educational perception and behaviour.
- They should be more resource-oriented than deficiency-oriented: a new orientation towards promoting strengths instead of counting weaknesses is needed. So far, in many countries bureaucratically determined school administration has concentrated on avoiding mistakes, on controlling, detecting, and eliminating weaknesses instead of – as would be desirable from an educational point of view – concentrating on the positive aspects, reinforcing strengths, and supporting cooperation; it should be about 'treasure hunting instead of uncovering defiencies'.
- They should follow the 'logic of trusting oneself and others': it is necessary to have trust in one's own abilities and as well as in those of the staff and others so that empowerment, true delegation, and independent actions can be facilitated. Then, mistakes can be addressed more openly.
- They should act according to the principle of 'collegiality in spite of hierarchy': individual and mutual responsibilities have to be respected and appreciated although special emphasis is placed on a shared collegial obligation regarding the shared goals.

Rosenbusch (ibid.) states, as consequences for 'schools' as organisations, that the structural conditions need to be designed in a way that they facilitate opportunities for self-determination, independence and cooperation. This requires:

- a flat hierarchy of school management (with one or two levels),
- replacing linear decision-making processes with circular ones (by searching for specific solutions for the individual school cooperatively across the levels of hierarchy),
- immediate bottom-up-introduction of experiences gathered at schools into the decision-making processes of superior authorities,
- opportunities for the individual schools to create their individual profiles, for innovations and fast adjustment to general and regional social, economic and cultural developments,
- a change in stipulations: instead of unnecessary regulations, more simplification and liberalisation,
- extended training, counselling, professional exchange, and support, focussing on the development of the individual school,
- introducing self-evaluations of schools in addition to external evaluations in order to enhance the professional feedback culture,
- establishing professional networks and learning communities: in addition to cooperation among teachers, also cooperation between schools and other educational institutions and others within the community,

- clearly defined standards, competences, and responsibilities, as well as democratic principles and transparency.

Therefore, the leadership concept of 'organisational-educational management' assumes a definition of 'educational' which not only incorporates teaching and education processes with pupils, but also with adults, as well as organisational learning. Organisational-educational management is committed to educational values, which are supposed to determine the interaction with pupils and the cooperation with staff as well. Administrative aspects fulfil a clearly defined function as instruments for reaching genuinely educational goals. These goals should determine the school as an organisation and thereby change it so that it becomes a deliberately designed, educationally significant reality for all. Leadership action also needs to be a model for what the school seeks to teach and preach, that is, it should shape a model-like social space for experiences for all the stakeholders by realising educational goals to the benefit of the organisation and the individual.

Consequently, the core principle of leadership action is 'democracy' and 'cooperation', both as an aim and a method. Due to the complex hierarchy within the school, democracy and cooperation represent an adequate rationale for actions concerning the intrinsic willingness and motivation of staff and the pupils for co-designing the individual school. However, cooperation is not only valuable as a means for reaching goals, it is a decisive educational goal in itself.

As far as 'cooperation' is concerned, following Wunderer and Grunwald (1980), Liebel (1992) defines 'cooperative leadership' as (1) exerting goal-oriented social influence for performing shared tasks or duties (goal-achievement aspect) (2) in/with a structured working environment (organisational aspect) (3) in the context of mutual, symmetric exertion of influence (participative aspect) and (4) designing the work and social relationships in a way that enables a general consensus (pro-social aspect). Here, an organisational and a cooperative perspective are combined.

Developing these ideas would result in a broad distribution of leadership responsibility, that is in a 'community of leaders' within the school (see Grace, 1995). This view is also taken by West and Jackson (2000), in their depiction of 'post-transformational leadership'. If the school is supposed to become a learning organisation, this implies the active, co-determining and collaborative participation of all (see also 'distributed leadership'). The old distinction between the position of the teachers on the one hand and the learners on the other cannot be sustained, nor can the separation between leaders and followers. Therefore, leadership is no longer statically connected to the hierarchical status of an individual person but allows for the participation in different fields by as many persons from staff as possible. This also extends to the active participation of the pupils in leadership tasks.

The delegation of decision-making power should not occur, however, in order to 'bribe' the stakeholders into showing motivation, but for the sake of a real democratisation of school. Therefore, cooperation or 'cooperative leadership' is not just a leadership style (like 'consultative leadership', 'delegative leadership' or 'participative leadership') but reflects a fundamental leadership conception as a general attitude. This can also be named 'democratic leadership'.

Overall, this has decisive consequences for teachers' actions and for school leadership actions; it also needs to be reflected in the preparation and qualification of those working in schools. Not only will the training of teachers benefit from this –

they also need to be trained for working within an organisation, whereas teacher training most often in many countries only focuses on how to teach the chosen subjects – but this will also affect the selection and development of the educational leadership personnel of the future.

2. Aim and Scope of this International Comparison

The starting point of the research project 'Preparing School Leaders for the 21st Century: An International Comparison of Development Program in 15 Countries' is the understanding of the central importance of school leadership for the quality and development of schools. The efficiency and the success of school improvement reforms and initiatives are highly dependent on school leadership. Schools – and also their leadership – are faced with new demands considering the quick change in society. Moreover, educational-policy stipulations such as, for example, local school management with increasing self-responsibility and accountability of schools, additionally put more pressure on school leadership. Therefore, a profound qualification of school leaders is necessary. Different countries have reacted to this need in different manners. Hence, an international comparison contributing to this topic by looking at the various approaches, and at their commonalities but also at their differences seems to be particularly necessary and urgent.

The central research question is: 'how are (aspiring) school leaders qualified in different countries of the world, what kind of training and development opportunities for school leaders are offered?'

The following questions are relevant in this context:

1. Provider(s): Which institution(s) provide training for school leaders?
2. Target group, timing, nature of participation, professional validity: Who is the target group of the qualification? At what point in time do they take part in training and development programs preparing them for their new job? What is the nature of participation? What is the relevance to their careers?
3. Aims: What are the aims and objectives of the program?
4. Contents: What content, topics, issues are covered?
5. Methods: What teaching strategies and learning methods are used?
6. Pattern: How is the training for these aspiring or newly appointed school leaders structured with regard to time?

What this study does not aim at, is to investigate quantitatively the effectiveness and empirical efficiency of the training and development programs. In contrast, this research study is supposed to be a necessary basis for later research into effectiveness and efficiency. However, it wants to identify best and promising practice from a normative standpoint by comparing the various approaches step by step, using a set of dimensions and criteria. Then, these findings will be confronted with theoretical and empirical research findings described in chapter I.1.

3. Methodology and Methods

The overall approach in this study is one of 'international comparative educational research'. In the appendix, the methodology of comparative research is reflected upon in general, and the methods used in the study are carefully elaborated. There, too, the approach of the study is discussed, and, subsequently, methodical implications are described in detail: criteria of the selection of the sample countries and their training and development programs, and the methods of data collection are described (international literature research, expert meetings, systematic collection of documents, a systematic questionnaire developed for this study, open questioning, and additional research via telephone and email). A depiction of the individual steps of the evaluation and analysis of the data is included in the appendix, too. Here, it should only be stressed that the analysis resulted in country or program descriptions, which in turn were modified and validated in several steps. An additional validation of the country and program descriptions by renewed feedback from the experts from the different countries was regarded as very important. Therefore, the reports were sent to the respective validating expert of each country. As the reports were originally written in German, they got translated into English, either here in Bamberg or by the validating expert. This expert was invited to be the co-author for the country report. Hence, a specific methodological step has been achieved that way, namely the combination of an independent external expert, who assures the consistent approach in the country studies, and an internal expert, who has specific context knowledge. The structure of how the results are presented in the country reports is explained at the beginning of Part IV.

II. Comparison and Discussion of Findings

Stephan Huber

This second part looks at the countries included in the study and their training and development programs according to certain classification criteria. The following dimensions are used for comparison: 'provider', 'target group, timing, nature of participation, and professional validity', 'aims', 'contents', 'methods', and 'pattern'. 'Certification', another criterion used for comparison, is covered in the context of 'provider' and 'target group, timing, nature of participation, professional validity'. A comparison featuring a discussion of 'costs' (as they are indicated separately in the outlines of the country reports) is not dealt with here.

By means of comparative statements, similarities are grouped together and differences are emphasised. The distinction between comparison and discussion sometimes cannot be – as far as some topics are concerned – easily made. In these cases an exact distinction was sometimes dispensed with for the sake of readability.

1. Provider: Monopoly or Market, Centralised or Decentralised Planning and Implementation of Programs

One dimension of the international comparison of school leadership development models is, who are the providers: who is responsible for the program, who has developed it and who carries it out?

1.1. Comparison

When compared internationally, a wide range of possibilities can be found. In some countries, development programs are either provided by universities alone, or by institutes linked to universities, or only by state-run institutes for continuous professional development of teachers, or by specific regional institutions. Programs are offered either by institutions of the same kind or even by one single institution for the whole country. In some countries, there are various cooperative arrangements among different providers, so, for example, among the respective departments of education (or rather offices of the departments) or universities or state-run institutions of continuous professional development (CPD) or among the departments of education and regional institutions and independent providers. Additionally, there are different kinds of cooperation among several universities, among institutions for adult learning and the profession itself (see table below). It can clearly be observed that the question of who provides school leadership development is responded to very differently in each of the countries.

Table 1: Different kind of providers

One provider	
Ministry	Hong Kong, China
State-run institute of teacher training (of the federal state or canton)	Germany; Switzerland (with differences among the cantons); Austria (according to state guidelines)
Independent provider commissioned by the state	South Tyrol, Italy (similar in other provinces of Italy)
Various providers of the same kind	
Universities or institutes linked to university	Ontario, Canada (accredited by the OCT); Washington, New Jersey, USA (according to CCSSO/ISLLC standards accepted by 36 states)
Cooperative arrangements/partnerships	
Ministry and state-run institutes (academies)	France
Ministry and regional institutions	New South Wales, Australia
Ministry and university	Singapore
Ministry and universities and communities	Sweden
Central institution and regional providers	England
Universities and regional/independent providers	the Netherlands
Cooperation of several universities	the Netherlands
College of education and regional teacher training institutes	Denmark; the Netherlands
Universities and state education authorities and individual schools and school districts	Washington, New Jersey, USA (similar to other states)
Institutions of adult learning and professional associations	Central Switzerland; the Netherlands
Different providers: 'diversity and choice'	
Universities and institutes linked to university, regional/ independent providers, professional associations, advisory boards etc.	Denmark; the Netherlands; New Zealand

1.2 Discussion

Development Approach and Context of Educational Policy

When reflecting upon the variety of providers, a number of relevant issues emerge. At first the different degrees of decentralisation and centralisation attract attention, that is on the one hand the degree of freedom of choice among various programs and/or various providers for potential participants, and on the other hand the amount of steering and control by the government as regards the school system in general. Here, the interrelationship between qualification and educational policy context, or the context of the school system, is interesting. Two central questions emerge: to what degree is the respective school system centralised, and how completely is the qualification of school leaders organised and carried through centrally? The countries may be divided into different groups according to their degree of centralisation or decentralisation (see table below).

Table 2: Centralization and decentralization of school systems and school leadership training and development approaches

		Approach to school leader development			
		predominantly centralised or using standards or guidelines		entrepreneurial	
Level of central control over school management	predominantly centralised	A	France; South Tyrol; Austria; Germany; Hong Kong; Singapore	B	
	substantially devolved	C	Sweden; England; Switzerland; NSW, Australia; Ontario, Canada; USA*	D	Denmark; the Netherlands; New Zealand; USA*

** Double listing due to differences in the approaches of the different states.*

Three groups emerge:

Group A: On the one hand, there are countries with rather centralised school systems and centrally regulated programs by one central provider. The programs are standardised, and there is (in case they are conducted regionally) a rather small degree of regional variability, and the state, specifically the department of education, exercises control over the school leader development. In Germany, for example, each federal state has its own qualification program, planned and implemented by the state-run institute of teacher training. In Austria, the respective Paedagogische

Institut of each federal state offers the qualification program, however, this is offered in agreement with across-state, national standards. In South Tyrol, the government has commissioned an independent provider, the EURAC, to carry out the central quasi-mandatory program. In France, there is a qualification program, which is mandatory for all school leaders. It has a standardised approach and is centrally organised, however it is carried out in regional academies. In Hong Kong, the Department of Education is directly responsible for the qualification. In Singapore, the Department cooperates with the university.

Group D: On the other hand, there are countries which have rather decentralised school systems, having a broad variety of development and training programs by very different providers, the choice among these is up to the individual participant. In this case it is not always easy for the potential participant to understand the quality of the programs that are available on the market. This is the case, for example, in the Netherlands, where there are cooperations among universities (for example the Nederlandse School for Onderwijsmanagement) as well as among universities and regional teacher training institutes, school advisory centres and also various other arrangements of providers The government abstains from mandating any preconditions or prerequisites[1]. Another country that does not have any central standards or preconditions set by the government is Denmark. In Denmark, programs offered in the communities by their training institutes can be found besides those made available by the educational faculties of the universities, and those developed by independent providers (for example an institute called Praxis). New Zealand, too, abstains from any central standards or conditions for licensing. There are a great number of competitive providers, among them institutes of, or linked to, university.

Group C: In addition to countries of group A and D, there are countries with rather decentralised school systems, which follow a rather centralised approach in school leader development for the sake of quality assurance. This control is exercised by different institutions. In some of these countries the responsibility for the development programs lies exclusively with universities, for example, in Ontario in Canada, and in many states of the USA, such as Washington and New Jersey. The educational departments or faculties of the universities in these North American countries, however, are not completely independent in developing their programs, but have to follow central guidelines and standards approved by each state or Province. In the USA there is a nation-wide catalogue of standards, which was set up by the Interstate School Leaders Licensure Consortium (ISLLC) and accepted by the Council of Chief State School Officers (CCSSO). Thirty-six US states have so far accepted these standards. In Ontario the guidelines were set up by the Ontario College of Teachers (OCT), that is the self-regulatory body of the teaching profession. Until now, only universities have been accredited by this body and offer this program (however, other providers will be allowed to offer the official Principal's Qualification Program, too). These rather decentrally organised countries aim at securing a high level of quality by controlling school leadership development centrally. A 'higher-ranking' institution develops nation-wide guidelines and is even

[1] In the Netherlands, some influence is exerted by the government, however, in the field of qualification programs for primary school leaders, as a number of programs are financed by the government by taking over the participants' fees.

partly responsible for the accreditation of the providers and the certification of the school leaders, as is the case in Ontario. In the USA, for example, the States of Washington and New Jersey present state certificates to the participants on completing the university-based preparatory courses. In England, there is now a central national institution, the National College for School Leadership (NCSL), which will be responsible for developing and conducting all state-run qualification programs for school leaders. Such a central institution can, for example, work out national guidelines or standards, be responsible for the accreditation of the regional providers, can carry out the programs itself, qualify the trainers and teams of trainers, train them, select the participants, choose success criteria, test the candidates against these, and award the qualification. Hence it can also be responsible for the certification of school leaders. Moreover, it can evaluate the various qualification programs and assure their quality, which certainly includes carrying out research projects on its own in the area of school leadership and school leadership development. Up to now the preparatory National Professional Qualification for Headship has been offered by specific regional centres subordinated to the Teacher Training Agency or later the Department for Education and Employment and now the NCSL. The program itself, however, has a fixed content and structure as well as integrated assessment procedures. A unique combination can be found in Sweden, where the government is responsible for a national basic training program carried out by universities, and the communities, as employers of school leaders, are expected to offer further training and development opportunities. In Switzerland, a nation-wide accreditation of providers is currently on the agenda.

Group B: Countries with rather centralised school systems, yet offering various programs by different providers do not exist (as may have been expected).

Generally it can be stated that this research project, when describing countries with a rather centrally organised school system, focuses on the major state-run and mandatory programs. As a matter of course in these countries independent providers can additionally be found, as is the case, for example, in Germany, where publishing houses, academics, or professional associations offer additional training and development opportunities. However, these events are typically smaller scale and serve as supplementary to the main program.

Interconnected Responsibilities

Regarding the fact that the state-run training institutes in the German states own a type of monopoly in school leader development, the interconnected responsibilities in many other countries appear interesting and instructive. Although empirical judgements about the effectiveness of individual programs cannot be made, the interaction of central preconditions and prerequisites and a decentralised realization of the programs, as well as the cooperation of the various providers are significant approaches.

Particularly interesting are the countries of group C, that is England, New South Wales, Sweden, Switzerland (as a country-wide accreditation of the providers is intended), Ontario, Washington, and New Jersey as their approaches are particularly progressive and point the way ahead. What these countries have in common is that they qualify their school leaders according to central preconditions or guidelines without planning the programs in detail in advance and standardizing them to a high degree. Teachers who want to qualify for a leadership position at school can choose among various providers. Whatever the criteria for their choice, be it proximity, personal development needs, or the reputation of the provider, the participants can be assured about the quality of the program and that it will be accepted by the state and by potential employment committees.

A Central Institution and Regional Providers

A favourable environment for the development of qualification programs for school leaders appears to exist in countries where a nation-wide initiative for making school leadership more professional has caused extensive efforts for the establishment of qualification opportunities. Qualification programs could then be developed with adequate financial and public support. The responsibility for this was transferred to a central institution, which set up guidelines and standards for the further development of the programs. Examples for this are the initiative of the Danforth Foundation or the CCSSO/ISLLC in the USA, and in Ontario the Ontario College of Teachers. In Sweden such an initiative was launched by the parliament, and in England with its National College for School Leadership, by the government. In France, the Centre Condorcet in Paris is centrally responsible.

The implementation of the programs, however, was not up to this central institution, but was decentralised, carried out by different regional providers. The advantages of a program that is carried out in a decentralised way are the geographical proximity for the potential participants and adequate flexibility to take their needs into account due to – in most cases – an easily manageable number of participants. Also regional particularities can be taken into consideration and cooperation with the regional and local schools and education authorities most often result in increased practical relevance.

If there are many different regional providers, it is particularly necessary for the suppliers to take the needs of the customers, or 'clients', into account, in order to maintain their hold on the market. Their offering has to be tailored to the participant and has to be needs-based. In general, there is an expectation that independent providers are like this. In order not to give in to the danger of only serving the status quo of current school leadership practice, the programs and the providers have to show future-sightedness. It is universities, however, that, in general, are believed to be capable of innovative approaches since, on the one hand, they have to follow political imperatives to a lesser degree than state-run or state-accredited institutions do, and they can focus less on implementing state guidelines. Thus, they are free to follow system-divergent and innovative approaches and may not be as subject to market pressures as independent providers may be, and thus can avoid having to concentrate on prescriptive pragmatic offers.

The overall concern, whether the local provider is an independent provider or a university, remains the assurance of quality. In a regional market of providers you can find offers of quite different quality. A fundamental assurance of quality is of high importance for the participants as well as a nation-wide acceptance of the program.

Both, quality assurance and national acceptance seem to be guaranteed in countries where a central institution sets up guidelines and standards for the various regional offers. In the US this is done by the standards of the CCSSO/ISSLLC, in Ontario and England by the accreditation of the providers and in France by specific state-run regional academies.

It appears that of particularly high value is if the profession itself accredits the providers, as is the case in Ontario. Switzerland, as another example, involves the profession itself in the organisation and implementation of the development programs by cooperating with the teachers' association representatives.

Summing up, an interesting solution seems to be the combination of a central institution working out and setting up universal frameworks and guidelines and regional providers carrying out the programs decentrally.

Staff: Trainers and Teams of Trainers

Who is in charge of the program and who delivers it, is of the utmost importance for the quality of the program. The training staff is decisively responsible for the implementation of the development concept and for the genuine core processes, namely the teaching-learning processes. The recruitment and selection of trainers are done by the provider and this process is – sometimes explicitly, in most cases however implicitly – part of the program concept.

What can we observe in this international comparison? One similarity of many qualification models is that the trainers mostly make up a team with very different professional backgrounds. At the University of York in Ontario, for example, there are teams of trainers that include people from a school background and also people from university, completed by guest speakers (from various other areas). At the University of Washington the teams consist of academics, school leaders, organizational consultants, as well as mentor school leaders at the internship schools. The program in New Jersey has members of the faculty and guest speakers who are experts in certain fields (from Law to Neuro-Science) as well as experienced school leaders themselves. In Sweden the majority are academics and experienced school leaders. In England the teams of trainers include academics, school leaders, people from teacher training institutions, from the education authority, as well as from economy and industry. In France there is a team of trainers at each academy including experienced school leaders and additionally there is a school leader-tutor and a school leader-mentor at the internship school for each participant. In the Netherlands in the NSO program one third of the trainers are school leaders themselves, or former school leaders, one third are organizational consultants and one third are academics from university. In Singapore they are members of the faculty and mentor school leaders. In New South Wales the regional Inter-District School Leadership Groups consist of a representative of the educational authority and one other member of superintendence, one school leader, some teachers from different school levels, and additionally one representative of the university. Finally, this model includes a cooperative arrangement with the respective school leader association.

Among the training staff there is also a specific group of professional teacher development or adult learning trainers. They often have a school background, but due to their long-term activity in the continuous professional development arena they consider themselves trainers or adult learning facilitators rather than ex-teachers. Besides, there are trainers who spent a lot of years at school being teachers and sometimes even, albeit rarely, being school leaders. During the time of their employment at the continuous professional development institute they have given up their position at school. Sometimes, one actually meets trainers who at the same time are still working at a school or even are school leaders in post (for example in England and in North America).

A completely different professional biographical background is that of the academics working in school leader development programs, most of them are educationalists. Among them, the background of being in a leadership position at school is rare. Sometimes they don't even have a license for teaching. Sometimes they come out of the field of change management or business management (as, for example, in England or the USA).

A problem may arise, however not necessarily, if the trainers lack school teacher and/or leadership experiences, since they may make too few connections to the everyday problems of school leadership, or show too little commitment to these problems while at the same time have a lecturing or moralizing approach. Then, their credibility is at risk and will be questioned. School leaders in general expect that those being responsible for their qualification have on the one hand grounded knowledge and on the other hand an understanding for the themes and burdens of the daily activities in school leadership. One conclusion is that the participants most readily accept those trainers who have experiences at school, and above all in school leadership, if not as a teacher or a school leader, then through coaching or other forms of professional consultancy work or counselling.

However, trainers without any such professional background, who are experts in certain areas and at the same time are able to establish clear references to school and school leadership, are judged very positively by participants and are well accepted by them. Some providers even claim that the participants hold school-external experts, in particular, in high esteem, above all, as it is highly inspiring to exchange information and experiences, and discuss general insights or particular problems and hot issues with them.

The demand that to a large extent the profession itself should play an important part in the training and development of their colleagues may even be judged critically.

> A successful principal may not necessarily be very analytical about describing the reasons for his success. One must be wary of the assumption that such persons are the best for developing the training assignment. In some cases, the most useful principals to have are those who have had training themselves and have had the opportunity to order the ideas gained from their experience into a conceptual framework provided by theory, but moderated by actual practice. (Hopes, 1982, p. 5)

Many providers in the countries included in this study try to provide the link between practical experience and theoretical input, which enables reflection on experiences, by establishing teams of trainers. New partnership arrangements have emerged. Typically, the partners will include representatives from the employing organisations (whether national, state or local level), from educationalists in the university sector and, increasingly, from professional associations that represent school leaders themselves.

Our study suggests that much of the coherence that has been brought to school leader training and development programs arises from the interaction of these groups, and the different perspectives they can bring to resolving the key issues of curriculum content and structure, of teaching strategies and learning methods and also of the timing and sequencing of the programs. The idea is not, however, to have trainers with different professional background to teach in turns, but rather to bring them together right from the conception of the program, and include them all in the design and implementation as a team. Such teams, comprising academics from university as well as school leaders have proven to be very worthwhile. In the US programs members of the respective educational faculty and practitioners from the cooperating school districts and schools work together. It is this new form of cooperations that makes a program such as the Danforth Leadership Program in Washington with its extensive internships rewarding for the participants.

In some programs such a team continuously works with one and the same group or cohort of participants. The team, however, additionally invites experts from other academic areas and fields outside of education as guest speakers, for example, from the area of organizational learning. Thus, representatives of sociology, economics, psychology, communication, and law are invited. One prerequisite, however, is that all people involved should have a high standard of self-reflection, verbalisation skills, and a rich variety of teaching strategies used for adult learning.

It is not only content knowledge but also the methodological competence that counts, particularly in the field of education (but also elsewhere in management training). Across the countries, there is still too little attention to the needs of trainers for adequate preparation. Thus, for the trainers 'train-the-trainer' modules should be offered. Moreover, it is certainly true that trainers must be able to analyse the aims of the school leader development program, to know adequate methods, and to develop adequate material. The programs in Sweden and France have integrated regular training opportunities for the teams of trainers. In both these countries, for example, the facilitators are trained in such a way that they meet with teams from different regions of the country and get input from the central coordinating institution. In less centralised countries or systems, the regional teams may also organise their own training locally. Thus, the team members may improve their competences and cooperate in developing course material together. It is relevant that trainers again and again get the opportunity to broaden their knowledge base and improve their competences. Also, experiences from other countries could provide important input here: the trainers should integrate international experiences and insights in their work.

Particularly in those programs whose explicit core goal is the development and improvement of schools in addition to the qualification of the participants, additional consultants play an important part. Here credibility is essential as well as a non-directive and non-judgmental style. Theirs is the role of a catalyser, they initiate options of acting, however, they do not prescribe them. Emphasis must be placed on competences in communication, active listening, creating an open and constructive atmosphere, creating a foundation of confidence, as well as expertise in group dynamics, problem solving and dealing with conflicts. Having discretion is a basic condition for all consultants.

2. Target Group, Timing, Nature of Participation, and Professional Validity: Multiple Approaches

2.1. Comparison

Target Group and Timing

Another dimension of the comparison refers to the target group of the development programs. Who is qualified and at what stage in their career are they qualified? Is there training and development provided after taking over or before taking over (in the time between appointment and starting the post) or even before applying for a leadership position?

Here, the survey shows clear differences among the countries. For example, nearly all German states qualify their school leaders after appointing them, and in most cases even after their taking over the leadership position. Orientation or preparation courses are the exception and, if offered at all, then they are voluntary. Qualification only after taking over the leadership position can be observed also in Hong Kong and in some European countries such as Sweden, Denmark, Switzerland, Austria, and South Tyrol (see table below). In other countries, however, a preparatory training program can be found. Qualification opportunities before taking over a leadership position are offered in England, the Netherlands, France, Singapore, New South Wales, New Zealand, Ontario, and the USA.

Table 3: Timing in participants' career

Preparatory	England; the Netherlands; France; Singapore; New South Wales, Australia; New Zealand; Ontario, Canada; Washington, New Jersey, California, USA
Induction	Sweden; Denmark; Germany; Switzerland; Austria; South Tyrol, Italy; Hong Kong, China

Nature of Participation

The question of timing should not be viewed without looking at the nature of participation, that is whether participation is mandatory or optional. It certainly makes a difference whether an orientation or preparation opportunity may be attended voluntarily or whether a training is mandated.

Table 4: Nature of participation

Optional	Sweden; Denmark; England; the Netherlands; Germany* Switzerland* ; New South Wales, Australia; New Zealand
Mandatory	France; Germany*; Switzerland*; Austria; South Tyrol, Italy; Singapore; Hong Kong, China; Ontario, Canada; Washington, New Jersey, California, USA

** Multiple entries result from the differences of the various programs included in the individual countries.*

A mandatory preparatory program (see table above) has been established in France, Hong Kong, Singapore, Ontario, and the USA,. In South Tyrol and in some states of Germany a school leader has to qualify after taking over head- or principalship. A voluntary qualification program can be attended by aspiring school leaders in England (where participation will soon be obligatory), the Netherlands, New South Wales, and New Zealand. In these countries the local, or regional employing committees generally expect a preparatory qualification, however such a qualification has varying levels of importance. A voluntary qualification after taking over a leadership position again is expected in Denmark, Sweden and Switzerland (in Switzerland it has not yet been decided whether it will become obligatory or not).

Professional Validity

The dimensions of timing and nature of participation have consequences for the planning of the professional career of the (aspiring) school leaders in the various countries. Actually, it plays an important part whether someone – in order to become a school leader or to remain a school leader – has to attend a qualification program or not (for the relation of both variables see table below).

Table 5: Timing in participants' career and nature of participation

		Preparatory		Induction
Mandatory	A	France; Singapore; Ontario, Canada; Washington, New Jersey, California, USA	B	Germany*; Switzerland* ; Austria; South Tyrol, Italy; Hong Kong, China
Optional	C	England; the Netherlands; New South Wales, Australia; New Zealand	D	Sweden; Denmark; Germany*; Switzerland*

** Double listing due to differences in the approaches of the German 'Laender' or Swiss 'Kantone'.*

Four groups emerge:

Group A: In the countries with a mandatory preparatory qualification (see quadrant A), participation actually has an impact on being appointed a school leader. There, it is a relevant criterion for selection. France is unique in having a close link between selection, qualification, and employment. Here, only the successful passing of the 'Concours' gives the right to participate in the state-financed qualification, which again is a pre-condition for obtaining a leadership position after having successfully passed it. Again, school leaders can only remain in this position if they have successfully gone through the mandatory second stage of qualification, the 'Formation d'encompagnement'. In Singapore, due to a fixed promotion procedure, being ranked in a higher career position is possible only after the state-financed full-time qualification. The situation for aspiring school leaders is much less secure in the North American countries. Here, a preparatory training is a precondition for the application, yet not automatically a guarantee for the individual's success. This is particularly interesting as the candidates have to commit themselves to investing quite a lot of time, money and work.

Group B: In case of a mandatory participation after appointment or even after taking over the leadership position, this qualification naturally had no consequences for recruitment. The qualification itself and its success could no longer be used as criteria for selection and appointment. It may, however, under certain circumstances, even if earned after appointment, play an important part for remaining in the leadership position. This is the case, for example, in Austria. Here, it is an explicit pre-condition for continuing after the first four years. In South Tyrol the leadership position is secure, but participation at the Führungskräfteschulung is a condition for a higher ranking.

Groups C and D: In the countries that have a voluntary preparatory or induction program, the professional validity is dependent on the employing committee and consequently differs regionally. However, there is a tendency towards the employing committees expecting an adequate preparatory training (such as in England, the Netherlands, New South Wales, and New Zealand) or participation in an in-service or induction program (like in Denmark, Sweden, and Switzerland).

In summary, it can be stated that in all countries included in this study, the qualification of school leadership personnel has a higher professional validity than ever before. Thereby, the acceptance of the qualification by the profession itself as well as the expectation and acceptance by employing committees is of particular importance.

2.2. Discussion

What Timing Is Best?

As stated above, timing and the nature of participation have professional consequences for the school leaders. The question what timing is best is answered differently by different countries and may raise some controversy. In general, arguments point to the costs, the effectiveness and the efficiency, but also to potential expectations and possible legal consequences.

There are arguments in favour of solely qualifying school leaders who are already appointed and have taken over their leadership position:

The first argument is that it saves resources, in terms of money and time. Qualifying before taking over principal- or headship is not a guarantee that the money invested is a reasonable investment. People may be qualified who never become school leaders (as we see in England, for example), maybe because of personal reasons (they do not apply), or because of a lack of competence (their application is not accepted by the employing committee).

Moreover, 'on the job'-training may also be seen as more effective, because learning goes hand in hand with experiences made in the workplace and may therefore respond to immediate needs.

In addition to this, there are further considerations of efficiency that can be determined by doing a cost-benefit analysis. It could be said that there is higher efficiency and effectiveness when qualifying after taking over the headship since there is a cost saving by not participating in a preparation program. Though there is the assumption that the greatest effectiveness can only be achieved by combining the two, a cost-benefit analysis may perhaps be against such a preparatory qualification (particularly, if it is not linked with selection and appointment procedures). The costs of training and development opportunities are particularly taken into account in those countries where the qualification is not only within the responsibility of the government, but is also state-financed, as, for example, in the German states, Austria, Switzerland, and South Tyrol. This is particularly relevant for the countries in which qualification is mandatory since these must have large-scale development programs. In such cases the investment seems to be more certain when spent for school leaders already in post.

Another argument against a preparatory qualification often put forward in Germany is that after this preparation high expectancies arise on the part of the participants to be appointed school leaders. There is fear that due to the preparatory qualification a quasi-legal claim for being appointed may be made and may be legally enforceable.

After a look at the international context, however, this view proves not to be convincing. In Ontario, Washington and New Jersey, for example, the participants who took part in extensive preparatory programs before applying for leadership positions (often self financed), know clearly that such a qualification may substantially improve their chances for promotion, but they must not make any legal claims for it. Why then should this be possible in Germany or elsewhere?

On the other hand there are some convincing arguments in favour of a preparatory qualification, which can prove the contra-arguments wrong:

Firstly, the qualification must correspond to the essential importance of school leadership. Recently insights and empirical evidence about the pivotal role of school leadership for the quality and development of a school has grown (see Part I), therefore efforts must be extensive enough to achieve the best results possible.

Secondly, even though in France and Singapore there is a mandatory state-financed preparation before taking over the leadership post, the concern to prevent bad investment is met by a rigorous application and selection process that has to be gone through successfully by the participants. To select the participants prior to training them seems to avoid spending resources inadequately.

Thirdly, a simple superficial cost-benefit analysis would not go far enough. From an education-economic view the cost-benefit analysis would have to be much more complex and long-term. As school leadership can be regarded a new profession, which requires a change in perspective in many respects and requires new and more comprehensive competences, it is hard to understand why such individuals should not be properly prepared through an adequate training and development program. For example, in Germany, people who hold other positions at school are qualified much more extensively than the school leaders themselves. Rosenbusch (1997b) points, for example, at the career advisors or the school psychologists. Even though it must be admitted that a school psychologist needs more specific knowledge, a qualification for the function of school leadership who may be regarded a generalist, is equally demanding. All this speaks in favour of a preparatory qualification.

Fourthly, adequate preparation may reduce the 'practice shock' experienced by new entrants to the role, when they change from 'teaching' to 'management' (Storath, 1995), they are confronted with various forms of stress. Additionally, there is the danger that new school leaders might fulfil their roles in very unfavourable ways if they are not adequately prepared. Psychology has described the well-known phenomenon of the 'first impression' as well as certain effects of labelling by staff. These problems can be prevented or, at least, reduced by a preparatory qualification.

Fifthly, the assumption that on-the-job-training is the most effective way has not been adequately researched and seems to be – when formulated on its own – rather doubtful. The line of argument that the best response to development and training needs is the interrelation of learning and reflection with practical experiences at school speaks in favour of on-the-job-training, but not necessarily of restricting development opportunities only to this. Too much stress, as it is most often experienced by newly appointed school leaders, increases the danger of searching for easily accessible recipes for action and short-term promises for improvement. It is especially important that leaders of organisations in the educational sector get opportunities for development in terms of long-term growth processes. Such opportunities would provide the chance to develop a new perspective when changing from teaching to managing and even leading a school, which obviously demands a mental shift, and to reflect thoroughly on one's own role, on the goals and objectives of one's own behaviour in the school and on the effects of the organisational actions taken. An adequate preparatory training and development seems to be essential for all these reasons. School leaders have to develop a 'big picture' of school and schooling before taking over their first leadership position in order to be able to tell what is genuinely important from what merely seems to be urgent.

Sixthly, a pre-service training offers the chance of assessing one's own interests and strengths. This may help candidates to make career decisions more consciously.

Seventhly, international experiences indicate that the provision of pre-service preparation may stimulate the number of women applicants to educational leadership positions. Women may be more self-critical, and may also be less connected to promotion networks. Obviously, training and development programs are encouraging.

Eighthly, experience shows that even participants who may not get into a leadership position in the end may extend the leadership resources of their schools in terms of empowerment and spreading of leadership and management responsibilities in line with the idea of building up a leadership capacity at each school.

All these arguments clearly favour orientation and preparation opportunities. Even more extensive are approaches to make these elements part of initial teacher training in order to identify and foster potential for leadership at the earliest possible stage. This has been done recently by the Australian State of Victoria. In Canada, too, long-term promotion is intended by a portfolio-system, and in Sweden, there is a project that offers enrolment in a school management course during initial teacher training. To provide such an opportunity during the first phase of teacher training was also aimed at by the Centre for School Development and Management at the University of Bamberg.

The Selection of Participants

It can be quite instructive for countries whose priority is in-service development for established school leaders to look at some countries with preparatory programs and participant selection procedures. This is the case with the university providers in the USA, above all in the course of the programs initiated by the Danforth Foundation (such as at the University of Washington). Right from the start of the Danforth programs, assessment procedures were used to select suitable candidates. At the same time these selection procedures were used for planning the individual course programs. Applicants had to meet relatively strict academic access standards, and the faculties conducted their own assessments. These comprised formal interviews by the members of the steering committees, academic assignments and the presentation of topics selected by the steering committee. Some universities commissioned special assessment centres with this procedure, for example, the National Association of Secondary School Principal Assessment Center (NASSP). It was argued that self-selected candidates might already identify with the status quo to such a degree that they might have built up a barrier against theoretical knowledge and against re-learning. Thus, the overall level of excellence might be reduced. Besides, emphasis was put on increasing the number of women and of candidates from ethnic minorities. Moreover, teachers were encouraged to name colleagues whom they thought particularly suitable for candidacy. The application procedures and the conditions of access to the programs of the University of Washington and the William Paterson University with their particular selection procedures show a strong effort to select the participants on the basis of their leadership qualities. By this selection process the qualification gains an even higher reputation.

In other countries similar procedures can be found, as in the Netherlands in the program of the NSO.

In England the procedure has two phases. There is a selection and later a needs assessment. The applicants have to prove experiences in school management and a well-founded professional knowledge base by providing references from their present school leaders or the school authority. Then, suitable candidates are tracked in a selection process. In the 'needs assessment' an analysis of one's personal development needs is used as a basis for setting up an individual qualification plan.

Even more extensive and costly is the access marathon that applicants for the school leader training program in France have to undergo. It is part of the recruiting procedure and meant to protect the state from investing considerable sums of money in candidates who prove to be unsuitable.

The Grouping of Participants

By taking a look at the grouping of participants, different ways can be found. Induction programs are sometimes decidedly school-type-oriented: for example only for school leaders from the primary sector or only for school leaders from the secondary sector, others, however, are explicitly cross-school-type-oriented. For the preparatory programs the same holds true. We can find a lot of diverse differentiations. Among the examples of state-run programs there are strictly school-type related ones, such as in France and cross-school-type related ones as in Singapore. In some programs the target group explicitly comprises not only potential applicants for a school leadership position, but also teachers aiming at enlarging their leadership competences generally without directly aspiring to a school leadership position (such as in New Jersey). The School Leadership Preparation Program in New South Wales, which is the first phase of the School Leadership Strategy conducted there, addresses a target group of potential school leaders as well as teachers interested in various subject or educational leadership positions at school (as subject leaders, heads of year, etc.). It is only in the second phase that the participants aiming at different careers are qualified separately. Other programs address applicants for school leadership positions as well as school leaders already in post, such as Foundation II in California and the program of the Nederlandse School voor Onderwijsmanagement, which additionally regards deputy school leaders explicitly as a target group. The same holds true for the program of the Danmarks Laererhojskole.

To group the participants according to their professional experience and to focus on one distinct target group, however, seems quite sensible for certain phases of the program. Proceeding like this ensures a unified focus and the necessary homogeneity of the cohort. This homogeneity, again, is helpful, as it facilitates concentrating on specific contents and teaching strategies, which support a better transfer and more direct applicability. It certainly makes a difference, whether or not the participants are able to translate what they have learnt into action parallel to taking part in the program. In other words, it obviously makes a difference whether or not they already hold a leadership position including the possibilities to implement certain innovations. Hence, their position definitely has an effect on the teaching and learning methods applied in the program, as the workplace can then be made use of in different ways, for projects or other kinds of homework (see chapter II.5.).

Four of the programs included in the study open up to target groups beyond the school personnel themselves. They set up groups of participants comprised of school leadership personnel and staff from other educational institutions. The course Master of Educational Leadership of the University of Waikato in New Zealand and the program of the NSO in the Netherlands are aimed at educational leaders and people holding leadership positions in extremely varied areas of the education system. The program Leadership in Development of the Danish provider Praxis addresses among others personnel of school boards and school authorities and advisory centres. In Sweden, in the part of the training that is offered by the local communities, school leadership staff and leadership personnel of local administration bodies and other community organizations take part in management courses together. The idea is that this might be of advantage for the cooperation of these organizations within the community.

Individual or Team? Competences for the School Leader or Competences for School Leadership Teams and School Development?

Particularly instructive when comparing various definitions of target groups is the explicit differentiation between programs focusing on the individual school leader and on equipping him or her with competences needed for school leadership tasks and those linking school leadership development closely to school development processes at the particular school, regardless of whether the target group includes potential school leaders (i.e. teachers aspiring to leadership) or those already holding a position. If school development is the decided goal, often whole school leadership teams (as in the Danish program Leadership in Development), or teams of staff members are included. Those teams sometimes even comprise representatives of the parents or the political community. Such teams comprising teachers, school leaders, parents etc. are the target group of the in-service program School Leadership Teams of the California Leadership Academy. However, only a few programs are designed explicitly for teams. Yet, it can generally be stated that even programs whose target group is individual school leaders put more and more emphasis on team aspects and on learning from colleagues and collegial problem solving. Knowing about the relevance of cooperation, there are efforts to make cooperative methods a principal feature of the programs.

Coming from a notion of the school as a learning organisation and viewing development and change in individual schools and in the school system systemically, it must be considered that not only staff within the schools are needed as change agents in various functions, but that there must be change agents on different hierarchical levels that initiate and promote development processes. Demanding a distribution of leadership responsibility results in an increase in successfully promoting change. Discussions about building up a change capacity within the school in school development literature means that more and more people recognise the potential of other team members, promote it, support it and thus give a stimulus for genuine grounded changes. It would be logical then, to qualify personnel from different hierarchical levels together, beyond the individual school. This might include the education authority or school administration together with school leaders, deputy school leaders, persons from steering groups, and perhaps even 'identified' informal leaders from the staff. Yet, none of the programs described here goes that far. Qualifying personnel from different levels, however, would result in a more profound understanding for each other. Possible reservations about each other's roles or even prejudices might be diminished. Hence, traditionally difficult relationships might be overcome or at least improved. Much more relevant, however, would be that such programs would create a shared knowledge about the concerns for school quality and school development. This shared knowledge can also improve communication though a common language. And this again might enhance cooperation and collaboration towards a common aim.

A big problem in many countries seems to be that the school system has become decentralised, which has strengthened the self-responsibility of the school, but the people involved go on thinking and acting on a hierarchical basis. This is the case in ministries, used to mandate regulations and to control, as well as among the teaching profession used to being passive and waiting for the general conditions to change. Chances for self-initiatives and for realising newly gained freedom for educational action remain unidentified and consequently unrealised. A comprehensive training and development approach that spreads across the various levels of the school system might be very useful here and would enhance professionalisation.

Part of such an approach is that continuous professional development would no longer take place solely away from school at an external institute, but rather would take place at the individual school, and focus on the initiation, implementation, and institutionalisation of school development processes.

3. Aims: Different Goals, Orientations, and Conceptions

The decision whether development opportunities focus on the individual school leader or the whole school implies that there must be a clear definition of the program's aims. We think that without stating clear aims, it is not possible to have a convincing conceptually based program. Hence, the program aims are a particularly instructive area when comparing training and development models.

3.1. Comparison

Most of the programs have an explicitly formulated set of aims. However, it is obvious by comparison that there is a broad range of aims among the various programs. They vary in terms of differentiation and degree of abstraction. Some providers state the general function of the program, namely – quite tautologically – to qualify the (aspiring) school leaders for their leadership tasks. Others quite pragmatically focus on the preparation for concrete tasks. Various providers refer to their visions, guidelines, or frameworks. Others start from a vision of school and/or of leadership, and others from a specific leadership conception. From these descriptions, the aims of their program are derived. Others go into the country-specific educational, political, and occasionally even the social situation. Some put their emphasis on the moral aspects of an understanding of leadership in a broader sense.

On the basis of an analysis of their foci, the goals can be differentiated according to their particular emphasis, that is, to the extent to which they:

- take into account the demands of the government (functional orientation),
- start from a quite pragmatic preparation for the different tasks of school leadership (task orientation),

- aim explicitly to develop the competences of the individual participant (competence orientation),
- explicitly focus on the development of the individual school (school development orientation),
- aim explicitly at the change or development of mental concepts of the participants (cognitions orientation),
- build explicitly on a vision of leadership, a conception of leadership, or on a vision of school (vision orientation),
- are distinctly oriented towards values (value orientation).

A clear grouping of the programs to a single criterion, however, is very rarely possible, most of the programs incorporate multiple foci, for example, some are task- and competence-oriented, or others are vision- and school development-oriented at the same time.

Three programs describe their goals with a clear functional orientation:
Directly related to the political or educational-policy context, the provider of South Tyrol states: the participants are to be prepared for the leadership tasks under the conditions of greater school autonomy, that is reflect upon the new range of action of the school, get the necessary abilities and methods to realise this autonomy in depth and develop their own ideas and understandings of the new regulations. Based on the aims formulated in the Ministerial Decree of August 5th 1998 and the suggestions made by the Regional Association of School Leaders, the following results for the participants are to be achieved: 1) They are aware of the radius of action of an autonomous school. 2) They know of the tasks and the responsibilities of a school leader of an autonomous school. 3) Building on an analysis of the actual present situation, they have determined the necessary capabilities for leadership of an autonomous school and have articulated the corresponding need for further training. 4) They have become aware of the necessary areas of responsibility and have had training practice in seminars. 5) They have become acquainted with the instruments needed for the realisation of autonomy in pilot projects and have taken steps to introduce such instruments. 6) They have newly-defined the professional image of 'head of school'.

For the qualification program in France, three goals have been explicitly formulated: 1) to provide insights into and experience of the leadership role, 2) to help to develop an appropriate range of skills, competences and knowledge which will enable the candidate to function effectively as the representative of the state, 3) to broaden and extend the candidates' cultural awareness, pedagogic skills and understanding in order to enable them to exercise a leading role in the school.

In Sweden, an aim for the school leadership qualification is a profound understanding and knowledge of school leaders of the Swedish education system, of the national education aims of school as well as of the role of school in society and in its particular community. The participants' knowledge of the function of leadership in an education system that is guided by goals is to be enhanced, as is their ability to plan, to implement, to evaluate and to develop school activities. The school leader is to develop the capacity to analyse the results of the activities of his school, to draw conclusions from this analysis and to present them to the committees in charge. Additionally competences to cooperate with the members of the school and the community and to represent the school outside are to be expanded.

The tasks of school leadership are the starting point of the 'working assumptions' of the program of the University of Washington, which comprise guidelines for school leadership activities. Beyond concrete everyday leadership activities, basic principles are enumerated here: equity and excellence, improvement of the quality of the school, educationally-oriented leadership, development of the organization, cooperation, personal enhancement of knowledge and reflection as a part of educational responsibility. These basic ideas can be found in the contents of the seminars and in the methods applied. The 'working assumptions' are: 1) Equity and Excellence – Public schooling in a pluralistic, democratic society demands that a quality education be received by every pupil under the care of educators in schools. Educational leaders are morally obligated to ensure that the twin goals of equity and excellence are the number-one priority of educators in their schools. 2) Leadership – Educational leadership is the exercise of significant and responsible influence. This includes, but is not limited to, the skills of management. Significant leadership requires the articulation, justification, and protection of the core set of educational values that underlie the purposes and functions of schools. 3) Organizational Change – Schools must be dynamic and renewing educational organizations. Educational leaders must nurture and sustain the process of dialogue, decision-making, action, and evaluation that lead to the improvement of schools as places for teaching and learning. 4) Collaboration – Educational Leadership is crucial at all levels of schooling. Administrators must create and support opportunities for the authentic involvement of all educators in organizational change and school improvement. 5) Inquiry and Reflective Practice – Educational Leaders must be committed to the importance and use of knowledge. Effective leaders must have the ability to reflect critically on practices in their schools and promote the importance of inquiry as a professional responsibility of all educators.

Some of the programs pursue a competence-oriented approach. Here, the individuality of the school leader is in the centre, she or he is to develop competences that are considered necessary: Switzerland aims at qualifying the participants for their tasks by imparting a compilation of the following competences: leadership competence, social competence, personal competence, educational competence, school development competence, administrative-organizational competence. Quite decidedly based on competences, too, is the program of the Nederlandse School voor Onderwijsmanagement. Before developing their program, there was a consultation that investigated necessary competences. The program's goal is to develop these competences for leadership functions in schools or other educational institutions. The following groups of competences are enumerated and put into practice (see contents): pedagogical, educational, and institutional competences, controller competences, leadership competences, organisational competences, consultation competences. The school management course in Austria (here, Salzburg) aims at promoting and developing the educational, leadership-related, social and personal competences of leadership personnel and, hence, at enhancing the quality of the institutions of education. Thus its primary focus is the preparation for educational tasks as well as administrative tasks and the ability to design medium and long-term change processes with the aim to continuously enhance quality of a locally based school. Here, competence- and task-oriented terms are used. The English qualification program NPQH aims at imparting to the participants leadership and management competences as preparation for school leadership tasks, whereby the core purpose of school leadership is formulated such that leadership is to ensure the success and the improvement of a school as well as a qualitatively high level of education for the pupils and an improvement of their achievements through professional leadership. The qualification is based on the National Standards, which divide school leadership activities into tasks that must be completed and the competences necessary to accomplish the tasks. There are five guidelines for conducting the program. Here again an explicit task and competence orientation can be found. The qualification 1) is rooted in school improvement and draws on the best leadership and management practice inside and outside education, 2) is based on national standards for headteachers, 3) signals readiness for headship, but does not replace the selection process, 4) is rigorous enough to ensure that those ready for headship gain the qualification, while being sufficiently flexible to take account of candidates' previous achievements and proven skills and the range of contexts in which they have been applied, 5) provides a baseline from which newly appointed headteachers can subsequently, in the context of their schools, continue to develop their leadership and management abilities.

The Canadian providers combine task, competence, and value orientation: in the guidelines of the Ontario College of Teachers, it is stated for the Principal's Qualification Program that it is to make the participants able to lead efficiently and effectively within a context characterised by change and complexity. Hereby, the political, economic and social influences on the school in Ontario are to be taken into account. Its goal is to develop the necessary competences, which, as in England, are explicitly enumerated. The University of York formulates its program on the basis of guidelines which include a philosophy, or vision, in a manner typical of the North American programs, to which moral principles belong. The program aims to prepare school leaders to act in cooperative and relationship-oriented ways in the every day life of the school. These actions take place amidst social communities and pursue the principles of social justice and equality.

A distinct school development orientation can also be found in most of the program goals, partly included in a functional, a task, or a competence orientation: the qualification in Singapore aims at making the participants effective and efficient professionals and creative leaders so that they can create the condition for a learning organization, motivate their staff, react to changes, initiate improvements (above all such improvements which are in accordance with the national goals), administer resources and implement information technology (and new media). The proposed goal of the Diploma in Educational Administration is to create 'thinking schools' and 'learning communities'.

A strictly cognitions-oriented approach is pursued by two programs. In both, the development of the abilities of the participants is focused upon, that is to analyse their own mental structures and mental models and to gain insight in the organizational structure and the organizational processes of the concrete school, out of which, then, change and development measures are to result. The cognitions-oriented approach is linked to a systemic one. The Danish Leadership in Development program of the independent provider Praxis formulates as its goal that the participants are to develop insights and competences by extensive reflection, intensive analysis and critical questioning. Of central importance is to detect, and to understand relevant patterns and relationships within a constantly changing environment and to react on this basis. A program which starts from a specific understanding of leadership and derives its goals from this understanding is Meesters in leiddingsseven of the GCO fryslân together with the Katholieke Unversiteit Nijmegen in the Netherlands. This conception of leadership, integral leadership, is about overcoming the contradiction of educational action on the one hand and administrative action on the other hand, instead the two are integrated. To develop this understanding of conceptualising the organizational unit of the school, and to act according to this mental picture, this is the aim of the program.

Some programs already start from a definite leadership conception or from a specific image of the profession: besides the cognitions-oriented Meesters in Leidingsseven, there is, for example, the program in New Zealand. The Educational Leadership Centre of the University of Waikato regards school leaders, despite all their manifold strains by market orientation and evolvement of administrative tasks, primarily as educational leaders, that is as leaders with genuinely educational tasks and with an educational mission. The program is to help them to further develop their ability to reflect, their interpersonal competences, and their fundamental educational values as basic conditions and foundations for their profession. The College of Education of the William Paterson University of New Jersey has based its program on a conception of transformational leadership. The participants are enabled to develop a personal vision and a personal leadership competence and to set up an understanding of how fundamental and continuous changes can be initiated and implemented within complex organizational structures. The orientation towards a specific leadership conception or a vision of school and school leadership also can be found in the Danish Laererhojskole (now Danish Paedagogiske Universitet). It aims to develop competences in the fields of management and leadership within the country's specific context, that is a strongly decentralised school system. The program is based on the vision of a democratic and reflective style of school leadership, linked to the central activity of school, that is to impart democratic knowledge. Hereby school leadership has to orient its own leadership activities towards the key goals of the organization. Among these are to develop and secure a democratic self-definition of leadership and leadership activities.

The development of a specific vision of school as their goal is quoted by the following programs: quite explicitly this is done in Foundation II of the California Leadership Academy. Here the vision is pupil- respectively learner-centred schools, in which successful and committed learning is possible. The prime task of the school leader is to help develop the school towards this aim. All the contents of this program are to support this. The contents and themes offered always bear the label 'for the sake of effective learning'. A concept of school as a learning community is the goal and aim of the School Leadership Preparation Program in New South Wales (this is the case for all phases of the entire qualification program School Leadership Strategy). This fundamental idea dominates the macro-didactic considerations and can be found again in the contents (a broadly defined target group, a multi-phase and modularised curriculum and so on).

These programs from Denmark, the Netherlands, from New South Wales, New Zealand, Washington, New Jersey, and California take very clearly into account what Rosenbusch (e.g. 1997b) has demanded again and again, that is an organisational-educational perspective, which means to always start from the core activity of school. Moreover, these conceptual leadership approaches are very much elaborated as transformational leadership, democratic leadership, or integral leadership.

3.2. Discussion

As the comparison shows, the aims of the programs differ widely. Some aspects are particularly striking.

Explicitly and Extensively Formulated Aims

In almost all countries – or programs – included in the study, the aims are rather extensively formulated. Apparently, the central role of school leadership is linked not only to the relevance which is attributed to an adequate training and development program, but also to the clear conception of aims. Of course, an alternative explanation for this could be that if extensive qualification programs are to be implemented, a convincing explanation and legitimatisation must be considered as well, no matter if they are paid for by the state or by the participants themselves. A part of this legitimisation of costs is the setting of transparent goals and objectives. Another desirable matter is to operationalise the aims so that the program can be evaluated. Sometimes the expected outcomes of the training and development program are formulated explicitly and in a concrete way. Unfortunately, sometimes they are only implicit.

Orientation Towards the Individual School Leaders Versus Orientation Towards School Leadership Teams and School Development

Internationally, two basically different approaches to the setting of aims can be observed: some programs focus upon the individual school leader and aim at imparting to her or him the relevant competences, respectively letting the person develop these competences for effectively leading a school. Other models, however, link school leader development closely to school development and put improving or restructuring the individual school into their centre. Both approaches have consequences for macro- and micro-didactic considerations. If school development measures are a relevant and essential part of the school leader qualification, this enhances the opportunity to learn from real situations and to use the reality of school not only as a starting point and target, but also as an intensive and extremely complex learning environment, embedding the school leadership development in the concrete working environment of the individual school.

A More Broadly Defined Understanding of Leadership with an Orientation Towards Values

Much more than it is the case in Germany, some of the programs worldwide have an orientation towards a specific value-based attitude. A more broadly defined understanding of leadership includes moral and political dimensions of leadership in a democracy. Leadership in a democratic society is embedded in democratic values, such as equality, justice, fairness, welfare and a careful use of power.

Leadership always implies some influence on others. It is essential to make the participants of development programs sensitive to that. They are to cultivate some awareness for the importance of dealing carefully and responsibly with power. Their educational aim has to be (regarding the pupils) that the people 'led' and 'influenced' will develop to become independent thinking, self-responsible and socially responsible, mature citizens who grow beyond being led. Principles such as self-autonomy, respect of oneself and of others, and cooperation play an important part, as they also do in adult learning processes and in leadership in general. Only a few of the programs, however, make this an explicit theme, such as the one of the University of Washington:

> Quality leadership preparation programs must be organized around, and guided by, an explicit set of values expressed in the program philosophy and working assumptions. (see Sirotnik & Kimball, 1996, p. 191)

New Paradigms for Leadership

Due to changes in the situation of schools, school leadership has to take over a new role. It seems clear that this new role can hardly be filled with old leadership conceptions. Therefore, some programs focus upon rather specific leadership concepts. If it is no longer central to see school as a stable system in which the existing structures have to be managed, conceptions like transformational leadership are referred to. Transformational leaders are not only to manage structures and tasks, but also to concentrate on the people who work there and the relationships they have in order to win their cooperation and commitment. They are to attempt to have an active influence on the culture of the respective school so that more cooperation, more cohesion, more self-reliable, and independent learning and working takes place. Leadership defined like this is supposed to be particularly successful in school improvement processes.

If the school is to become a learning organization, this implies the active empowerment and cooperative commitment of all stakeholders. The previous division between the positions of teachers on the one hand and learners on the other hand cannot be maintained, nor can the division between leaders and followers. Leadership is no longer statically linked to the hierarchical status of an individual person, but empowers as many staff members as possible as partners in various parts of the organisation.

Integral leadership, as suggested by the Dutch Meesters in Leidingsseven, emphasises the integrated perspective, which overcomes the classic divide between management and leadership. Both areas are to serve the educational aims to which school leadership activities are obligated. Through working with the cognitive maps that school leaders have of their schools, such an integrated perspective is to be developed.

Adjusting of Aims

In this study, a decisive fundamental consideration has been found in some programs, it is the idea of new conceptions of school. The old idea of school as an 'administered addition of lessons', as the 'lowest body' in the school system's hierarchy which has to carry out and implement the stipulations of superior bodies and is controlled by these is no longer up to date. Since there are so many changes in society, economy, etc., but also in the school system, the individual school has to become a 'learning school' in a twofold sense. Besides promoting the learning processes of individuals, the whole school, as an autonomous unit, has to learn, which means that it has to flexibly adapt to social, economic, and cultural developments, sometimes even precede them, but also counteract problems resulting from them.

Therefore it would be quite important, before taking over a leadership position, to make the aims of school in our society and possible goals of schools of the future central themes. These should then be discussed in a training and development program, debated, and adapted into a thinking framework for taking action towards these goals.

These reflections on leadership activities, the school, its role and function are explicitly made in some countries. Increasingly, there is more conceptual elaboration about the aims towards which schools are to be developed. Questions are being asked in more and more countries, such as: what is the aim of current education in schools? How can this aim be realised through teaching and through the communicative everyday practice in schools and the culture of a school? What must the organisation and administration of school be like in order to reach this aim? And: what is the role of school leadership in this context? This means that leadership activities like decision making processes, dealing with conflicts, problem solving, interpretations of regulations and instructions, as well as the everyday routines at school have to be brought in line with these fundamental premises.

The principle that 'school has to be a model of what it teaches and preaches' (Rosenbusch, 1997b) thus has consequences for school leader development. Training and development of school leaders has to be based on a clear conception of the aims of education and teaching-learning processes at school. This idea has to shape the programs with regards to contents, methods, patterns in terms of timetabling, etc.

The demand made by Rosenbusch and Huber (2001) for the German context that school leadership personnel must be trained as professionals and experts in school education and organizational-educational leadership and management seems to have already been put in practice in other countries, which are much more advanced in terms of leadership development.

Internationally, some programs provide a model that quite decisively aims at a new conception of school: changeable and adaptable organisations, developing as a learning community, a pupil- respectively learner-centred school. A leadership conception such as transformational leadership, too, applies an idea of school as a culturally independent organism, which is able to develop and has to develop. The efforts to combine a cognitive approach and a systemic approach also regard school as an independent organizational unit. Here, approaches in Denmark, the Netherlands, New South Wales, Washington, New Jersey, or California can be referred to.

The aims formulated by the provider should, if they are not merely rhetoric, have an impact on the selection of contents as well as on methods and structure of the program. It is, however, beyond the scope of this study to see how far this has been realised in the programs. Moreover, this study gives no evidence of how well the aims stated have been reached. For this purpose, effectiveness studies would have to be conducted. In order to get to international comparisons of the effectiveness of programs, meta-studies would be necessary. However, this study is a starting point and provides a basis for this kind of research.

As a whole it can be stated that it is difficult to establish a convincing training or development program without a clear conception of aims. When designing, planning, and conducting a program, it is essential to have a vision for the program. From such a vision, a concrete set of aims can be derived. Thus, coherence and sustainability are assured.

4. Contents: What Is Important?

The aims of the program should, as stated above, have an impact on the contents chosen. What topics are dealt with?

4.1. Comparison

Comparing the contents of the programs of the individual countries, instructive insights in the conception of school leadership and of the tasks allocated to school leaders can be gained. In Germany for example, some years ago, a clear emphasis on legal and organizational issues could be found, whereas today communication-oriented contents are focused upon. Looking comparatively at the contents of the programs, it is easy to notice that the contents reflect different foci. The present legal framework for school and schooling of the particular countries sets up a kind of basis of orientation, however more or less distinct. Besides the international phenomenon of centralisation and decentralisation, and as a consequence giving the individual school more decision-making power, but also holding it more accountable, new leadership conceptions have gained importance. Moreover, everyday leadership tasks and to a different extent also the concrete interests of the participants play a role in the curriculum of the programs. That means that the following areas of interest are taken into account (here they have been generated via an additive-grouping procedure, the coding categories are supplemented by a compilation of example topics and themes from the programs):

Table 6: Contents of the training and development programs

The role of school leadership	
Educational Leadership in General	Leadership in Times of Change; Leadership in a Team; Leadership: Leading the School; Guidance and Leadership; Leadership according to a Role Model; Guidance and Leadership in Learning Communities; Values; Ethics and the Legitimisation of Leadership; Dealing with Power; Ethics and Legitimisation of Leadership; Moral and Political Dimensions of School and Pedagogic Leadership within a Democracy; Developing a Joint Role Model; Leadership Shared with Others in the Context of Successful Learning; etc.
The Concept of Leadership	Leadership Conceptions; Strategic Leadership; Instructional Leadership; Transformational Leadership; Integral Leadership; Shared Leadership; Leadership Styles; Values; Ideals; etc.
Personal Vision, Personal and Professional Development	Self-Expectations and Expectations By Others; Professionalising School Leaders; Developing One's Individual Professional Personality; Reflections on the Experiences from Everyday School Life; Developing Skills and Basic Values; Further Development of the Ability to Reflect, of Interpersonal Competences, of Pedagogical and Educational Competences; Emotional Intelligence; Time and Self-Management; Organising One's Tasks; Work Techniques; Cognitive Mapping; etc.
Acquiring Knowledge	Psychology; Methodology; Controlling; Law; Action Research; Organisational Theory; Qualitative and Quantitative Research Designs; Know-How of Information Technology; etc.
Tasks of school leaders	
Communication and Cooperation	Presentation Techniques; Designing Conferences; Conducting Conversations; Rhetoric; Solving Problems and Conflicts; Conflict Management; Dealing with Conflicts and Micropolitics; Delegating Responsibility; Communication and Leadership; Professional Dialogues; Conference Techniques – Effective Meetings; Decision-Making; Group Processes; Creating Structures of Relationship and Communication in the Context of Successful Learning; Communication and Perception; Working in (Regional/National) Networks; Cooperation with Authorities in Administration and Politics; Cooperation with Associations; etc.
Organisation Development	Models of Pedagogic Organisations; School as a Learning Organisation; Organisational Theory; Organisational Culture; Organisational Psychology; Working in the Learning Organisation; Quality and Quality Development of Schools; Quality Management; Instructional Leadership; Integral Leadership; Transformational Leadership; Introducing Innovations; Creating a Role Model of School and its Practical Implementation; Vision of School; Pedagogic Development of School; Managing School Programs; Initiate Change – Organise Change; School Improvement Projects; Project Management; Organisational Learning; School Development and School Culture; The Concept of School Culture; Systemic Thinking for the Benefit of Successful Learning; Corporate-Identity-Models; TQM-Techniques; etc.
Staff Development	Leading and Developing Staff; Recruiting, Selecting and Leading New Teachers; Human Resource Management; Employment of Staff; Teamwork and Team Development; Training and Professionalisation of the Faculty; Staff Development and Supervision of Teachers; Counselling; Counselling and Supporting the Faculty; Dealing with Difficult Staff Situations (Mobbing, Alcohol Abuse etc.) ; etc.

Classroom Development (Work with and for Pupils)	Curriculum and Lessons; Structuring Learning Topics; New Curriculum – the Role of the School Leader; Developing the Syllabus; Creating a Favourable Learning Environment; Individualised Promotion of Learning Processes; Multicultural Education; Providing Lessons for Foreign Pupils; School Leadership and Promoting Special-Needs Children; Designing a Selection of Subjects in the Context of Successful Learning; New Forms of School and Learning (e.g. Learner-Centered Lessons; Working Independently; Computer-Assisted Learning; Cooperative Learning; Project Work); Project Work Days; School Cafeteria and Extracurricular Activities; Social Work at Schools (Day Care, Taking Care of Pupils during Lunch or Afternoons; Mentors; Tutors; Internal Social Worker); Pupils' Board; Preventing Violence; Drug Prevention; Environmental Education; Disciplinary Measures; etc.
Teacher Evaluation	Evaluation in Practice; Monitoring Lessons-Analysis-Counselling-Evaluation; Supporting and Evaluating Lessons; Evaluation of Teacher Achievements; etc.
School Evaluation	Evaluation Methods; Internal and External Evaluation; Pupil Achievement and Assessment; Achievement and Assessment Standards; Standardised Test Procedures; Evaluation Practice; Evaluation and Rating in the Context of Promoting Successful Learning; Accountability towards the Parents, the Government, the Public; Action Research and Evaluation; Organisational Learning and Evaluation; Supervision and Evaluation; etc.
Administrative Tasks	School Management; (School) Administration; Administration Processes; Budgeting; Budget Tasks; Controlling Skills; Financing; Managing Human and Financial Resources; Concluding Contracts; Insurance; Managing Information Systems and Technologies; Managing Buildings and Organisations; Selected Problems of Managing Schools; etc.
School in the Context of (Educational) Policy and Society	Topical Issues Concerning (Educational) Policy; The School's Mission in a National Context; State Administration; the Consequences of Actions Taken by the Schools; School in the Context of Educational Policy, Society and International Affairs; Changes in Educational Policy and their Effects on School; Changes in Society – Changes in Schools: Developments and Challenges; International School Systems / International School Innovations; Topical Issues for Schools and Society; Guidelines Issued by the Ministry of Education; School Supervision; Models of Cooperation Between Supervisory Institutions and Schools; Models of School-Based Management; Keyword 'School Autonomy'; etc.
School and the Public	Designing School-External Relations; The Work of Parents and Boards; School and the Community; School Leadership and the Community; The Social Context; Cooperation with Local Institutions (Youth Welfare Office, Police, Counselling Centres etc.); Exchange Programs for Schools; Marketing and PR (Working with the Press, Representing Schools, Marketing Schools, Schools Online); etc.
Laws and regulations	
	Constitutional Background; School Laws; Regulations; Concluding Contracts; Labour Legislation; Salary Regulations; Budget Regulations; etc.

4.2. Discussion

Legal and Administrative Issues

The comparison shows that the state and its stipulations not surprisingly influence the selection of the contents of the programs particularly in those countries, where school leader development is organised and carried out centrally. In France, the modules of the preparatory training put an emphasis on themes and issues that help to make school leadership more supervisable by the government. Among them are administration, school law, management techniques, and budget issues. In Austria, too, the guidelines of the ministry of education comprise, among others, school law, employment law, budget law, and administration. South Tyrol also ranks school management and school law issues as essential modules. Also Hong Kong and Singapore's programs underscore law and administration themes. In the programs of these countries it becomes apparent that school leaders have to perform reliably and effectively according to the decisions and stipulations of the state's policy.

In the programs of the other countries included in the study, school law, if mentioned at all, ranks much lower (for example in Washington, New Jersey, and the Netherlands). Undoubtedly, no trustworthy school leadership worldwide can do completely without legal knowledge, especially since its range of responsibility is increasing. Decisions made at school level have to be legally correct as parents increasingly tend to sue the school for their true or supposed rights. Abstaining from legal knowledge would undoubtedly be inexcusable and an evidence of naivety. Nevertheless, legal issues in general are no longer a dominating content of the programs.

Similar to legal topics, issues of administration are also on the decrease. As administrative tasks however belong to the everyday work of school leaders and even have increased particularly in countries with more school-based management, these themes regain some importance through the back door, so to speak, particularly regarding administration of the budget and administration of personnel resources. Examples for this are the programs of the NSO (comprising themes such as personnel management, management of information systems and IT, buildings and equipment, financing and budgeting), and of New Zealand (comprising themes such as administration of resources and issues of school administration). The US programs (covering topics like administration of the budget, school management and so on), or the preparatory program of England (with themes like deployment of personnel and financial resources) round out the picture.

School-Based Leadership

The contents of the programs in nearly all countries show a distinctive orientation towards the new tasks of school leadership when school-based management was introduced. To a high degree, the present reality of school leadership is seen as the starting point and the point of reference for the choice of contents and of methods applied in the programs (see also chapter II.5.).

In most programs, 'leadership' is an explicit topic, more often than 'management'. Often the term is explicitly conceptualised, for example, as instructional leadership, transformational leadership, integral leadership, or leadership in a learning community. There are titles of modules such as leadership strategies, leadership according to a vision, developing a shared vision, shared leadership in teams, etc. Increasingly, as in the program of Danmarks Laererhojskole and that in New Zealand (both without any state guidelines), but also in the standards-oriented program of Washington, or the qualification model of Singapore, the role of values, ethics and morals, the question of power, and how to legitimate leadership in a democracy and leadership for social justice are explicitly made central themes.

In the focus of programs, there is the idea of an educational leader, whose central task is the development of the school in cooperation with all the stakeholders. School is no longer regarded as something static, but seen systemically as a learning organization, with its own specific rules and conditions, with a focus on education and teaching–learning-processes. Thus, leading a school no longer means only to maintain the status quo, but above all to develop this learning organization. School leaders should be the driving force for effective change processes. Consequently, the focus can no longer be on maintenance but has to be on development. Yet both have to be coordinated: what is worthwhile has to be maintained and should not be lost due to a blind actionism, and at the same time structures that can be improved, should be improved. Reasonable achievements, again, have to be maintained, have to be institutionalised. This paradigm shift from managing schools to leading and developing schools is mirrored in the lists of contents themes and sub-themes of a lot of development programs.

The program Leadership in Development by the independent provider Praxis in Denmark and Ontario's Principal's Qualification Program link educational leadership explicitly to 'change' and 'development', and regard the school leader as a first class change agent. Others, too, like South Tyrol emphasise school development and development planning as it has become crucial in the course of restructuring the school system. Even if change is not explicitly indicated, none of the programs does without the field of school development (the only exemption is the brief induction program in Hong Kong, but the second program in Hong Kong focuses explicitly on school development).

Further themes within this field are: development of an organization, staff management, evaluation, classroom development, and educational work with and for pupils.

Due to the clearly broadened range of tasks of school leaders in many countries, the assessment of staff, and the staff's employment and, if necessary, dismissal are part of the school leaders' duties. Consequently, these topics are dealt with in some of the programs as well (see, for example, England).

Another range of topics in the efforts to qualify school leaders according to the new steering of schools, particularly in more decentralised countries, is school and the public. In Denmark it is called leading the relations to the community. In Ontario there are themes like the school and its community. In New Zealand it is called school leadership and the community.

An explicit focus on PR-work in the sense of advertising for the school and of representing the school to the public astonishingly is mostly missing, even in countries whose education system is strongly based on market principles. Marketing and PR as a topic on its own is only mentioned in the program of the NSO.

In England, putting accountability of the school leader to the government and the public at local, regional, and national level on the development agenda, and making it an obligatory module, can also only be understood against the background of transferring responsibility to the individual school and its leadership.

Some programs explicitly include the function of the school within educational policy and society, and make current educational policy a theme (for example in New Jersey), or deal with the social mission of school, which is, for example, in Denmark and in Sweden understood as developing and maintaining democratic awareness.

Communication and Cooperation

In spite of the increasing strain on school leaders due to additional administrative tasks – particularly in countries with more devolved systems – school leader development has not become dominated by administrative topics. Nowhere is school leadership reduced to purely administrative or executive tasks. The overall focus of school leader development has shifted to contents like 'working with and for people'. It has been recognised that communication and cooperation play an essential role. Contents like communication and leadership, motivation, coordination, gaining the cooperation of all stakeholders are – in various forms and with different emphasis – part of all programs. That is mirrored in the prevailing module titles in the context of 'communication', such as moderation, leading conferences and meetings, leading conversation, problem- and conflict-solving, conflict management and micro-politics, empowerment, staff meetings, effective meetings, creating structures of relation and communication. In the context of 'cooperation', it is about creating a shared vision, a shared school program, group processes in general, shared leader-ship in the sense of shared decision-making and empowerment, team work, and team development.

Hence, internationally a paradigm shift can be stated. Instead of qualifying school leaders to administer their schools, they are qualified according to a new understanding of school leadership as experts in communication and cooperation. Besides the communicative shift, a cooperative shift has occurred.

> During the 90s there has been a paradigm shift towards communicative and cooperative competences of the school leader, and towards school development. [...] This 'communicative and cooperative shift' shows the effort to cope with the new roles and responsibilities of school leaders. As a result, the contents of the curriculum are characterised by the effort to combine necessary administrative aspects and legal topics on the one hand and aspects of internal and external communication and cooperation on the other hand. The emphasis is clearly given to the notion of 'leading by communicating and cooperating'. (Huber, 2001, S. 17)

Role and Self-Concept

School leader qualification seen against the background of a changing job profile can no longer start from a fixed role that the participants are trained to fulfil. It implies that the (aspiring) school leaders are motivated to reflect upon their own conceptions of the role of a school leader and to modify them if necessary. For this purpose, in some countries, the programs include components such as personal vision, personal and professional development, development of fundamental values, development of one's ability to reflect, time and self management, designing a mental picture of the organizational processes in the school and of one's leadership actions, mentally processing the experiences of everyday school life and of internship. Some programs put the individual school leader in the centre by focusing on his or her development of competences, and deal with contents that refer to the competences aimed at.

Theoretical Knowledge and Academic Orientation

The contents of the programs also comprise of theoretical knowledge. On the one hand these are to extend the participants' existing knowledge and to impart new knowledge. On the other hand offering a kind of background through theoretical input and through imparting academic knowledge, they should reflect on practice beyond their well-worn subjective everyday theories and automaticised personal interpretation patterns. Core questions in this context are: how much knowledge should be delivered? What should school leaders know? From which disciplines should the knowledge come from?

Considering the first question, an adequate knowledge base has to be guaranteed and an overload has to be prevented. For the latter many US programs were blamed in the past (see Murphy, 1992). To have an academic input only for the sake of gaining the status of a university program is not helpful at all. Sound academic value on the other hand is very useful in terms of professionalisation of school leadership. Pivotal in the discussion about the amount and the kind of theory and academic contents in the programs, however, has to be how far a link to workplace practice is successfully made. The participants should not have to make this theory-practice-link painstakingly entirely on their own after they have suffered from the well-known practice-shock. Friction due to transfer should be avoided for many reasons, also because otherwise very soon an aversion to all theoretical knowledge may emerge and much of the benefit that this knowledge may offer is lost. To solve this problem, however, is predominantly up to the use of appropriate methods.

It would be beyond this survey to completely follow the considerations of the individual providers, nor can it verify theoretical and academic foundations of the content offered. It only can clarify, which knowledge bases have been brought in, that is which explicitly appear in the programs.

We find, besides education:

- Psychology (psychology of learning, social psychology, psychology of organisations),
- Sociology (sociology of organisations),
- Communication,

- Administration,
- Jurisprudence,
- Economic science,
- Political sciences,
- Statistics,
- Philosophy and ethics.

Huber and West (2002) state that school leaders themselves seem to show a strong preference for what they describe as 'practical training', and that 'theory' is not always seen as valuable to the practitioner. However, we also see (West et al., 2000) that school leaders find it much easier to generalise from their experience and repeat effective behaviours when they have a conceptual framework underpinning the decisions they are making. Theory and practice need one another, and need to be developed together.

5. Methods: Learning in Lectures, Seminars, Workshops, from Colleagues, and in the Workplace

For imparting the contents there are manifold teaching strategies and learning methods applied in all countries. It is important to note beforehand that the methodical implementation of development programs has to take macro-didactic and micro-didactic considerations into account. Among the macro-didactic elements are the organisational conditions such as the provider (discussed above) or the pattern (e.g. number of training days, scheduling etc., discussed below). Such conceptual considerations are didactically extremely important when designing and conducting development programs. The focus of the following chapter, however, is the methods applied relevant from a micro-didactic perspective.

5.1. Comparison

Analysing and comparing the methods used, it is worthwhile looking at the variety of types of program events, the different ways of learning of the participants, and the study material available to them.

Types of Program Events

How many participants are on one single course? And what types of events do the programs comprise? The range of program events includes:

- plenary sessions for up to 150 participants (as, for example, in South Tyrol, where all the participants of the Führungskräfteschulung come together),
- lectures, plenary discussions, and so on for 20 to 40 participants (this number appears frequently since it is probably the maximum size and very efficient economically),
- small group events for 5 to 15 people to work cooperatively and intensively on certain topics,

- small team events for supervisions, coaching, learning from colleagues, etc.,
- one-to-one counselling, supervision, or coaching,
- events for teams comprising the school leader and one or several members of the school leadership team or staff, for groups of teachers, parents and pupils of a particular school, or even comprising representatives of the community.

Besides the lectures, seminars, and the team meetings, there are hospitations and internships at schools and non-educational institutions (such as in France, England, in the USA or Singapore). Very rare are international study trips.

Hence, the comparison indicates that a broad variety of events are chosen. In order to come to an evaluation or assessment, participating observations as well as interviews with the participants would be necessary. Therefore, judgements are not possible here.

Ways of Learning

As far as the choice of concrete methods is concerned, we also find a broad variety. Obviously, there is no single best method. The methods coexist and do not exclude each other. Among them are cognitive-theoretical ways of teaching (such as lectures), which are to impart information and knowledge, cooperative and communicative and process-oriented methods (group activities of different kinds), and also interactive methods like role playing, which strongly aims at the development of application competences. Internships, shadowing, mentoring, or coaching as practice-oriented methods also aim at changing thinking patterns and patterns of behaviour. Particularly interesting are cognitive-reflective methods aimed at changing mental models, patterns of thinking and patterns of interpretation.

The following teaching and learning methods can be found in almost all of the countries:

- lectures, introducing new themes or illustrating an overall coherence, introducing a particular theme with a theoretical overview or presenting a special theme,
- input lectures, offering stimuli or introducing a topic in a brief summarising fashion,
- plenary discussions after lectures giving participants the chance to ask questions and to raise issues,
- group discussions and seminar discussions with strongly process-oriented components, for elaborating questions for a plenary discussion or extending the discussion of a plenary,
- various forms of group activities, to work on particular contents and to elaborate problems taken from the everyday life of the participants,
- role playing or simulation exercises, sometimes also change games, allowing for experimenting on different ways of behaviour.

Moreover, methods are applied that may be particularly instructive. Among them are combinations of traditional ways of learning such as literature studies and discussion groups. In the meantime these are also organised in web-based learning communities via the Internet, which are established by the training providers and supported by experienced tutors. Besides, the personal and the professional development can be fostered through working on academic assignments, documentation on projects, and, maybe even more effectively, through reflective writing (for example in Ontario or Sweden), and through learning journals (for example in the Netherlands, the USA, and Ontario). New learning experiences are to be reflected and made fruitful against the background of one's own biography and the theoretical framework, which perhaps has been gained in the development program. A very innovative method, such as cognitive mapping, attempts at enabling new perspectives through a distinct process and reflection orientation.

Certainly attractive are excursions abroad, for example, to particularly innovative schools, which are offered in a few countries. Such international study trips built into a part of the program of the University of Waikato. In Germany they can be found, too.

Interesting – and certainly relevant particularly against the backdrop of the infrastructure of a large country such as Australia – are the opportunities to exchange leadership positions, which are integrated in the development opportunities in New South Wales. School leaders of different schools (city schools, rural schools, small schools, large schools) exchange their positions for a limited time span and thus can make valuable experiences. In some countries, a change of school leader positions is expected at regular intervals, as is the case in Ontario, for example, where the school leaders are assigned according to their strengths and weaknesses as well as to the needs of the individual schools. The change of positions becomes part of their professional development, as their competences grow.

In addition to the methods of learning indicated above, several other experience- and participant-oriented methods are employed. These include problem-based learning, projects at one's own or a different school, internships, shadowing, coaching, and mentoring. These will be elaborated below. The programs investigated, however, differ greatly in their reliance on these possibilities of learning-in-practice.

As Huber and West (2002) show, the training provision can be conceptualised as being spread across two continua of course-based and experience-based learning opportunities. Hence, it is possible to distribute the programs worldwide according to the relative emphasis given to these two strategies (see table below).

However, one has to be aware that by grouping the programs according to their relative emphasis on experiential versus course-based learning, one risks simplification. For example, it is not taken into account whether the offers are made to teachers aspiring to leadership or to school leaders already experienced in their role. Again, the different emphasis could be viewed in reference to the total amount or length of training available; since offering experiential learning opportunities inevitably means expanding the program accordingly. Nevertheless, mapping the emphasis of training methods is useful, since it tells us something about the ways both the content and processes of school leadership are conceptualised locally.

Table 7: Emphasis of learning opportunities within school leadership training and development programs

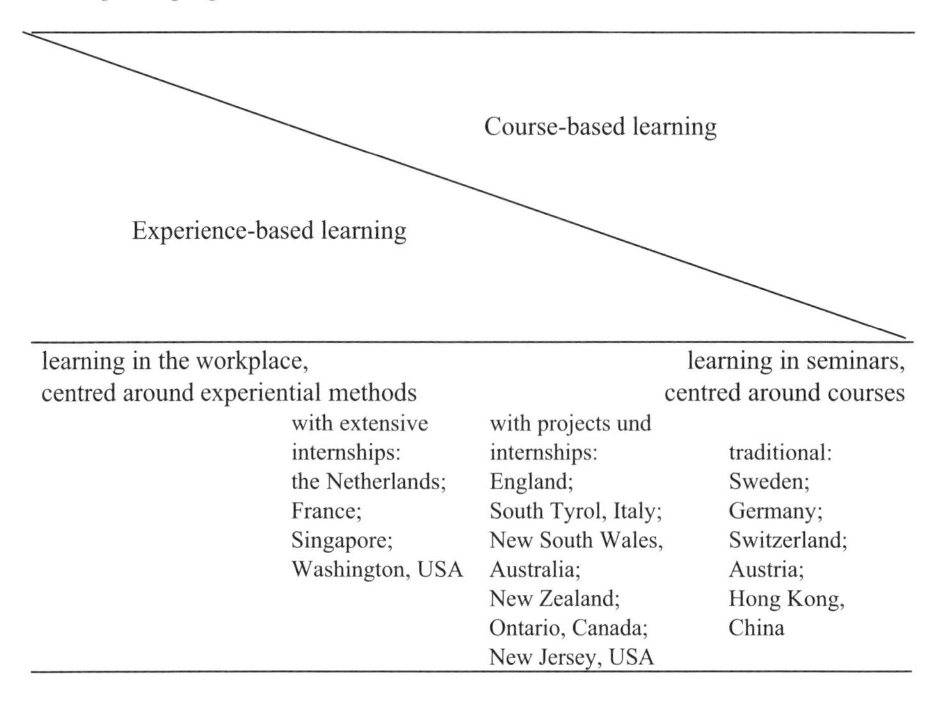

with extensive internships:	with projects und internships:	traditional:
the Netherlands; France; Singapore; Washington, USA	England; South Tyrol, Italy; New South Wales, Australia; New Zealand; Ontario, Canada; New Jersey, USA	Sweden; Germany; Switzerland; Austria; Hong Kong, China

Project work and/or internships are included, for example, in the National Professional Qualification for Headship in England, in the Management- en Organisatieopleidingen of the Nederlandse School voor Onderwijsmanagement, in the Master program in Educational Leadership at the William Paterson University of New Jersey, in the Principal's Qualification Program in Ontario, and particularly extensively in the central program in France, in the Diploma in Educational Administration in Singapore, and in the Danforth Educational Leadership Program at the University of Washington. However, countries which still favour more or less an approach to leadership development, which is centred around courses, also indicate that certain modifications are under consideration.

Hence, it is obvious that development programs in many countries move from learning solely centred around courses towards learning centred around experiential methods. Increasingly, programs are centred around experiential methods.

Moreover, the didactically useful application and the quality of the implementation of these methods have not been controlled. Thus, nothing can be said about the concrete realisation of the development and training, or about the emphasis of the individual methods within the respective programs. The competence of the provider, or of the team of trainers, seems to be decisive, which means that there is quite a broad range of quality in the training and development.

Study Material

As far as the study material is concerned, the statements in the questionnaires are quite brief and scarce. The material used consists of academic literature, individually compiled texts, collections of texts, school laws and administrative stipulations, curricula, school programs from various schools, inspection minutes, evaluation reports, documentations of school development projects (case studies), descriptions of problems, video documentations of school programs, TV records about topics linked to educational policy and to school development, audio records of interviews with experienced school leaders, and the like. There is much effort for a practise-relevance and for a certain relevance to the current situation.

5.2. Discussion

Considering Adult Learning Principles

Generally, there is consensus among providers and participants that the methods applied have to take the learning needs and the abilities of adults into account. Hence, it is important to consider basic andragogic principles. We will summarise such fundamental characteristics of adult learning briefly (for a further account see, for example, Kidd, 1975; Knowles, 1980; Corrigan, 1980; Blum & Butler, 1989; Siebert, 1996; Harteis et al., 2000; Gruber, 2000; Mandl & Gerstenmaier, 2000; and others).

Characteristics of adult learning
From a constructivist point of view, learning is a self-referential, reflective process: experience is built on previous experiences, new knowledge emerges out of existing knowledge. This previous knowledge is necessary on the one hand, to be able to interpret new information, on the other hand it is changed and modified by new information itself. The previous knowledge is embedded and maintained in the newly constructed knowledge (see Ebner, 2000). To a high degree adults bring their personal and professional experiences, their knowledge and their own way of seeing themselves to bear in the learning process. While among children learning something new prevails, the learning biographies of adults have the effect that their own learning is above all follow-up learning (see Siebert, 1996). Their individual experiences always have a subliminal influence on the new information and at the same time are the foundations on which something new can be learned. Themes that cannot be linked to previously existing cognitive systems are very much up in the air, so to speak, and mostly are quickly forgotten (ibid.). Hence it is preferable to refer the new information explicitly to the experiences and anchor them there. The reality and the experiences of the participants, their needs and problems, should be the starting point and the point of reference for the selection of contents and of methods applied.

For adults, there is no reduction of the ability to learn, however, there may be a certain decrease in the learning speed. The learning capacities of adults should therefore not be underestimated, but they should be provided with sufficient time, especially if their last intensive training phase dates back a long time and certain learning and training techniques have not been applied in their working repertory for quite some time.

The everyday life of adult learners usually takes place outside of the learning context, both in terms of its place and its organisation. The workplace, however, is their source of motivation for learning. The subjective meaningfulness of the content of what is learned decisively influences the achievement of adult learners and the transfer of something newly learned into professional action (see Gruber, 2000). This holds true, above all, when their everyday life surrounding is changing, which releases a great potential for motivation. Adults expect that the contents correspond to the changes that they have experienced, and that these are referred to, that the knowledge and understanding gained is a tool that can be applied in the specific and extremely complex work situation, with as little loss as possible due to the transfer. Inert knowledge (see Whitehead, 1929; Bransford et al., 1991; Renkl, 1996) that cannot be adequately made use of in real work life situations should not be produced.

Moreover, adult learners select what they learn, they filter information, consciously or subconsciously. Thereby they proceed much more problem-oriented than theme-centred and the effects of learning are more sustainable when there is the possibility to apply in practice what they have learned. According to Gruber (2000), gaining experience for professional competences means learning in complex application-relevant and practice-relevant situations (see also Joyce & Showers, 1980). New competences are mostly gained by practice followed by feedback and reflection. However, sufficient theoretical foundations should be imparted as well so that a reflection of practice beyond the well-worn subjective everyday life theories can take place.

All this undoubtedly has concrete consequences for the teaching strategies and learning methods applied, there is no single top-priority strategy or method, but it is obvious that a big spectrum of different strategies and methods is most successful. It is advisable to choose a variety of methods that help the individual learners to accept new information, not only on the cognitive level, but to motivate them to call outdated patterns of thinking, patterns of interpretation, and mental maps into question, and maybe to give up well-worn patterns of behaviour, which might have been useful when mastering some previous task, but would be an obstacle when acting in the new areas of responsibility (see Antal, 1997).

Case-based learning (see Kolodner, 1983, 1997) plays a particularly important role in this context. When real cases are used as study material, it is possible to represent knowledge of objects, situations, events and actions in the form of schemes which are dynamic in terms of being adapted in a way specific to a situation, and hence dependent on experience, and are constructed according to the aims, intentions, and plans of the learners (see Gruber, 2000). Experiential knowledge builds up. In this episodic experiential knowledge, on the one hand general episodes are represented, which have been generalised by the learner from a number of experienced single episodes. On the other hand, divergences and differences of particular single episodes from the general episodes are represented as well. By the flexible way of representation, declarative as well as procedural and conditional knowledge is created[2].

Adults should be taken seriously as self-confident, independent partners in the learning process, as subjects, not as objects in this process. It is most suitable if they can function as self-reliant, self-motivated, active partners of the learning processes and take over responsibility for their learning themselves. Thereby, a non-threatening learning environment characterised by mutual confidence and mutual respect is supportive. It is decisive that the learning goals to be achieved cover personal, self-defined aims as much as possible.

In fact, in the programs there is a clear tendency towards experience-oriented and application-oriented methods instead of theoretical methods, and the methodology of learning and processing of information are themes as well. There is a shift of emphasis in school leader development away from theory-oriented learning towards practice-oriented development and imparting of competences. This can be seen in the application of chances for profound group experiences and of bringing in practical work experiences from the schools. The cognition- and reflection-oriented methods (such as learning journals, discussion groups, working with mental images) too, which are coming into fashion in school leader development, are linked to authentic experiences in the school reality.

Concrete consequences for the seminars are drawn from the insight that learning processes that are made together with other participants and that come from experiences offer greater chances to be intensive. From a constructivist point of view, the learner constructs knowledge by linking new information to previous knowledge. For this to occur, firstly old knowledge has to be made conscious by reflection. This reflection process can be fostered, by thinking, but also by a discourse with other people (see Vygotsky, 1978). When this process occurs in an exchange, the big advantage is that the dialogue can help to eliminate uncertainties, to bring clarity, to take up alternative interpretations, and experiences and insights from the other participants, to have a critical look at them and to combine them with one's own knowledge. During such a communicatively moderated acquisition of knowledge, that is a kind of co-construction of knowledge (see Roschelle, 1992), learning becomes a kind of bi-product of communicative interaction (see Henninger and Mandl, 2000). It is essential for the intensity of the learning processes, however, that this interaction occurs consciously and reflectively.

Additionally, the participants get the chance to make concrete experiences in cooperating in a team and in leading a team, which is valuable for any leadership task later on. How should the demanding interpersonal competences, which are hard to impart (such as creating and maintaining relationships, shared decision-making processes, negotiating, mediating, moderating conferences and professional dialogues of every kind, dealing with conflicts, solving of conflicts) be effectively developed and trained, if not by the participants' exercising those tasks and getting feedback during the program as well.

---

[2] See also the 'Cognitive Apprenticeship' approach by Collins et al. (1989) with its principles of 'authenticity' and 'reflection'. This approach prefers 'situated learning', that is learning in discourse with others, using examples given by authentic situations. Acquiring and applying knowledge basically coincide here. Transferring knowledge to future situations where it is needed is thereby facilitated tremendously.

Hence, in seminars methods should also be used that enable process-oriented experiences. Here a combination of methods seems to be more effective than the isolated use of a single method (see Seljelid, 1982). Lecture-oriented methods alone do not enable group experiences or social learning, and process-oriented methods alone often lack theoretical clarity and thoroughness. In concrete terms, a training and development event may be structured like this: after introducing a topic (such as moderation of conferences, decision processes, working with teams) by a short lecture giving an impulse, the participants themselves are involved in activities, for example, in group discussions or various other forms of exercises giving them the opportunity for profound critical thinking while going through the learning process. Then, experiences and insights should be integrated in a more abstract conceptualisation. For this, a theoretical reflection and discourse might be helpful, and further information should be provided.

Ideas like these partly go back to Kolb and his model of experiential learning (see Kolb et al. 1971; Kolb, 1984). Kolb defines four stages in a learning process: concrete experience, reflective observation, abstract conceptualisation, and active experimenting. Consequently, it would be useful to begin a theme, or a learning unit, by having a participant describe a concrete experience. This means that the learning process is not started by traditional methods, such as literature study or a lecture, but by the individual personal experience of a participant. This experience will then be the theme of the following analysis and discussion so that deep acquisition and conceptualisation are possible, which lead to conclusions and generalisations, which again can be applied to new experiences.

Participant, experience, and needs orientation

Adult learning must take the participants seriously as partners, and must take their experiences and needs into account. This basic attitude, which can also be called 'client orientation', however, does not cover everything that is meant by 'participant orientation'. Beyond this marketing principle, participant orientation means that the participants take an active part in their qualification program. It is about the role that they play as far as the trainers, the provider, the given framework and so on are concerned, it is about which decisions they can make and they should make. The aim of participant-oriented didactics is to find a balance between the learning demands and the participants' experiences. This balance has to take into account the relation of content-logic and psycho-logic as well as the relation of teaching behaviour and learning behaviour, but also the relation of participants' motivation and the setting of the event. In order to achieve this balance, there is the need of not only establishing an anticipation of the participants' motivation and knowledge (by the trainers, instructors, facilitators etc.), but also of enabling active participant-participation, understood as didactic-methodical co-determination by participants.

This participation, however, is harder to achieve than may be expected at first sight, respectively than the educational reform efforts of the 1960s and early 1970s might make one believe. On the one hand, expecting from participants, who are about to learn something new, that they co-decide immediately upon contents and methods of learning, is unrealistic: how should they always know what and how to learn? They should, however, be able to reflect on their previous knowledge and their learning needs. On the other hand, some participants may not want to decide by themselves, but prefer leaving the decisions to the trainers and have new knowledge presented to them. Siebert (1996) regards this as a 'didactical dilemma': shall an educationally based participant orientation be realised even against the wishes of the participants themselves? Moreover, the needs and demands of the individual participants may differ greatly and even the needs of a single individual may differ according to the circumstances. Doerry et al. (1981) are in favour of a concept of 'distributed responsibility', which is split among the participants and the trainers.

From a constructivist perspective, participant orientation means to recognise the different constructions of reality, to formulate them, to take them seriously, to accept differences, to make the individuals aware of them, to acknowledge that people in different situations regard different solutions as feasible, and to do this without obliging answers for all, but to allow for variety.

In connection with participant orientation the term experience orientation is used. Sometimes the terms are used interchangeably, as the learning of adults mainly comprises of linking new knowledge to previous knowledge out of experience. At the end of the 1960s the term was often connected with an emancipatory impetus (see Negt, 1968). In the meantime this has been looked at more sceptically, the value of experiences has been seen in relative terms, the ambivalence of experience orientation has been realised. Experiences may be important as stabilizing factors. Being conservative and reaffirming traditional ways of thinking, however, they may also be an obstacle to learning, barring new approaches of problem solving. Nowadays, a fine distinction is made between experiences that happen to people without further influencing knowledge and such experiences that lead to increased knowledge. If the communication within a group remains a mere exchange of anecdotes, this may be emotionally relieving, new learning experience, however, will not be achieved this way, rather it results in an affirmation of old problem-solving strategies and interpretation patterns. Here, it is crucial that the trainers do not leave the learning cohorts to themselves, but open up new opportunities of experiences to them, for example, by a suitable input taken from academic research. The danger of merely sticking to one's own status quo of knowledge has to be avoided.

Frequently Used Teaching Strategies and Methods of Learning

Lectures and plenary sessions

Lectures certainly represent a time-efficient method of adult learning for imparting information and knowledge. Plenary sessions after the lectures often seem of little use to the participants, the bigger the number of participants, the less they are favoured as in such plenary events only a limited number of participants can actively take part at all. Their use can only be for further clarification of what has been talked about in the lecture. In an English-speaking context, plenary events are often compared with staff meetings: "They should be short and only important issues should be discussed", is one of the demands.

It is seen as important that lectures are made as vivid and stimulating as possible, even provocative if the topic allows for that, and that the participants can go on working on the topic in group discussions or workshops after the lecture. Vice-versa a lecture can also be prepared for by a discussion of a case example or the elaboration of an individual problem. Hence, lectures are often used at the end of a content unit in order to clarify, sum up, and accentuate the theme. This, however, requires a very high quality level of the lecture, a high degree of abstractness, and a good deal of competence of the speaker, so that the lecture clearly goes beyond a mere repetition of what has been elaborated before, and succeeds in getting through to the heart of the essential aspects introducing the complexity without risking too much simplification. An audience of teachers and school leaders (all of them experts in teaching) is often said to be particularly critical. This implies high demands on the speakers, facilitators, instructors, trainers, and so on.

Reflective writing

Quite a number of programs have integrated reflection-oriented methods. Developing reflection-competence is closely linked to enlarging action-competence, both competences are interdependent. Moreover, linking new information to previous experiences and integrating it into existing knowledge improves the learning process of adults, as their own learning is above all follow-up learning. Methods such as writing learning journals and reflective texts on particular topics taken from everyday work at school, but also project reports and written assignments are instructive examples.

Group work

Group work fulfils an important function in the programs. It can precede plenary discussions or be the platform for exchange for the participants, for example, for those conducting similar school projects during their internships. This method is widely known and often favoured by the participants. A very important factor here is the size of the group. The smaller the group, the more it offers a non-threatening platform for statements by individuals about their experiences and views. If it is too small in number, too few contributions can be made, or it means too much work for the individual. The 'optimum' size seems to be between five and ten participants, mostly there are seven. Groups of this size develop an individual atmosphere, a special team climate, and this often results in a kind of team spirit.

Besides the size, the combination of these groups is important and should be organised very carefully. In many countries teams are favoured whose members represent a broad spectrum of schools (type, size, context, etc.). However, sometimes it may be equally important to build homogeneous groups, or, in case not only school leaders take part, differentiating according to the different backgrounds of the participants (deputy school leaders, members of the school leadership team, teachers in other leadership positions, and so on). Sometimes, heterogeneous groups lack a shared understanding (managing a big school versus managing a small school, rural or urban, etc.)

Group work is particularly favoured by already established and experienced school leaders as these normally do not have the chance for continuing exchange with peers due to a certain isolation in everyday life. The security of a supporting and understanding group makes it possible to communicate uncertainties and fears, and enables shared reflection on problems. Professional-collegial relationships that have started here, often are maintained beyond the qualification program and enable further meetings or mutual visits. This is an opportunity for long-term mutual professional support (see also below the section about learning from colleagues).

Role playing and simulation exercises

Role playing and simulation exercises have the advantage of allowing the participants to approach concrete exemplarily selected problems in less complex circumstances than they would be required to do in the workplace reality. They can then be interrupted for the sake of reflection, which would hardly be possible during real situations. However, it must be stated, that this method is applied more rarely than might be supposed. Reasons for this could not be found in the questionnaire responses. It might be explained through a partial refusal of role playing by participants[3], through the high demand on preparation or the difficulty to find suitable problems[4]. Here, particularly, high competences of the trainers who prepare and supervise the role play are required, above all, psychological knowledge, empathy, and understanding for interpersonal interaction.

School Leaders Learn from their Colleagues – the Foundation of Learning Communities

Being the subjects in their own learning process, the participants can and should learn from one another. Such peer learning may be designed quite differently. Using peer assisted learning or critical friendships, the participants counsel each other; knowledge is generated through their dialogue that in this context-rich form could not have been provided otherwise. Thus, full use is made of the self-learning potential of each participant. The trainer only gives an input or feedback and at most offers a pattern of problem solving strategy, but no concrete directives or solutions. This, for example, is aimed at, too, in the cognitive mapping method of the program Meesters in Leidingsseven, but also in the Danish Leadership in Development program, and partly in the program of York University.

Another chance to learn from experienced colleagues is offered by methods like shadowing or mentoring. Shadowing means to accompany systematically an experienced school leader in her or his everyday work. The observations made by this are then reflected cooperatively. Mentoring means that the participants go beyond the role of observers and get into acting (either during a project at their own schools or when carrying out leadership tasks independently in their internship schools, or, in case they already are school leaders, in the course of everyday working life). The mentor observes the participant, and together they reflect upon the actions taken and decisions made.

A great deal of independent work can be found in the networks that are explicitly aimed at in some programs, which are initiated during the program and are meant to be maintained beyond it, for example, in South Tyrol, Denmark, the USA, New South Wales, etc.

---

[3] A rather negative attitude towards role plays has been confirmed by a survey of 300 teachers in three German federal states on the topic of teachers' continuous professional development needs (see Huber, 2001b), which was carried out for the Bertelsmann Foundation.

[4] The instruction for simulation exercises needs to be clear and motivating. When applied, the simulation is often prepared by a team, which combines different media (video, film, slides, role play). Moreover, the instructions have to be authentic by all means since (aspiring) school leaders – as already indicated – seem to have a sceptical attitude towards role play.

Particularly the US programs put an emphasis on establishing cohorts of participants, which remain unchanged during the entire course. Considerations like this are, of course, of no relevance in highly modularised programs, whose individual modules are not taken part in by all candidates (see chapter II.6.). Even if there is such a variety of individual training and development events to choose from, it seems important to encourage establishing cooperative learning teams or even to provide adequate structural conditions to help the participants establish those teams.

Learning cohorts offer a broad range of advantages. In addition to the support that members of learning cohorts can give each other during the qualification program, the development of professional friendships can be of much help after the program as well. These learning cohorts can directly be modified to a network of professional contacts. The members of such informal networks can support one another and exchange experiences. Here, the qualification shall achieve what Krainz-Dürr (2000) observes when reflecting on continuous professional development for teachers: it is to bring existing approaches and ideas into a professional discourse. Since establishing networks at first means investing in relationships, the learning cohorts are a good foundation for the social management required, because they give the opportunity of shared work experiences.

The positive impact of networks has also been described by Lieberman and McLaughlin (1992): networks prevent individuals from feeling isolated, they offer an opportunity for getting involved in a professional dialogue, and the shared foundation of values and aims guarantees a basis of mutual trust and psychological confidence. Individual differences and differences in context assure vividness and mutual stimulation. This combination of stimulation and challenge on the one hand and a secure team on the other hand is a good basis for personal growth and for the development of an organization.

This cooperation is made easier by the new communications technology, which has also become an intensively used part of various programs, for example, in moderated online groups, e-mail correspondence with the mentor or trainer, and interactive projects of small teams (among others in the program of York University in Ontario or the training program in South Tyrol, in New Zealand, and in the USA, but also in some approaches in Germany.

Learning in the Workshop: Problem-Based Learning

When thinking about the best possible way to guarantee participant orientation, practice-relevance, an easier transfer to the reality of work, and also about the possibility to counteract the phenomena of inert knowledge (see Whitehead, 1929; Bransford et al., 1991; Renkl, 1996), there basically are different but not mutually exclusive approaches.

In many countries programs use workshops, and confront participants with modelled situations of school leadership work life and carefully constructed cases, and involve them in a cooperative problem-solving process, such as the case method or, even more consequently, the problem-based learning (PBL) approach, for example at the University of Washington.

PBL has been applied for many years in management training by companies or in training medical staff. In the educational sector, especially concerning the qualification of educational leaders from school boards and schools, PBL has been developed predominantly by Bridges and Hallinger at Vanderbilt University in the USA (see Bridges, 1992; Bridges & Hallinger, 1995, 1997). The PBL-approach uses concrete and complex problems, as they are experienced in everyday practice by school leaders, as a starting point in order to involve the learners in a cooperative problem-solving process and to find solutions interactively.

In the setting of a development program using the PBL-approach, a problem serves as an introduction to a project. Thereby, the participants receive a written case description (see an example in Bridges, 1992, pp. 144-159). Alternative presentation forms could include a video film, a computer simulation, a scene from a role play, etc. Additionally, the participants can be provided with ample material. In the cited example, these are statistical information on the school district, a brochure the school uses to inform the public, a description of the planned center for new, Spanish-speaking citizens, a list of references on topics such as bilingual education, material about teaching foreign languages, the situation of Spanish-speaking immigrants, the legal basis, a guideline for designing and presenting conferences, etc.

The task in the presented case consists of assembling the relevant information for the advisory board as it is indicated in the case description. This information package should incorporate an indication of competence, a formulation of goals to be achieved, and a suggestion for a task schedule with concrete individual steps. Additionally, an agenda for the first meeting and a information paper with all the necessary background facts for the board members have to be included.

The participants work in project groups, which have sole responsibility and decide independently on how the knowledge from the suggested literature and other resources are to be employed. Each team consists of three to twelve members. The group seize should be in accordance with the complexity; to make sure that the distribution of the assignments make sense for the individual. Smaller groups of about five persons seem to prove well – especially in the opinion of the participants.

One of the team members functions as the team speaker. She or he coordinates the group process and the various single tasks. The others assume certain alternating tasks of presentation and documentation, or acquire and process information for the next team meeting. Documentation material of the meetings is compiled regularly. The instructors keep a low profile during these meetings and confine themselves to occasionally giving impulses and suggestions, answering questions and provide the participants with feedback. They do not, however, intervene when the team decides to strike a wrong or unfavourable path since mistakes are considered to be learning opportunities and often deliver insights into the problem, the problem solving process, the group function, and the ways of thinking of the participants. Certainly, the instructors can also play the role of a facilitator, who can give extensive information when necessary. The range of variation can extend as far as cognitive apprenticeship, where an expert gives a model solution to the problem and, thereby, explains the partial stages. Here, the expert must have reflected on her or his solution, and the method of solving the problem has to be made clear to the participants.

The participants usually organise their work independently. At first they have to agree on ambiguous terms in the description of their case. They have to debate on the information they need for solving the problem, on who is in charge of the different tasks and in which form the information has to be processed. They have to agree on systematics and act out different hypotheses, problem solving strategies and perspectives. After an exchange of arguments, solutions have to be made (here especially the dynamic development of the group in terms of communication and cooperation is interesting to be looked at).

Upon finishing this project, the team will present their result to a group of experienced school leaders and individuals who are involved with similar problems or tasks in their daily work. Here, unanswered questions can be discussed and the participants receive concrete, case-related, and general feedback concerning the solution and the problem-solving process. This forms the basis for a formative evaluation. Each participant receives feedback by the members of the project group, the instructor, as well as by experienced school leaders and others who are familiar with the topic of the project. To ensure a better transfer of the newly acquired knowledge, the participants create a summarizing essay on their own learning experiences. This is achieved by reflecting on the contents and the potential transfer of knowledge to their future role as a school leader.

The particular feature of PBL is that the usual separation of learning on the one hand and application on the other is revoked and particular disciplines, which often are isolated from each other, are brought together. The idea is to link learning to real situations. It is no longer about isolated knowledge of facts, but about combining specific knowledge bases of different subjects to a real context.

Rather traditionally oriented development programs may still be based on the idea that teaching means imparting knowledge in terms of fixed contents and that learning means acquiring these fixed knowledge bases. They assume that knowledge shall be relevant for the future function and that the learners will later on realise when and how they can transfer what they have learned into their work. Application is considered to be a relatively easy and linear process. It is of no importance in which context the knowledge once has been acquired.

The PBL-approach, on the contrary, considers learning as creating knowledge by the participants themselves. Here, learning is cooperative, interactive, and participative. More consistently than the case studies and simulations often applied in qualification programs, the PBL approach starts with real-life experiences and then looks for supportive knowledge as a tool and takes it from various academic fields, from experiences of practitioners, from the political context of the respective school board, and from the previous knowledge of the participants. The slogan here is: "First the problem, then the content." (Bridges and Hallinger, 1995, p. 8) Everyday life problems are seen as stimulus for learning and as offering a learning situation as well as learning material and not only as the future context to apply what has previously been learned.

The PBL approach distinguishes itself in some central aspects from the many variants of 'case methods' adopted in seminars: Usually in the 'case method' theoretical material is given, the 'case' then serves as the field of application for this theoretical knowledge. In PBL the theoretical material only has auxiliary function. While the instructor in the 'case method' administers the work much more, in PBL the team structures and accomplishes its project work independently, whereas the instructor keeps in the background. Furthermore, the kind of assessment varies: PBL goes beyond the usual written analysis of the case and the likewise presentation of solutions in written form, by presenting the result of the project work to experienced experts in order to get feedback of its feasibility and efficiency.

Special attention must be drawn to the experience of working as a team, getting know-how by learning through interaction in a group and, therefore, attaining solutions in an interactive-participatory way. This especially seems to be important for leaders in school systems, where the single school already has a high degree of autonomy and the range of responsibilities of the school leader has already expanded. This leads to a stronger distribution of leadership tasks and increases the need for cooperation. PBL is meant to introduce this new way of problem solving. In addition, working in a group as in PBL offers a kind of shelter, where experiments can take place.

The supporters of the PBL approach formulate the following premises (see, for example, Bridges and Hallinger, 1995):

- PBL is based upon the assumption that learning always implies knowledge as well as action. Both, acquiring knowledge and developing competences, has the same relevance for the process of learning itself as for the results of learning.
- Every learner brings valuable knowledge to bear in the learning process.
- Learning is the more successful the better certain preconditions are given. Among them are that the previous knowledge of the participants is activated and they are encouraged to integrate new knowledge in this previous knowledge. Besides, there should be a rather broad offer of opportunities to apply this knowledge in practice. Important hereby is that the new knowledge gets imbedded in a context that is as similar as possible to the context of application later on. This context of application will contain signals or stimuli that motivate to retrieve and apply what has been learned in the training situation.
- The assessment should be based on the performance of tasks that are as closely in accordance with the reality of school leadership as possible.

The following aims should be involved in a PBL-program:

- to assist the participant in acquiring knowledge, by giving them the opportunity to practise their knowledge in handling real, practical problems,
- to train the ability of application,
- to train problem-solving strategies, as future problems cannot be foreseen regarding their content,
- to support the development of leadership competences, so to speak to inspire the ability of teamwork, the forming and leading of teams, of solving problems and finding consensus, of planning and organization of projects,
- to support the development of a number of affective abilities, such as commitment for teamwork, patience, tolerance towards frustration, and self-confidence,
- to initiate the development of competences for self-determined learning, to critically realise one's own deficits in knowledge and to repeatedly make accessible suitable resources of different kind.

When developing a program based on the problem-based learning approach, a number of closely linked components have to be taken into account (see Bridges,

1992; Bridges & Hallinger, 1995; Gräsel, 1997; Gräsel & Mandl, 1999; Renkl, 1996, 2001):

- Learners have to be confronted with a problem.
- The problem has to be real and authentic (it should be in accordance with the context of application).
- The problem has to be of relevance for the learner.
- The problem has to be complex.
- The problem has to be solved according to the context of application or the situation.
- When solving the problem, different offers of assistance should be given (according to the complexity of the problem and the methodical competence of the learners).
- Problem-oriented learning will usually be arranged in a participative, co-operative, interactive, and to a certain degree self-organised way (the degree is defined by the facilitator, mentor, coach, tutor, etc.; exception: 'cognitive apprenticeship', which more likely resembles 'shadowing'.
- The problem should be considered from different perspectives (through cooperation or role playing).
- The problem should ideally be suitable for multiple contexts (see also below 'transfer-transfer-problem').

There are different methods to assist problem-oriented learning. In overview they are:

- Case method: the application of the acquired knowledge is demonstrated by means of case studies, which give impulses for learning and stimulate motivation and meaning.
- Real cases: not invented examples or constructed cases, but real cases from practice are provided.
- Life cases: authentic cases, which actually happen in a real organization, parallel to the development program, are being worked at, with the opportunity for the participants to act as external advisors and get adequate feedback.
- Simulation: simplified presentation of a sector of reality through role playing (see below), complex context games (see below) or through a medium such as a suitable software on the computer (see below).
- Role playing: didactical-dramatically arranged situations, which put participants into another context for a short time in order to allow different perspectives and to behave differently; role playing bears the opportunity of experimenting; it also stimulates reflection and trying out a certain behaviour.
- Context games: didactical-dramatically arranged complex circumstances, either as a complex longer lasting role play (e.g. AT Car & Company), or in written form (e.g. The Change Game).
- Cognitive mapping: learners are prompted to visualise the problem graphically. Of special interest, besides the elements, are the relations among them, which shall be displayed as well. Cognitive mapping motivates the learners to reflect on their work and, when applied cooperatively, to use certain terms more precisely.

The following processes are also associated with PBL, however, they are better applied in procedures where the place of learning has changed from a seminar or workshop to the real working environment (see next chapter). While the above mentioned are practice-oriented, the following can be called practice-integrating:

- Mentoring: the novice will be coached by an experienced colleague (similar to tutoring).
- Shadowing: the novice observes an experienced colleague or expert, sits in with him and accompanies him during his work days. He takes notes of his observations and gets an authentic insight into the complexity and variety of school leadership activities and in the leadership style of the particular school leader. Any kind of potentially artificial behaviour shown by the school leader who is shadowed will stop after some time, as well as any awkwardness at the beginning or the attempt to specifically 'control' situations by selection of interactions. Both sides profit from the shared reflection on the observations collected.
- Cognitive apprenticeship: the novice is watching an experienced colleague or expert, who explains relevant steps and working activities; the many comments allow additional insights into patterns of thinking and behaviour; the expert models a specific behaviour.
- Peer-assisted learning / collegial case discussion / critical friendships: colleagues exchange experiences and insights in specific questions (see also below peer coaching).
- Coaching: the participant gets intensive offers of support and attendance; it can be differentiated between individual and group-coaching, between internal and external coaching, as well as between peer coaching (on the same hierarchical level) and senior coaching (by an executive or a person higher in hierarchy).

A categorisation of these methods is possible within two dimensions, one as classifying them as practice-oriented versus practice-integrated, the other as preparatory, induction, and continuous development.

To sum it up, the method of problem-based learning has become a consistent part of a number of training and development programs for school leaders. It is meant to offer practical relevance through concrete problems and to involve the participants into a cooperative process of finding solutions. However, when PBL sticks to made up and constructed problems, the necessary complexity of the real situation in school may be missing. Nevertheless, there might still be exemplary learning effects, which may reduce transfer interferences.

Going one step further means using genuine, real cases that are taken from the real life of a school. Thus, the cases are grounded more concretely and more authentically in a school situation. One precondition, however, is that the trainers have enough contact with school leaders who can deliver these case examples from their own schools. These school leaders then can also function as mentors in the trainer team and, due to their experiences, can give feedback when the participants present the solutions to the problem. Hence, the participants of the project group become external consultants for the school leaders. Through this interaction, both parties may gain something.

This method can also be used in online mentoring schemes. For example, experienced school leaders present a problem taken from their work life to the group they are in charge of in seven to ten-day intervals. The participants then work on this problem by investigating into it or interviewing their own school leaders. After that, the participants put on the web their suggestions on how to act, and then write their comments about the suggestions of the others. After the ten days the school leader who has brought the problem up for discussion reports about how she or he has acted and whether it has been successful.

However, a 'transfer-transfer-problem' can show up. The more concrete the context of application, the smaller the problem of transferring specifically designed knowledge to such a situation. The more specific, however, the designed knowledge is for the concrete situation, the bigger gets the problem of transferring it to another context of application. Thus, when choosing a problem, its concreteness, complexity, and specificity have to be adapted to the aims of the learning process. The desired degree of abstraction influences the choice of the method.

The big advantages of PBL are its application orientation, the development of (social) competences, the acquisition of problem solving strategies as well as of cooperative and interactive working and learning techniques, and the development of an attitude towards lifelong learning, which is indispensable in a leadership position, due to an immense increase in demands.

From the Workshop to the Workplace

Some programs take this one step further and leave the workshop and turn to the authentic workplace using it as a clinical faculty. It is argued that only the authentic working context can assure adequate complexity and authenticity leading to the learning processes required. For the participants of pre-service school leader development, internships at one school or several schools are organised parallel to the training. In these internships they can observe the school leader by shadowing him or her, can partially take over leadership tasks themselves and can carry out projects independently. In this way the school leaders at the internship schools function as mentors or supervisors. Here, certainly the best possible practice relevance is created: exemplary learning processes take place in the reality of school. Mentoring, however, is not only useful for aspiring school leaders, but can also support those already in office. In the following two sections, internships and mentoring will be described.

Learning in the Workplace: Internships

The internships, or school attachments, in one's own or another school differ with regard to their duration and their emphasis. In some countries the participants spend many hours observing the school leader in their school in order to experience school leadership work and take over leadership and management tasks on their own. Other internships, however, aim at applying knowledge directly to the workplace by making the interns plan and carry out innovative projects. These projects, again, are meant to be useful to the internship school. In general, there is a lot of effort needed to create a balance of observation and active working on one's own project.

In the Führungskräfteschulung in South Tyrol conducting a project at one's own school is included. For this, all participants have to design a project to meet the needs of their own schools. Then small teams are built according to the themes of the projects. Each team has one tutor, who supports the school leaders in working on project relevant aspects, testing their applicability in the specific working context, and adapting them to the needs of this context. This coaching approach is intended to help carry out the projects successfully.

In the National Professional Qualification for Headship in England practical application is offered parallel to partaking in the respective module. The mandatory module Strategic Leadership and Accountability, for example, requires 180 hours, of which 50 hours are estimated for shared course activities and the remaining 120 hours for individual studies and conducting a project at a school.

In the Principals' Qualification Programs in Ontario the participants have a 60 hours internship between the two parts of the program. As it is easier to organise, they have these 60 hours during a longer time period of ten weeks at their own schools. They accompany and observe their school leader, try to bring their knowledge (e.g. legal knowledge, administrative knowledge, research findings and theoretical concepts) to bear by taking over school leadership tasks and carrying out a project independently. Additional visits to other schools are to give them a broader insight into school leadership and management.

In the second part of the second course year of the NSO program, the participants carry out a project in a school which explicitly is not their own school. Its aim is to coordinate all training objectives by making pairs of participants work on one question from the school leadership area and evaluate their own action. Notes are taken on the projects and the results are presented. For example, an aim may be to introduce an alternative pupil assessment system, or a new way of budgeting, or in-service development opportunities for staff at the internship school. The trainers and team leaders support the pairs as coaches.

In Singapore the participants of the development program have two four-week internships alternating with the seminars. During these internships (taking place at the same school) they can watch their mentor school leader in a shadowing procedure in everyday work life and can take over leadership tasks independently. They get important feedback and counselling for their own activities, and the school profits from the innovations that the participants, who are additionally counselled by a faculty member, bring in during their internships. Further insights are gained through visits to other schools.

In France, during the first basic part of the development program, before taking over the first school leadership post, there are various internships at different institutions. As a rule, each participant has about a twelve-week internship at a school (alternating with seminars), accompanied by a school leader as her or his mentor who sets up a schedule suited to the participant. Additionally, the participant leads her or his own school project. In further internships, one of them in a company (four to six weeks) and one in a community or regional administrative board or office (two weeks), the aspiring school leader can gain insights into organisational structures and processes outside of the school system with a chance for helpful analogies and conclusions as to decision making processes, procedures of communication, and so on. The aspiring school leader gets in touch with the institutions independently. These institutions have to fulfil a set of criteria given by the trainer team of the academy.

With 720 hours, the internship of the University of Washington is particularly extensive. The participants of the qualification program have to spend 16 hours a week at minimum at their internship schools. Hence, they have to work part time in their own schools in order to have enough time left. The internship schools should be as different from their own schools as possible. They can also go to a type of school different from their own one, for example, as a primary teacher to go to a secondary school. Then, one of the school leadership team, either the school leader her or himself, or another experienced teacher functions as the mentor. Mentor and intern work cooperatively with a representative of the university, that is a university facilitator. Hence, there is a team of two people responsible for the activities of the respective intern. Due to the fact that the internship is continuously spread throughout the school year, these activities can be well-coordinated with typical leadership tasks during the school year, starting with the preparation of a new term until its conclusion in June of the following year.

In the post-graduate program of the William Paterson University of New Jersey, too, the internship experience is continuous and parallel to the development program. In every semester the participants have to do 30 hours of field experience. The school is regarded, parallel to the university faculty, as a clinical faculty. The two universities, Washington and William Paterson, consider spreading the internship across a whole school year as much better than concentrating it at one point in time, as is the case in many other post-graduate programs in the USA.

In the medical and legal professions application orientation through internships has been practised in training programs for quite some time. For the qualification of educational leadership personnel, internships were suggested in the USA for the first time in 1947 at a session of the National Conference of Professors in Educational Administration. In the following decades the number of universities in the US that integrated internships in their training programs increased continuously and rose to about 87% of the university programs up to 1987 (see Milstein et al., 1991). A particular strength of internships lies in the balance they create between knowledge and skills, "between learning about and learning how, rooted in a solid foundation of learning why" (Milstein, 1990, p. 122).

The extensive prevailingly American literature enumerates a number of advantages of such learning in the workplace. For the participants themselves it offers the possibility to get to know concrete school leadership activities. Moreover, the participatory observation of school leaders in a shadowing process fosters individual reflection on the experiences set against the background of one's own attitude, values, and previous experiences. This 'knowing in practice' respectively the 'reflection in action' (see Schön, 1984) also enables them to examine critically and if necessary to modify their own intuitive understanding of the experiences made. Besides, the internship offers concrete orientation regarding one's own decision whether or not to actually apply for a leadership position. Additionally, even when participants eventually do not aim at a leadership position, they develop new competences for the benefit of their schools. Another advantage is that the school board and school involved gain the chance of influencing the qualification of their future leadership personnel. They also get the opportunity to get to know and to be able to assess the candidates for school leadership positions before employing them, which is particularly interesting in countries with a decentralised school system. For the universities responsible for conducting the qualification program, manifold chances for cooperation with practitioners in schools and school boards emerge. Hence, internships are regarded as profitable for all sides involved.

Within the whole qualification program the internship is considered as an integrating factor, bringing about a synthesis of what has been learned. This makes a more holistic learning process possible. It is necessary to motivate meta-cognitive processes enabling 'reflection on action' and helping to work out the meaning of experiences made retrospectively: 'Making the meaning out of experiences' (see Schön, 1983; Kolb, 1984; Greenfield, 1985). The aim is, in the end, to improve the competence for acting in future situations. In particular, the participants should get the chance to develop technical skills, human skills, and conceptual skills, but also cooperative and communicative competences for managing group processes, dealing with conflicts, decision making, problem solving, as well as the ability to keep the balance between the big picture and detailed knowledge and to develop an understanding of mutual links and inter-dependencies in complex situations.

Anderson (1989) describes ten knowledge domains (see table below), which the participants should approach by working in different schools with different staff and by reflecting upon these settings.

The internships take place in some instances in the participants' own schools and sometimes in one or several other schools. Working in one's own school has the advantage of being familiar with the staff, this may mean at the same time that the intern is often only seen in her or his old role. Other schools, on the contrary, offer the opportunity to get to know different school cultures and different school styles, but also bring with it the necessity to become accepted by staff and not to be seen as a kind of intruder.

The internships may be embedded in a training and development program in various ways (as the programs included in the study show):

- parallel to the seminars, projects at one's own school are conducted (e.g. England and South Tyrol),
- parallel to the seminars, an extensive project at an internship school is conducted (e.g. Nederlandse School for Onderwijsmanagement and Ontario),
- several-week internships alternate with several-week seminars (e.g. France and Singapore),
- a one-year internship parallel to seminars (e.g. Washington and New Jersey, USA).

Table 8: Knowledge domains to be developed in internships

1. Local knowledge: this is knowledge that is specific to the settings, the community, and the organization.
2. Tacit knowledge: it is described as much like intuition, known but difficult to express. Tacit knowledge in administrative practice involves both cultural and craft knowledge. If interns are encouraged to introspect on their actions and the actions of others, evidence of tacit knowledge emerges.
3. Cultural knowledge: this domain may be defined as knowledge about the basic assumptions and beliefs that are shared by organizational members.
4. Moral/Ethical knowledge: moral/ethical knowledge affects decision-making situations in which moral dilemmas occur. Educational administrators must examine both individual and codified systems of ethics.
5. Research knowledge: research data influence the practitioners either directly or indirectly, affecting the ways in which educational organizations are conceptualised by those involved in organizational change.
6. Personal knowledge: personal experiences, abilities, preferences, and styles have an impact on messages sent and received, decisions made, and knowledge acquired. In short, person-environment fit is critically important and must be considered to be a part of the knowledge base.
7. Theoretical knowledge: 'theories-in-use' (e.g., Argyris and Schön, 1974) as well as 'espoused theories' from the behavioural sciences act as templates for the development of individual beliefs, attitudes, and values about educational systems and the environments in which administrators function. Administrative interns must be given opportunities to examine behaviors relative to espoused theories.
8. Critical knowledge: this domain may also be referred to as 'critical reflection'. This knowledge involves viewing administration from the perspective of the larger social system and the structures embedded in educational institutions.
9. Political knowledge: this area of knowledge is comprised of the awareness/knowledge of the power structures and networks of influence within the larger community as well as the educational organization.
10. Craft knowledge: craft knowledge is based on professional experience: it is the repertoire of the practitioner. This domain goes beyond that of technical core operations to include reflective practice as well.

(according to Anderson, 1989; from: Milstein et al., 1991)

Milstein (1990) describes four ideal-type variations in the USA, namely:

- The linear approach: this rather traditional approach is one in which campus-based learnings are pursued first and clinical experiences comprise of the culminating program experiences. The assumption is that students must first master a knowledge base before they can maximise their learning in the field.
- The dialectic approach: this approach assumes that learning takes place best when it combines acquisition of a knowledge base with application in real-world settings. Internship programs based on this assumption involve students in field experiences at the same time they take campus-based seminars.

- The reflective practice approach: this approach is based on the belief that theory is best learned if it is based on experience. This implies that practice must come early and that theory building becomes a possibility only as students gain an experiential foundation.
- The developmental approach: this approach assumes the need for a foundation of knowledge and theory but also assumes that field experiences must begin early in the program so that these theories can be tested in practice. The implication is that campus-based activities will initially be the primary focus but that, over time, there will be an increasing level of field-based activities and a comparable reduction of academic seminar activities.

So far there have been no insights into the effectiveness of the different structures, neither of Milstein's ideal types nor of those found in the study. However, an essential factor for the success of the concept is that the activities of the participants in the internships and those in the seminars are carefully coordinated, both complement each other. The workplace environment offers chances through getting to know and experience the everyday life components of school leadership. Writing a learning journal, in which experiences and reflections are continuously collected, is demanded in most programs. It serves as a basis for feedback sessions for the participant with the mentor, but also for discussion in the seminar. The technique of writing journals itself is a sensible method to reflect on experiences and to anchor what has been learned, which often has been unknown to the participants before. An important condition, however, are seminars linked closely to the internship to avoid the danger of a pure on-the-job-training. In the seminar the participants exchange their experiences, structure them, process them, reflect on them, and receive further information and impulses for reflections. Thus, theoretical contents receive an immediate and a direct relevance for practical action. The seminar group offers the chance of partnerships, for exchange and cooperation and for setting up a network lasting beyond the time of the program.

During the projects, a critical problem may emerge, when a school leader needs the particular school project to successfully fulfil the qualification requirements. Then, the staff may feel used and may not be ready to identify with the project. This seems to be partly the case in South Tyrol, as has been stated in some comments by representatives of professional associations (see Amplatz, 2000). It is different, however, if not the school leader alone, but a team of the school takes part in the qualification program and then initiates and implements school improvement projects, as it is the case in California (above all in the program School Leadership Teams) or in New South Wales. Fundamentally, the school must not become a stage for the actions of an individual, and the projects must not become productions of a stage play but have to be taken seriously by putting the school and not the school leader at the centre and making the school perceive itself as being of central importance. This problem has to be taken into consideration when school leader qualifications and school improvement initiatives are combined.

Besides the individual demand on the participants and the demand on the organisational capacity of the provider, one specific problem of qualification programs comprising long internships is how to finance them. Somebody has to stand in for the participants at their own schools when they are absent. Mentors have to be selected, trained, and then because of the time factor, be granted leave or be additionally paid. It would be ideal to release the participants completely from their professional obligations, as in France or Singapore. In these countries they obtain full pay during their full-time qualification. The opposite extreme means participants keep their positions at their schools, additionally take part in the whole qualification program and even pay for their own qualification themselves. The creativity and imaginative ideas of the providers and the persons organising the programs are called upon to find alternative ways for at least partly financing participation, for example, through regional subsidies, grants, and scholarships.

Mentoring: Support during Internships or after Taking over a Leadership Position

Mentoring can be found in the training and development programs of several countries, for example, France, England, Singapore, and the USA. Mentoring on the one hand is an essential part of internships, on the other hand it is used as a professional support when taking over a leadership positions. Compared with shadowing, the participant is active at the workplace. Mentoring can be described as a complex interactive process between persons with different levels of experience and expertise which stimulates interpersonal and psycho-social development. The relationship between mentor and protégé or mentee is dynamic and has different stages (see below).

Essential preconditions for the success of a mentoring program are described by Thody (1993):

1. Selection of the mentors: the mentors should be experienced school leaders who fulfil certain conditions. Selection criteria are stated by Milstein et al. (1991):
   - a clearly shown interest in supporting others in developing their potential,
   - being continuously a model for lifelong learning,
   - the ability to ask questions, but also to give stimuli,
   - clearly shown leadership competences such as intelligence, ability to communicate orally and in writing, the readiness to develop alternative solutions for complex problems, a personal vision and the ability to communicate it, well-developed interpersonal competences,
   - an awareness of the political and social situation.
2. Preparation and qualification of mentors: all established mentoring programs train their mentors in a several-day course, which above all is to impart knowledge from methods taken from adult education.
3. Combination of suitable mentor-protégé-pairs: personal wishes (for example if male or female), the type of school, low distances to avoid unnecessary time for travelling, but sufficient distance (in order not to belong to a competing school) are mostly taken into consideration.
4. Duration of the process: to create a confident relationship and an open communication is essential. The relationship mentor-protégé is a key factor for the learning process. It develops in several ideal-type phases, first a formal stage, second a cautious stage, third a sharing stage, fourth an open stage and fifth the continuation of the confident collegial relationship even after finishing the program (see Walker & Stott, 1993). Not every collegial

partnership is able to exceed stage number two, which makes the learning effect much less intensive. The partners are free to decide (within a proposed frame work), which aims they focus upon and they make an explicit working agreement, which may be modified during the process, if necessary.

5. For the shadowing phase see above.
6. Regular reflection talks: in regular meetings, the observations during the shadowing phase and the experiences with the participants' own actions are reflected upon and analysed. This does not mean that simple 'recipes' for action are passed on to the future colleague nor that she or he takes over patterns of thinking and acting from the mentor. Nor is the mentor a kind of superior school leader at the school of his protégé (in case she or he is already in a leadership position). What is aimed at is a dialogue based on partnership, which ideally can become a cooperative mutual counselling, which is supportive for both.
7. Supervision and evaluation: the mentors themselves are supported by supervisors. Moreover, the whole process should be evaluated. Often, this is done by university personnel as is the training or qualification of the mentors.

The function that mentors have in school development programs can be summarised as follows (see Wasden & Muse, 1987; Daresh & Playko, 1989):

- Giving stimuli: being ready for questions and for passing on information.
- Assessing: formal and informal evaluation of the interns.
- Coaching: showing effective school leadership and providing opportunities for the participant to practise her or his own leadership actions in a non-threatening surrounding.
- Communicating: having an open communication with the participants, being open for dialogue.
- Counselling: providing emotional support and professional and personal help.
- Guiding: introducing the participants to the procedures and the culture of the organization, and supporting them.
- Being a role model: showing attitudes and behaviour which may model leadership.
- Motivating: encouraging the participants to pursue their goals.
- Protecting: creating a kind of shelter in which the participants can act without fear of failure or loss of self-confidence.
- Developing competences: supporting the participants to continuously develop competences for leadership.
- Promoting: using her or his own influence and contacts to school administration in order to promote the professional career of the participants.
- Supervising: giving the participants feedback.
- Teaching: teaching the participants knowledge.
- Validating: evaluating the achievement and the personal goals of the participants, helping to modify them if necessary, and supporting them.

An advantage of mentoring as one part of the qualification of school leaders is the possibility to learn at a different school without getting stuck in the culture of only one. The mentor, coming from another school, and the group meetings with other mentors and protégés bring in perspectives with which the aspiring or newly appointed school leader would not normally be confronted. The mentor is an experienced colleague, whose credibility is appreciated and whose competence and role are accepted. Mentoring does not lead to some copying of the behaviour of somebody else, as may be the case with shadowing, for the participants are active and their own actions are the central theme.

This procedure, however, carries with it the risk that – in unfavourable cases – it may turn out to be not much more than a kind of a low-level apprenticeship and a mere passing on of recipes. In this case, a purely affirmative tendency has to be feared: Traditional values and a kind of school leadership oriented towards maintaining the status quo are passed on to a new generation of school leadership personnel. Thody, too, recognises this danger: "It will result in principals cloning principals." (Thody, 1993, p. 74) In this unfavourable case, an increased readiness to question established traditions, to think critically, to favour change and development may not be sufficiently developed; academic inputs would be disregarded. The profession would isolate itself and a method that is basically meant to be innovative could indeed no longer be seen as increasing professionalisation.

Learning at the workplace – in internships or through mentoring – offers a genuine chance to reduce the 'practice shock' experienced by new entrants to the role (Storath, 1995). The methods outlined here, however, will usually be used only in programs with very small cohorts, due to the demands on time, the costs, and the organisation. In large-scale state-run programs they are very rarely found.

6. Pattern: One Size for All or Multi-Phase Designs and Modularisation

Selecting adequate teaching strategies and learning methods based on micro-didactic considerations correspond directly with macro-didactic considerations such as the planning of the time pattern. A training course will significantly differ from a Master's program not only in the number of course days or in the total number of training and development days but also in terms of scheduling the various parts of the program.

6.1. Comparison

Number of Training and Development Days

As to the number of qualification days, a direct comparison is difficult because the way time is calculated varies greatly. The North American university-based programs use credit hours, others indicate contact time in courses or seminars. Additionally, there are internships, the time spent on projects, self-study and literature study. These components were sometimes explicitly stated but otherwise not fully incorporated. This study uses the information of the providers and, if the contact time is not given in days but in hours, introduces the unit 'course day' for the sake of comparison and bases one course day on six hours. The time needed for internships etc. is not brought into line in the country reports, but converted into days in the same way for the comparison itself. By adding up course days and additional days, the total number of training and development days was defined, so that the time needed by the participants as a whole can be acknowledged more or less appropriately, knowing that the real effort and real time demands are not necessarily expressed adequately in this way. There might be slight distortions due to these conversions. The value of a purely numeric comparison should not be overestimated anyway. One should note the status and the professional validity that the particular program has for the participants.

However, the amount of time gives us some idea about how much importance is given to preparation and development in particular and therefore to the role of school leadership in general. Considering the various programs, astonishing differences become apparent (see table below). According to the information provided, the number of course days varies from nine days of the induction course in Hong Kong to a group of programs with 20 to 30 course days and a group of programs with 40 to 60 course days up to the more extensive models with 90 up to nearly 150 course days.

Table 9: Number of course days of the various programs

7 to 20 course days	Denmark*; the Netherlands*; Germany*; Hong Kong; New South Wales, Australia*
20 to 30 course days	Sweden; Denmark*; England; Germany*; Switzerland; Austria; South Tyrol, Italy; Ontario, Canada
40 to 60 course days	France; Singapore; New South Wales, Australia*; New Zealand
90 to 150 course days	the Netherlands*; France; Washington and New Jersey, USA

*The days have been adjusted. * Multiple entries result from the differences of the various programs included in the individual countries.*

Even this merely quantitative comparison implies different ways of understanding the relevance of school leadership and school leadership qualification. Here, the contradiction becomes apparent between the increasingly widespread notion of school leadership as a new profession in its own right when moving from teaching to management, and the particular qualification for this profession. Which demanding profession can be learned within a few days, or through in-service experiences only?

While the programs from the first group may be seen as kind of short programs adding only little to the competences of the participants, particularly the last group clearly shows that in these countries the training and development of school leaders is taken very seriously.

Undoubtedly the validity of the qualification and the professional understanding of school leadership cannot be separated from the actual systematic role that school leadership plays in the particular country. If school leaders are seen as administrators putting regulations given from super-ordinate levels of the educational system into practice, and if the function of school leaders is not acknowledged as a profession on its own with an individual qualification profile, the qualification, at least the mandatory one, will be limited to the transfer of an accumulated small knowledge base and be relatively short. The broader responsibility in the school system is extended and the more decentralised decision-making is, the more extensive the listing of tasks for school leadership will be and the broader and more profound the qualification should be. Examples such as those in the USA, Canada, Singapore, but also the voluntary qualification programs in the Netherlands, New Zealand, and England (still voluntary) show an awareness of this. The programs here much more clearly take the demands on school leadership and its relevance for securing the quality of schools into account.

For illustration, some examples from Europe, Asia, Australia/New Zealand, and North America are shown in the following table. It is important that all programs shown here are preparatory ones, that is they take place before participants are appointed to school leader positions (with the exception of the program from the Netherlands which is accessible for practising school leaders, too).

Figure 1: Number of course days of preparatory development programs from Europe, Asia, Australia/New Zealand, and North America chosen as examples

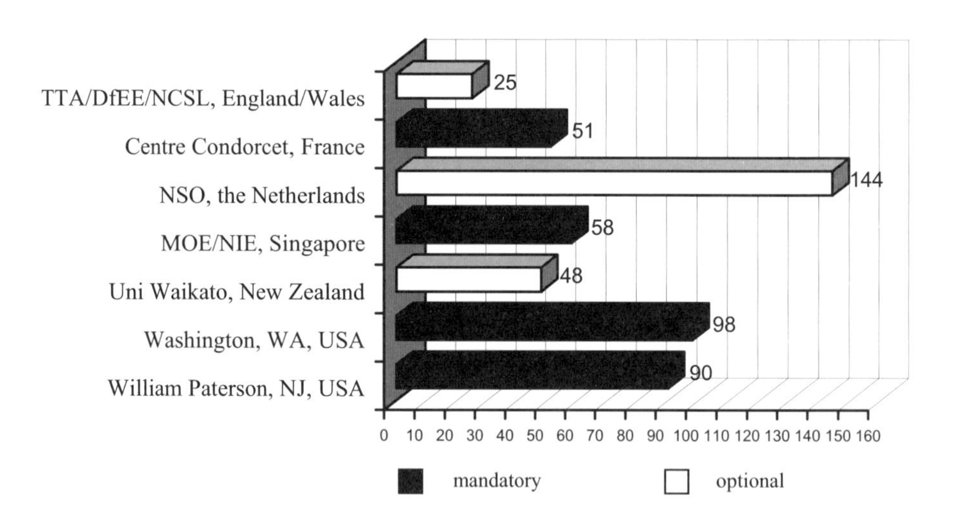

The real time needed, however, becomes apparent only when trying to additionally include the time required for internships, project work, and documentations next to the actual course days. Required literature study, too, has to be added sometimes if the provider does not give any information as to the time needed. In the last group of the table above with 90 to 150 course days it adds up to 110 to 240 training and development days altogether.

Figure 2: Number of total training and development days of preparatory development programs from Europe, Asia, Australia/New Zealand, and North America chosen as examples

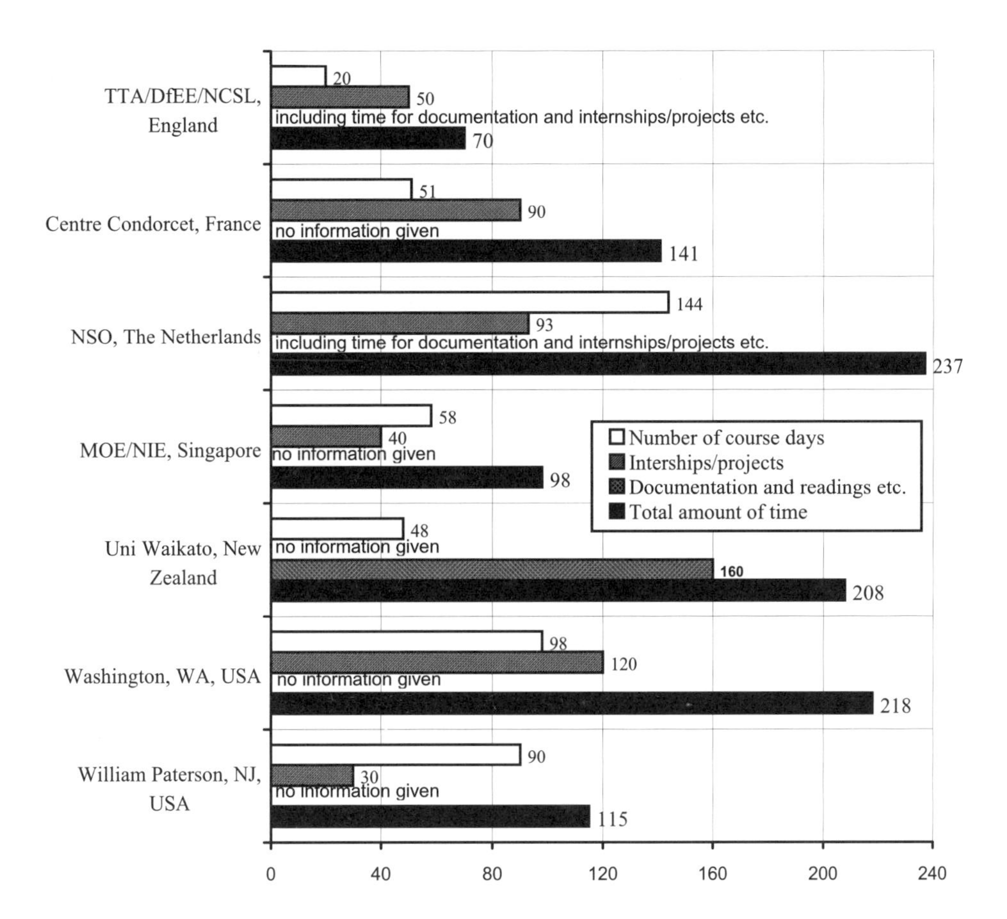

The program at the University of Washington takes 39 credit hours (assuming 15-week semesters) and 720 additional internship hours (i.e. 16 hours per week). In addition, the participants have to write assignments and learning journals, as well as read course literature. The program of the Nederlandse School voor Onderwijs-management comprises four semesters, each with 20 credit hours for seminars, 175 training session hours and up to 20 hours of consultation sessions plus 140 hours for internships per semester (in the first three semesters), literature studies and a written assignment. In the fourth semester, the realisation of a final project and the report, as well as a thesis account for an additional 140 hours. The University of Waikato offers a program which comprises 24 credit hours (assuming 12-week semesters) as well as 1 600 hours (this number of hours is assumed by the provider) for individual studies, participation in the email-forum, and for conducting school-based projects.

The high professional validity of the programs becomes even more apparent when one considers the fact that all the examples in the table are preparatory programs, partly even before appointment, and academic Master degrees can be obtained. This gives us the impression that the importance of school leadership and adequate professionalisation have actually been taken into account in those countries.

Scheduling

How are the training and development days scheduled? How are they structured with regard to time? Here again, the programs vary. The course days are scheduled in units of several weeks, units of one week, of half a week, or in sequences that last from one to several days, the timetabling naturally being closely linked to the number of training and development days as a whole.

The state-run obligatory full-time programs in France and Singapore set up units of several weeks at academies, or institutes, alternating with internships at schools and other internship places. The university programs in Canada, New Zealand, and the USA usually are organised in one-week units during the summer holidays and then in one-day or half-a-day sessions on Saturdays or in the workday afternoon or evening. Here, again, different extensive internships must be added, which are completed on several days per week, usually half-day at local schools. The course-like programs in Austria, Switzerland, South Tyrol, Sweden, and Denmark are scheduled in several-day (mostly two-to-four-day) sequences. This is similar to the way the programs are organised in many of the German federal states.

Time Span

The time within which the programs will be accomplished results from the number of days and the time scheduling. For the 9-day training and development program in Hong Kong, two to three weeks are planned. Quite differently, the course-oriented university programs of New Jersey and New Zealand as well as the NSO program, they usually take about two years, and in the case of part-time even up to four years. In Ontario and Washington the programs take one year (or two semesters) and even longer if there is a break between the two semesters. The state-financed preparatory programs in France and Singapore with their strict scheduling require six or nine months full-time. The in-service programs in central Europe (Austria, Switzerland, South Tyrol, Germany) can be finished in around one year. The preparatory program NPQH in England can be taken in one to three years. Sweden's basic training program covers two and a half years.

The overall duration that has to be taken into account is in general quite similar. However, the amount of seminars and sessions that have to be attended and the workload differ greatly. This is obvious when looking at the respective number of course days, the internships that have to be done within this duration, the project work and the documentations as well as the individual literature study that has to be accomplished. The energy and time required by the participants is to be estimated clearly higher in the countries of group 4 from the table above.

6.2. Discussion

For the overall time span and the scheduling of the course days, cost-benefit considerations are paramount. On the one hand, the participants should not have to take a lot of leave from the schools where they are working in order to keep the demand of personnel to replace them as small as possible. On the other hand, both for the providers and the participants the time-effect-ratio is essential. Only a carefully planned scheduling guarantees a continuous learning process. Too short phases (comprising few meetings over a short time span) admittedly may provide relevant information and may cause an increased awareness and sensitivity. However, learning in the sense of changing one's own behaviour and way of thinking cannot be achieved this way. Learning processes require time: "Learning is a lengthy process." (Leithwood et al., 1987, p. 25) But learning does not only occur during the course days themselves, but also to a great extent through reflections and activities on a day-to-day base at the work place particularly during supervised practice. In the time between the course days participants try things out, develop their own ideas, implement them, evaluate and analyse the success, etc., and thus learning takes place.

Regarding macro-didactic considerations such as the ones concerning scheduling, Hopes (1983) also states that it is necessary for certain topics (such as communication or social interaction, for example) to have some alternation of theoretical seminars and concrete opportunities to gather experiences over a longer time span. Individual learning processes as well as group processes have to be given a chance to develop; single, short training events alone cannot achieve this.

An Example of an Instructive Time Pattern

An example to illustrate various considerations regarding time span and scheduling is the program of the University of Washington (see Milstein, 1993). This university spreads its course over one year. An advantage of this is that the core themes can be adjusted to the range of demands and tasks of a school leader during one complete school year. This effect is even enhanced by the internship components. The master program starts with a one-week seminar in the summer (beginning in August). The participants are provided with the opportunity to develop into a learning team, being locally and organisationally away from the everyday work of school. This cohort remains in the same grouping throughout the program. In the course of the year the contents are divided into sessions of six hours each, which take place on Thursdays from 1 to 7 pm. Concerning content and methods, links to the experiences from the participants' own schools and to the internships are to be established, and the advantages of the exchange within the learning team are to be rewarding. Additionally, there are ten full-day sessions spread out over the year on Saturdays. The program finishes with a five-day seminar in the following summer, which is meant to facilitate an easier transition to the leadership position and to help the learning team systematically set up a collegial network of mutual support.

This example demonstrates a careful interrelation of overall time span and the distribution of seminar phases of various duration within this time, pursuing different intentions and aims. Regarding the time intensity of this program, naturally there is much effort to reduce the participants' absence from their school. Hence, as much of the work as possible is scheduled in the holidays and on weekends. This means that the demands on the participants are extremely high.

Multi-Phase Designs and Modularisation of Programs

The international comparison shows that school leader development is more and more regarded as a continuous process. This could be divided into several phases like orientation, preparation, induction, and continuous professional development. In more concrete words, it seems to be important to qualify teachers with special responsibilities other than school leadership, to provide teachers interested in leadership positions with the opportunity to reflect on the role and responsibilities of a school leader, considered in the context of their abilities and expectations, to prepare them prior to taking over a school leadership position or even before applying for it, to introduce them after taking over a leadership position, within the first three years, which provides opportunities to support the school leaders in their new position, to provide opportunities for continuous professional development for established school leaders, designed to meet their individual needs and the needs of their schools. Additionally, a kind of 'reflective phase' for very experienced school leaders might have a twofold effect, for themselves and for others. They continue to grow introspectively by being involved in development programs for others as coaches. This phase would be supported through methods like 'learning by teaching' and through supplemental train-the-trainer-programs.

Considering that raising the levels of knowledge and modifying the behaviour of participants requires a serious commitment of time, providers are increasingly moving towards several phases of development. This is resulting in the implementation of multi-phase development models. Multi-phase development in this sense does not merely mean that pre-service and in-service training options offered by the same provider exist, but that the different phases are well coordinated and match with each other; they are based on a coherent conceptual approach.

In England, for example, the development model is comprised of three phases. Firstly, the National Professional Qualification for Headship (NPQH) is a preparatory program for aspiring heads. Secondly, the Headteacher Leadership and Management Programme (HEADLAMP) addresses the needs of newly appointed school leaders. Thirdly, the Leadership Programme for Serving Headteachers (LPSH), is a program for school leaders who have served for more than six years. This overall conceptualisation of a three-phase training and development opportunity for school leaders as well as the content design within each phase represent good examples of the multi-phase model.

In the USA, the California School Leadership Academy offers a combination of programs, which fit in with the career cycle and at the same time have a systemic approach: the target group of Foundation II are teachers interested in leadership positions, School Leadership Teams is a continuous professional development program for teams including the school leader, and Ventures aims at experienced school leaders.

The Department of Education in New South Wales has designed the five-stage development program School Leadership Strategy, based on a conception of school as a learning community and aiming at the empowerment of staff and the distribution of leadership responsibility. Its modules cover three career stages. Hence, participation in one of the single modules depends on the professional position and also on the development needs of the individual participant, which implies that the modules need not be attended necessarily in a particular order, but may be selected quite individually.

In these examples, some emphasis is put on the possibility to continue the modules by taking up the support offers of the providers and by joining a network of (former) participants. The basic idea is that an adequate qualification cannot be completed in one go and then ticked off, but should cover the whole career cycle.

We also find professional development provided through a series of modules, as already noted above in the examples of England or New South Wales. This kind of modularisation takes two general forms. In the first form, the modules are conceived of as a mandatory sequence of rounded single programs. In the second form, there is no specific sequence for completing the modules. Rather participation in the modules depends on the professional position and development needs of the individual participant. The modules may be collected in a kind of personal portfolio. The individual school leader may well fall back upon them as support in crucial career phases.

Examples can also be found in the Netherlands and in Ontario, where in addition to the preparatory Principals' Qualification Program there are many different development offers by the school boards, which, again, can be collected systematically in an individual portfolio.

The table below illustrates six ideal-type models generated from the international comparison. They are meant to mirror the following issues:

1. There are single preparatory, induction, and in-service programs.
2. The programs vary in duration.
3. There are programs that are scheduled in different ways (for example extensive preparatory programs then followed by shorter in-service units).
4. There are programs whose individual events are attended by all participants or programs with a modular structure, allowing for an individual choice according to the personal needs and those of the individual school.

Figure 3: Ideal-type models of school leadership training and development

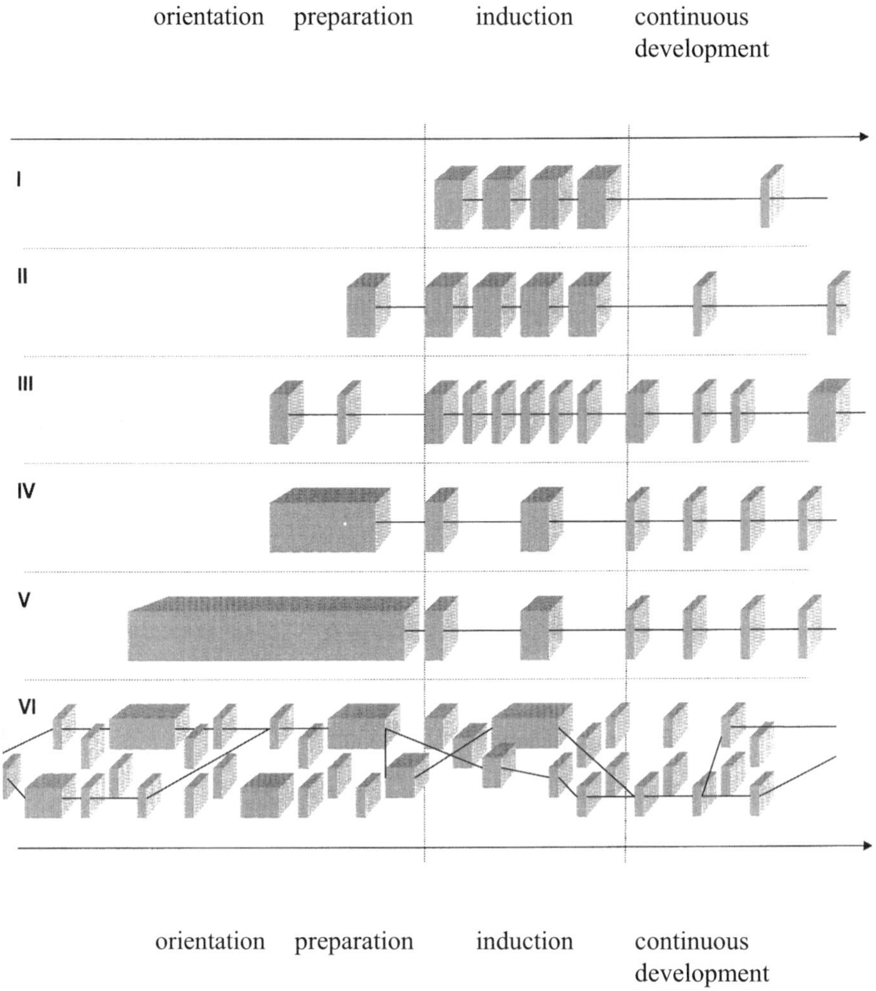

orientation preparation induction continuous development

I

II

III

IV

V

VI

orientation preparation induction continuous development

* *The individual squares represent courses; the black line signifies the sequence of participation.*

Explanation of the ideal-type models

Model I Several development programs for school leaders are designed exclusively as induction or in-service offerings after appointment or after taking over leadership. Often there is a sequence of development units of several days. Building on these units there are individual short events as continuous development, for example half-day or weekend sessions.

Model II Quite frequently, too, are programs that, similar to those of model I, offer developmental courses before appointment or taking over leadership, as a kind of short-term preparation or orientation offer, for example, as a compact seminar.

Model III There are qualification programs that offer more extensive weekend or one-week seminars for interested teachers, meant as orientation offers. Then, however, they differ from Model I and Model II as far as the induction or in-service programs are concerned. Here, they are not organised in one-week seminars, but in one- to two-day seminars with a higher frequency and shorter intervals. The conception of this qualification model is focused on transferring what has been learned in the seminars to the workplace and establishing collegial networks as well as stimulating interactive problem-solving processes. The participants are given theoretical inputs or stimuli for interactive problem-solving through case studies and then all this is linked to assignments or projects, integrated in practice in order to implement what has been discussed and reflected upon. In these countries quite frequently more extensive continuous development opportunities for school leaders can be found.

Model IV In some countries, there is a preparatory training which clearly exceeds a mere orientation program. The aim is to prepare aspiring school leaders before they actually suffer from the 'practice shock'. Additionally, it is to provide the necessary training and development in advance, since, pragmatically speaking, particularly in the first year of principal- or headship there is not sufficient time for being away from school taking part in seminars. In this ideal-type model, an induction or in-service program naturally plays a much less important role. The offerings made in the first years are carefully adjusted to the needs of the participants and their schools. In this model there are also more opportunities for continuous professional development.

Model V In this model, the focus is clearly on a preparatory training. These countries show an awareness of the important role of school leadership and of a conception of school leadership as a profession in his own right that has to be learned and that requires an adequate qualification. Training and developing school leaders is high on the agenda; this certified qualification has a high professional validity and is valued in the application procedure.

Model VI This ideal-type qualification model takes various professional stages into account and covers a preparation, induction and continuous development phase. Moreover, it is characterised by a distinct modularisation. In this context modularisation does not only mean spreading the content over several courses, but it means that individual participants can attend the individual qualification events in a needs-oriented way and need not do so in a given order. Additionally, this ideal-type model stands for a school leader development landscape in which there is no standardised, centrally recognised, or centrally controlled program, but a broad variety of offerings made by various providers. In other words, it is best described as 'diversity and choice'.

The following examples for these ideal-type models can be cited: model I and II can be found in some qualification programs in Germany, Model III in the Dutch program Meesters in leidingsseven. Model IV represents the situation in England with its preparatory NPQH, its induction in-service program HEADLAMP and its continuous development program LPSH. Model V reminds one of France, Singapore, or the US with qualification models ranging from half a year to two years (part time up to four years). An example for the ideal-type model VI is the situation as a whole in the Netherlands or New Zealand with their many providers and training and development offers.

III. Conclusions and Outlook

Stephan Huber

The third part outlines a number of current trends identified internationally in school leader training and development. Although some are mere changes in emphasis for the time being in terms of tendencies, others already have the character of paradigm shifts. Then, although evaluating statements concerning the effectiveness of the programs cannot be made, preliminary assessments are accomplished against the background of the aspects discussed in the previous chapters. Examples for best and promising practice are cited and the necessity of a multi-stage adjustment of aims in school leader development is emphasised. Additionally, 18 recommendations for designing and conducting development programs for school leaders are deduced. Finally, research desiderata are formulated as a general outlook.

1. Current Trends from a Global Perspective

This project on school leadership development was based on researching, analysing, contrasting, comparing, and discussing programs of 15 countries in Europe, Asia, Australia/New Zealand, and North America. A broad variety of school leadership development approaches and models became apparent. In spite of differences in cultural and institutional traditions, there are common tendencies and trends throughout these countries. While some of them may be viewed as differences in emphasis, others may be so significant as to represent paradigm shifts. The largest differences are evident in those countries with longer experiences in school leadership development and school leadership research.

Current trends and paradigm shifts in qualifying school leaders include:

- Central quality assurance and decentralised provision
- New forms of cooperation and partnership
- Dovetailing theory and practice
- Preparatory qualification
- Extensive and comprehensive programs
- Multi-phase designs and modularisation
- Personal development instead of training for a role
- The communicative and cooperative shift

- From administration and maintenance to leadership, change and continuous improvement
- Qualifying teams and developing the leadership capacity of schools
- From knowledge acquisition to creation and development of knowledge
- Experience and application orientation
- New ways of learning: workshops and the workplace
- Adjusting the program to explicit aims and objectives
- New paradigms of leadership
- Orientation towards the school's core purpose

Central quality assurance and decentralised provision

Regarding the provider or the mode of providing development opportunities, two major tendencies become apparent when comparing the historical development in the countries. On the one hand, the development of new, qualifying programs and suitable quality control measures are being more and more centrally implemented or handed over to a central (super-ordinate) institution. On the other hand, numerous decentralised providers, that are meant to meet regional needs, are then responsible for actually conducting the programs.

Centrally issued guidelines and standards are apt to provide fundamental quality assurance. Other instruments used for quality assurance include the accreditation of local providers and centralised participant certification. This provides teachers intending to qualify for a leadership position with certain advantages. They can choose from a variety of providers and, at the same time, expect certain uniform basic standards. This, in turn, ensures a certain quality connected with the program and the acceptance of the degree or credential obtained by the government and the educational authorities or the respective employing committees. In addition to the state taking on the major role in certifying school leaders, an accreditation by the professional associations seems to be valued by the participants.

Since the central guidelines do not account for all the details of the programs, adequate freedom in developing the actual design of programs is left to the local providers. This results in increasing flexibility towards the participants' needs and provides better opportunities for cooperation with the local school authorities and the individual schools.

New forms of cooperation and partnership

New arrangements concerning partnerships in numerous countries can be viewed as the second trend. These arrangements were created to conceive, implement, supervise, and evaluate school leader development programs. The most striking feature of this development, however, is the fact that representatives of the recruiting committees (either state or local), of the colleges of education at the universities, and more and more representatives of the profession itself (predominantly from professional organisations, but also from local schools) are now included as well.

It becomes apparent that much of the coherence that characterises the new programs in these countries is due to this cooperation. These groups contribute a variety of perspectives concerning the essential content of the programs, the teaching strategies and learning methods, and the organisational and chronological conception of the programs; that is to say, their conception on a macro- and micro-didactic level.

It is the cooperation of these groups, especially in collaboration with universities, school boards and particular schools within the region, which supports field-based projects and school internships, and enables the implementation of innovative approaches for adult learners.

These partnerships have also contributed to the creation of a pool of highly qualified and accredited or certified trainers in some countries. This, too, is important, since the credibility, the currency, and the current knowledge of trainers have been a matter of debate in several countries in the past, and the preparation of trainers is likely to become an increasingly important issue in the future.

Dovetailing theory and practice

Partnerships like these have also contributed to the next trend: the increasing combination of the theoretical and practical aspects of school leadership development, which is an important task that is difficult to achieve. It might sound axiomatically that theory has to be made accessible through practice and vice versa. Seemingly it has never been easy to achieve both at the same time. In many of the countries investigated, it was perceived that either development programs emphasising theory were developed from those focusing on a more practical approach or that courses evolved from being theoretically oriented to experiences for practitioners by practitioners. Both models therefore seem to suffer from one-sidedness, and do not seem to be attracting participants or leading to the expected increase in knowledge, understanding, skills, and abilities. Only a more balanced model leads to participant satisfaction and is a suitable method to meet the participants' needs. It is thus safe to assume – although this aspect has not yet been investigated sufficiently – that dovetailing theoretical and practical aspects is essential for designing effective development programs which aim at changes in one's behaviours and dispositions through the process of teaching and acquiring knowledge. Admittedly, school leaders themselves seem to prefer what they refer to as practical experience and, at times, regard theoretical and academically oriented topics as less useful. It can be seen (see West et al., 2000), however, that they find it much easier to deduce general knowledge from their experiences and to use effective strategies when they have a theoretical conceptual framework that underpins their decisions and actions.

Theory and practical experiences are interdependent and therefore have to be developed together. The partnerships indicated above appear to be a suitable starting point, since research is conducted alongside the development programs and can affect the development concepts. Hence, research-based training concepts are realised. This connection requires partnerships between the individuals working at schools and those who research and study schools. This will more effectively link the work carried out in both areas. Mutual respect and collaboration between both groups is essential for this to occur.

Preparatory qualification

Another shift observed in the international comparison concerns the target group and when training and development takes place. Many of the countries included in the study offer pre-service preparation, that is training scheduled before taking over a position of school leadership, instead of relying solely on in-service induction, that is training once one has been appointed to a leadership position.

In the countries that have mandatory preparatory qualifications, participation in the program is an important selection criterion for future employment as a school leader. However, successful completion of a preparation program does not automatically guarantee employment in a leadership position. In countries where preparation programs are optional, there is a growing tendency among employing bodies towards expecting some preparation for the position or requiring applicants to complete in-service training immediately after appointment but before taking over the leadership position. This tendency is certainly matched by an increasing understanding of the central importance of school leadership for effective schools.

Additionally, pre-service training offers the chance of self-evaluation and of assessing one's own interests and strengths. As a kind of orientation process, it may help to decide on one's next career stage more consciously. Participants who may not achieve a leadership position at the end of the program are then looked upon as resources for professional development and change agents in their schools and may be involved in both leadership and management activities, especially as shared or distributed leadership and management becomes inevitable.

More and more countries are thinking about offering preparatory courses in addition to their already existing induction programs as they move away from the concept that the school leader is nothing more than a teacher with a few extra responsibilities, a position which is associated with the phrase 'primus inter pares'. This mirrors a prevailing view of 'school leadership' as a profession in its own right that requires a shift of perspective in the knowledge, skills, and dispositions that school leaders need.

In many countries, school leaders must meet centralised qualifications and have preparatory training, as a conditio sine qua non. This may be regarded as a kind of paradigm shift in the view of school leadership and leadership development because it supports an increased recognition for the importance of specific (and often extended) training and its central part in adequately qualifying candidates for their new leadership role.

Extensive and comprehensive programs

This tendency to regard school leadership as a profession in its own right has implications for the depth and breadth of training and development programs. This comparison indicates a significant tendency towards more extensive training programs that are then able to explore many of the challenges connected to this new leadership role and its responsibilities.

Many of the countries that have, in recent years, gained more experience in the field of school leadership development originally started with short courses of a very practical nature. These courses often focused on fairly limited areas of interest, and were designed to provide answers rather than encourage reflection and development. The programs were then extended so that the courses might add up to a more comprehensive package, supported by a theoretical framework. These training and development opportunities have become quite extensive. Examples can be found in North America as well as in Europe, Asia, Australia and New Zealand. Since the extensive set of required activities is usually preparatory and often takes place before one applies for or before assuming positions of leadership responsibility, it is safe to state that the relevance of school leadership for the effectiveness and improvement of schools has been realised in many countries in the last years.

Multi-phase designs and modularisation

The international comparison shows that there is a tendency to move away from the idea that adequate preparation and development could be completed in a specific time frame using a standardised program. Instead, school leadership development is more and more regarded as a continuous, life-long process linked to the career cycle and to specific needs of the leader and the needs of her or his school.

This continuous process could be divided – ideally speaking – into the following phases:

1. A continuous development phase for teachers: this provides training and development for teachers in the fields of school effectiveness, school improvement, and school leadership.
2. An orientation phase: this provides the opportunity for teachers interested in leadership positions to reflect on the role of a school leader in respect to their own abilities and expectations.
3. A preparation phase: this occurs prior to taking over a school leadership position or even before applying for it.
4. An induction phase: after taking over a leadership position, development opportunities are provided to support the school leader in his or her new position.
5. A continuous professional development phase: this provides various training and development opportunities for established school leaders, best tailored to their individual needs and those of their schools.
6. A reflective phase: this provides the opportunity for experienced school leaders to continue to grow introspectively by being involved in development programs for others as coaches and to gain new experiences through learning by teaching, supplemental train-the-trainer-programs, and the exchange with the younger colleagues who participate in the programs.

Additionally, a 'reflective phase' might have a twofold effect, for themselves and for others. Experienced school leaders continue to grow introspectively by being involved in development programs for others as a coach. This phase would be supported through methods like 'learning by teaching' and through supplemental 'train-the-trainer-programs'.

Although it may be erroneous to state that this ideal model has currently been realised in some countries, tendencies towards developing in this direction are emerging. Instead of a standardised program for all participants that intends to 'teach' all the required competences at once, more and more countries provide professional development through multi-phase designs. These phases are, ideally, based on a coherent conceptual model.

It can also be observed internationally that programs become increasingly modularised. These modules tend to be organised according to one's individual needs that become evident during different stages of the school leader's career, but also to the needs of the school she or he is in charge of. For the modules, there is no mandatory sequence for completion. Moreover, the individual school leaders may well rely on these modules for support during crucial phases of their careers. They will be 'collected' and archived in a kind of personal portfolio.

Personal development instead of training for a role

As the role of school leaders is becoming more and more complex, it becomes more and more evident that it is no longer sufficient to train potential candidates or school leaders for a fixed role, whose model may be quickly outdated. Instead, aspiring school leaders must develop a vision within the context of their school and adapt their role and responsibilities to that context. To achieve successful adaptive leadership, the programs of some countries include components such as personal vision, personal and professional development, development of fundamental values and of one's ability to reflect, time and self management, developing mental models of the organisational structure, and activities in the school that mirror good leadership activities. Moreover, day-to-day school or internship experiences have become reflective activities that result in constant re-conceptualisation.

As far as qualifications are concerned, the emphasis has shifted from focusing on a specific role to a broader one that concentrates on personal learning and one's needs in the areas of knowledge, dispositions, and performances that would be useful in a more complex environment. Often, training for a management position has been replaced by offering professional development opportunities for one's leadership style. It is the personality of the (aspiring) school leader that becomes the focal point of any program.

The communicative and cooperative shift

In spite of the increasing stress on school leaders due to the complexity of the role – particularly in countries with more devolved systems – school leadership programs are not preoccupied with administrative topics. On the contrary, the overall focus of school leadership programs is no longer on administrative and legal topics as it used to be in earlier programs, but has shifted to topics that focus on communication and cooperation.

The image of school leaders as experts in administration has shifted to school leaders as experts in communication and cooperation. This trend has become another international paradigm shift. Topics such as communication, motivation, collaboration, collegiality, and cooperation are essential parts of all programs. Internationally, there is the recognition that understanding and effectively using these topics is essential for one to become a successful school leader.

Communication and cooperation as essential components in leadership development programs also play an important role as far as the methods applied in those programs are concerned. Realizing that learning processes that take place in groups provide participants with better opportunities for experiential learning, more programs are moving in the direction of small and large group interaction. The aim then becomes one of creating reflective practitioners and this will intensify the teaching-learning experiences. In addition to traditional seminars, 'collegial learning' – learning together with other colleagues – is being realised through a variety of strategies including peer-assisted learning, peer coaching, critical partnerships, acquiring knowledge from experienced peers by shadowing or through mentoring programs or collegial networks that were created (for example as a result of experiences from the cohorts that existed during other training programs). When one uses these strategies, learning evolves through mutual reflection and problem-solving processes; it is about learning with and from colleagues.

From administration and maintenance to leadership, change and continuous improvement

Throughout the countries that were involved in this study, an important paradigm shift has occurred: from a focus on managing schools with an emphasis on maintenance, to a focus on leading and improving schools. The aim is no longer to make the organisation function within a static or fixed framework, but it is considered essential that programs adequately respond to the challenges created by social, cultural, and economic changes. Schools are no longer static organisations, but must be considered learning organisations, each with their unique culture. Therefore, leading a school no longer means simply maintaining the status quo, but, above all, developing a changing learning organisation. Consequently, what is worthwhile has to be sustained and, at the same time, necessary changes have to be made, and after being successfully implemented, they have to be institutionalised.

This paradigm shift can be identified in the lists of themes that are in the courses of many school leadership development programs. They take into account that school leaders must be educational leaders and that is about initiating, supporting, and sustaining substantive and lasting change as well as continuous improvement in schools for the benefit of pupils. The focus is then on a collaborative and collegial style of leadership.

Qualifying teams and developing the leadership capacity of schools

One trend in development programs for school leaders is particularly interesting: attempts are made at linking one's qualification and development more directly to the improvement of individual schools. School leader development programs then become a means of school improvement. They intend to affect and impact directly on everyday activities at school.

Training and development providers, therefore, try to attract more and more teacher leaders to some of the preparation programs, thus broadening the target group. Rather than simply attracting aspiring school leaders, teachers who want to enhance their leadership competences are admitted to these programs. These applicants may not plan to apply for a school leadership position, but may be interested in other school-level leadership positions such as department head or head of year.

If school improvement is the explicit goal, whole school leadership teams or teams of staff members may participate in these programs, and this may sometimes include parental and community representatives. While the trend towards team-based training is only apparent in a few programs, an increasing number of providers indicate that they intend to focus on developing leadership teams in addition to focusing on school improvement. They believe that this approach is necessary in order to develop stronger leadership and enhance the leadership capacities within schools.

This shift in focus to enhancing the leadership capacity of the school (rather than qualifying 'just' one individual person) implies that the professional development activities no longer take place solely in an institute away from the school site, but at the individual school itself, where school improvement processes are initiated and implemented. Programs then are much more focused on content-specific topics that are generated at the individual school level.

From knowledge acquisition to creation and development of knowledge

In many development programs, two conceptual considerations appear to be taken into account. Firstly, at a time when swift changes in many areas, including education, are coupled with a worldwide explosion of information, it would not be sufficient to simply increase the quantity of declarative knowledge that aspiring school leaders must learn. The development programs must prepare the aspiring school leaders for new knowledge as information continues to expand. This is a shift away from imparting a seemingly fixed knowledge base towards the development of procedural and conditional knowledge. The notion of 'acquiring' knowledge will be replaced by the concept of 'developing' or 'creating' knowledge through information management. Participants should enhance their ability to learn and to question traditional thinking patterns and cognitive processes. They should acquire skills to be proactive in complex work environments. How to learn and to process information are therefore increasingly emphasised.

Secondly, in general there is consensus among the providers and the participants that the teaching strategies used in development programs have to meet the needs of adult learners. Hence, fundamental andragogic principles must be taken into account: while children learn new things, adult learning usually supplements what has been already learned. Their individual experiences always have a subliminal influence on the new information and, at the same time, represent the foundation upon which something new can be learned. Consequently, development programs increasingly create learning environments that offer the opportunity of deliberately linking and embedding new information in previous experiences. The reality and the experiences of the participants, their needs and problems, become the starting point as well as the point of reference for the selection of contents and learning methods that are used in these programs. The knowledge gained during the development programs should be directly transferable to one's specific working environment. Therefore, knowledge cannot simply be imparted but it has to be created and developed.

Experience and application orientation

In the programs studied, there is a clear tendency towards experience-oriented and application-oriented learning. A shift away from purely practice-driven or from purely theory-driven learning towards practice-with-reflection-oriented learning became evident in many programs. This becomes obvious as development models bring practical experiences from the schools into their programs. Case studies play a particularly important role in this context. Popular learning strategies including reflective practice such as learning journals, discussion groups, working with mental maps, etc. are linked to authentic school experiences.

New ways of learning: workshops and the workplace

With the aim of providing some orientation towards the participants' needs, being relevant to actual field practice and to be able to transfer learning into the world of work, the participants are often placed in workshops in which they role play school situations within the context of carefully constructed cases. Learning becomes team focused as these role plays and case studies unfold. Problem-based learning is a concept employed by many programs, although notably by those in North America.

Going one step further to bring theory to practice, genuine, authentic cases are taken from everyday school life. Thus, the cases are grounded more concretely and authentically in school situations. Many providers reported that real-life case studies are used more widely than ever before.

An increasingly high number of development programs take another step, leaving the workshop model and going into the actual workplace, using school as clinical faculty. For the participants of predominantly pre-service school leadership preparation programs, internships at one or several schools are organised within the preparation programs. They 'shadow' the principal or head of school, assist, or take on leadership tasks, and carry out school-wide projects independently. Here, the school leaders at the intern's school function as mentors or supervisors.

Adjusting the program to explicit aims and objectives

It becomes increasingly obvious that the process of developing school leaders is becoming more professional. This also includes explicitly stating the program's aims that aspiring leaders must achieve. Until now, programs were not necessarily developed with explicit goals or objectives, especially in the early stages of their development. Instead, generalised statements like 'school leader development aims at developing school leaders', were used. Content-wise, however, the aims postulated differ greatly at a higher level of explicitness. They can be classified according to their main focus: those with an explicit functional orientation and/or task orientation, those which are distinctly competence-oriented or cognitions-oriented, those with a definite orientation towards school improvement and some which are clearly vision- or value-oriented.

As new concepts of leadership and schools emerge, based on the values of society, they begin more and more to impact on the programs.

New paradigms of leadership

Preparation programs reflect more and more the new concepts of leadership. The school leader is often called the educational leader, an instructional leader or a visionary leader, and schools are no longer seen predominantly as static systems in which the existing structures have to be managed. Concepts like 'transformational leadership' are increasingly being advocated. Transformational leaders regard schools as culturally independent organisms, which have to continue to evolve. Hence, transformational leaders attempt to actively influence the school's culture so that it values collegiality, collaboration, cooperation, cohesion, and self-reliant learning and working. They are not only expected to manage structures and tasks, but to focus on establishing relationships with and within the staff and make an effort to influence their thinking towards a common vision and commitment. The application of this definition of leadership appears to be particularly successful in school development processes.

If schools are considered learning organisations, this implies the stakeholders are empowered and collaboratively work together. Leadership is no longer hierarchical and dependent upon one person, but is shared and empowers others as viable partners in leadership. The previous separation between leaders and followers, as well as between the teachers and learners, begin to blur. A new concept emerges, called 'post-transformational'. Another concept that emerged in the study, is 'integral leadership'. It aims at overcoming the classical division of management and leadership and emphasises an integrating perspective focused on the overall aims of the school.

Orientation towards the school's core purpose
Another trend that emerged seems particularly interesting: New concepts of schools are embedded in the programs. Schools are now seen as learning, problem-solving, creative, self-renewing, or self-managing organisations. This has an impact on the role of school leaders, and, on how training and development programs have been designed.

The schools' core purpose, namely teaching and learning, and the specific aims of schools within society today and in the future have increasingly become the starting point for designing school leadership development programs. These reflections on the school, its role and function, and – derived from this – on successful leadership have definitely influenced development programs in more and more countries. The principle that 'school has to be a model of what it teaches and preaches' (see Rosenbusch, 1997b) has become the implicit foundation of some leadership development programs.

The development models strive to create the vision of the school leader as educational leader whose focus is on improving the schools' teaching and learning processes and outcomes. Focusing on the school's primary goal is not only a reasonable means of guiding the school leaders' decisions, but also becomes a criterion for reflective inquiry into their ways of thinking and behaving.

2. Evaluation, Best Practice, and Multi-Stage Adjustment of Aims in School Leader Development

Evaluation, and Best and Promising Practice

An evaluation of the individual programs would be important considering the variety of substantially different approaches, conceptualisations, and macro- and micro-didactic realisations of development programs for school leaders in the countries included in this study. The focus would be to look at whether the measures taken by providers meet expectations. Do they reach the goals established by their providers, do they meet state requirements, the policy goals of educational organisations, and the expectations of school administration, the hiring boards, the participating school leaders and the profession as a whole, as well as the schools' wider community namely teachers, parents, and last but not least pupils? How effective are these programs? Are there substantive and lasting effects on the development of school leaders as a result of the programs?

The study presented here is in no position to answer these questions. There are several reasons for this:

Firstly, this study aimed at exploring a comparison of school leader development programs in different countries. A comparison like this is the conditio sine qua non for extended research into the effectiveness of such programs. Further studies concerning program effectiveness are to be carried out elsewhere (see chapter III.4.). However, such a study is going to be prepared at the moment.

Secondly, although the questionnaire used for data collection included an additional question about if and how the respective programs were evaluated, an examination of the data collected indicated that only very few countries had responded to this question. This may be due to the fact that some of the programs had been introduced so recently that an evaluation simply could not have been undertaken at this early stage. In some countries the data collection took place parallel to the first cycle of implementation but hasn't been analysed yet. In some other countries, especially those that are extensively market-oriented, it was found that the competitive providers are not willing to reveal existing evaluation data.

Thirdly, in cases where this information has been provided, it was largely unsystematic and informal, evaluation data concerning the level of satisfaction achieved as requested from the participants. Naturally, some quite interesting inferences can be deduced from this, although they are not suitable for systematic application. There are hardly any follow-up assessments or external evaluations (involving the teachers' perspectives, school advisory boards, or – as a classical method of school effectiveness – the correlation with school and pupil outcomes). Hence, we know little about what is achieved by the programs and its relationship to program effectiveness because such studies were rarely conducted.

Evaluation processes were found in the state-run basic training of the Skolverket in Sweden, the English National Professional Qualification for Headship (whose evaluation results are not available for publication), the School Leadership Preparation Program in New South Wales, the Danforth Educational Leadership Program at the University of Washington, and the Management- en Organisatie-opleidingen program offered by the Nederlandse School voor Onderwijs-management. An exemplary survey of school leadership development in the area of primary education has been carried out by order of the ministry in the Netherlands as well.

Generally speaking, studies that determine effectiveness continue to be something that is needed. Evaluations that consider a range of programs or even international, comparative studies of effectiveness are completely non-existent. This may be partly due to the fundamental problems that arise whenever attempts at evaluating the 'effectiveness' of the teaching and learning processes in development programs are made. However, considering the high costs that are often associated with development programs (also for the participating candidates), efforts at evaluating program effectiveness would be of major importance.

Notwithstanding, what can indeed be accomplished here is a preliminary assess-ment based on the aspects discussed in the previous parts. It seems feasible to examine the insights gained from the analysis, discussion, and comparison of the data retrieved. Additionally, the conceptual designs of development programs for school leadership may be examined within the context of theory starting from organisational education and the paradigms of school effectiveness and school improvement, as it was presented briefly in the first part. This will become the basis for such an assessment.

There surely will not be a single development program that could serve as a perfect model in this sense (even if the question of transferability to the contexts of other countries were ignored). However, it can be recognised that some programs feature obvious components of what could be characterised as best and promising practice. These examples have a high potential as incentives for others to consider when planning school leadership development programs.

One program that could be cited as an example for 'best and promising practice' is the Danforth Educational Leadership Program of the University of Washington. It appears that the conception of the program accommodates recent research, especially in its paradigms of 'school effectiveness' and 'school improvement'. The critical role of school leadership as a key factor for school effectiveness and as a change agent for school improvement processes has been realised and considered carefully in this program. In addition to that, the results of international school leadership research, particularly research concerning school leadership tasks and behaviours, as well as criteria for establishing a certain level of expertise, are also incorporated in this program. Another unique aspect of this program is its history: it emerged from analysing the pervious leadership development models in the USA in the 1980s and the criticism brought forward by educational researchers and the profession itself about these programs. The Danforth Foundation's attempted to improve school leadership development programs. In this program, quality assurance is guaranteed to a relatively large degree by complying with certain guidelines and standards and by awarding a state certificate for the participants. This prevents arbitrariness and restricts providers in a complex market from offering different programs that vary in quality in a decentralised education system like that in the USA. Nevertheless, the program preserves flexibility and variability.

This program is initiated before one applies for a position as a school leader. Since it is a prerequisite for such positions, it has a strong professional validity. Rather than focusing on mere on-the-job-training, long-term development of skills (which stands a bigger chance of resulting in a 'mental shift' for the new tasks) is emphasised. The rigorous selection process of participants aims at preventing those who are not yet qualified enough or who do not have the capacity to lead in the view of the school officials who grant financial support. Hence, it results in selecting candidates who are reasonably motivated as well as those who have a high academic ability.

The program's 'working assumptions' use 'outstanding school leadership' as a starting point. In this program, school leadership is focused on shaping and improving schools, as well as making prospective leaders skilful managers. Schools are regarded as dynamic, self-improving organisations. Therefore, school leadership must include a belief in these values. This broader understanding of leadership incorporates the moral aspects of leadership. Leadership in a democratic society is based upon values like equality and justice. Topics in this program include the role of values, ethics and morals, the question of power and the legitimacy of leadership. The participants are inspired to develop their own educational model within a leadership platform on which they can base their actions as school leaders. This is discussed by the school leaders in the context of the qualifications imposed on school leadership by the state of Washington in its 21 Performance Domains.

The Danforth Leadership Program includes a focus on the central role of collaboration as it relates to leading and guiding a school. The importance of collaboration is reflected in the topics included in the seminars. Regarding best practice, however, it seems especially interesting that communication and collaboration are not only included as far as the content is concerned, but are also skills for which the participants strive, and seek to apply in their leadership activities at the same time. The teaching strategies and learning methods correspond to these values and beliefs. During the development program, the learning groups are firmly established and they should continue beyond the program as collaborative networks. Collegial learning or learning through discourse with peers becomes actualised through variations of problem-based learning as a consistent way of learning in workshops based on case studies. Here, learning is collaborative and interactive, it emphasises participation, and it is partly self-organised and self-directed. The most important method in this preparation program, trying to integrate authenticity and real-life situations, is learning at the actual workplace. The participants required to take part at an internship during the school year which includes shadowing and mentoring as well as taking over school based leadership tasks.

Special consideration is given to the selection and professional development of the course instructors. The teams of facilitators, trainers, etc. consist of university faculty and school practitioners. Close collaborative relationships are maintained with school officials, and local schools become the focal point of the process. Additionally, these relationships are critical to establishing the internships.

As far as the scheduling is concerned, seminars vary in length and frequency. This flexibility attempts to achieve the various learning goals of the program.
In addition to the example of the University of Washington as cited above, another best and promising practice is particularly noteworthy to examine – the highly modularised, multi-staged program School Leadership Strategy in New South Wales. This program looks at school leadership as a continuous process which takes place in different phases. These phases match the various stages before and after assuming a leadership position. This example presents another interesting concept: the target group is not exclusively potential applicants for school leadership positions. It also includes teachers who would like to generally improve their leadership skills or attain other school positions that require additional competences. As a consequence of this idea, entire school leadership teams or even teams within the teaching staff, together with representatives of the parents or the community are included. This is also the case in the School Leadership Teams development program of the California Leadership Academy. Programs similar to this that incorporate leadership teams can be considered 'best practice', since teams increase the leadership capacity of schools as tasks and responsibilities are distributed. School leadership development then becomes a measure of school improvement and directly affects the entire school community, including the pupils. Leadership means that as many individuals from the staff participate as possible. This is clearly the goal intended here.

Multi-Stage Adjustment of Aims

If the goal described above is to be realised, school leaders have to be qualified to understand the complexity of the system along with the different individuals and groups involved as well as the interactive and collaborative relationships between them. Additionally, school leaders need to be able to develop influencing relationships and 'lead' proactively. Moreover, they need to be familiar with the potential 'stumbling blocks' and 'structurally disrupted conditions' that may exist and how these obstacles can become challenges that will be overcome. School leaders have to be qualified to intervene appropriately whenever situations like these occur. School leadership must shape the school in a way so that the teachers who work there can then ideally be more effective in achieving learning outcomes for pupils. The leader then becomes a facilitator of change and someone who supports teachers in their work with pupils. This requires reflection on the role, function, and goals of the school, and consequently on the role, function, and goals of appropriate leadership and management. Reflections on these topics have influenced the programs in this study to varying degrees. The aims of each program may differ greatly and are more or less pragmatic or theory-based. A number of programs feature the much-demanded adjustment of perspectives in their aims: to firmly establish school leadership within a comprehensive conception of school and schooling.

A consistent connection between educational and organisational action that takes into account both viewpoints would imply an even more stringent connection among aims. Development programs for school leaders therefore require a multi-stage adjusting of aims (see table below). The first question would be: what are the essential aims of education? From these, the corresponding aims for schools and schooling in general can then be derived: what is the purpose of school and what are the aims of the teaching and learning processes? Considering the perspective of the new field of 'organisational education', one should ask: how does the school organisation need to be designed and developed in order to create the best conditions possible so that the entire school becomes a deliberately designed, educationally meaningful environment? This in turn would enable teaching and learning to take place as well as multi-faceted and holistic educational processes that would lead to achieving the schools' aims.

This, again, leads to the essential concern of school leadership: what are the aims of school leadership regarding the school's purpose and the individual context of each school? To be even more precise: how do school leaders lead if it is the goal of leadership to shape school in a way that makes it possible to reach its aims?

Therefore, the aims of school leader development programs should answer questions like this: what is school and schooling about, and what is leadership and management about? What is the core purpose, what should be the aims? What kind of training and development opportunities are therefore needed to prepare and support (aspiring) school leaders in adjusting their perspectives, conceptualising their role and function, developing the necessary competences, and mastering the manifold tasks within the individual school in order to provide conditions and support staff so that effective and efficient teaching and learning takes place for the sake of the pupils.

This should be the essential or core goal for aligning and evaluating school leadership development programs.

Table 10: Multi-stage adjustment of aims of school leadership training and development

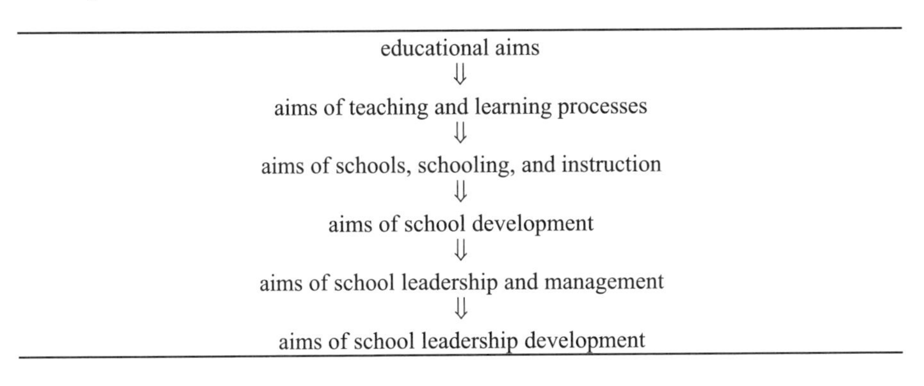

educational aims
⇓
aims of teaching and learning processes
⇓
aims of schools, schooling, and instruction
⇓
aims of school development
⇓
aims of school leadership and management
⇓
aims of school leadership development

The multi-stage adjustment of aims as indicated in the table above represents the basis for the overall concept of development program models for school leaders. This alignment must be apparent as one conceptualises and develops the basic approach to any program. The objectives developed from the aims in the multi-stage adjustment model should influence macro-didactic considerations that is defining the target group(s), the timing, the nature of participation, the professional validity, but also the pattern with the total number of training and development days, the time-span, the scheduling, etc. Moreover, it should affect micro-didactic reality such as the curriculum, teaching strategies and learning methods. As stated previously, additional studies concerning the effectiveness of the programs would have to be completed. Appropriate meta-studies based on this comparison could tell us as well whether the programs' aims have been realised. The multi-stage adjustment of aims should be a benchmark and a tool for the evaluation of school leadership development programs.

3. Recommendations for Designing and Conducting Training and Development Programs

On the basis of the international study in 15 countries worldwide as well as the comparative analysis and the discussion of the results, a number of basic principles could be deduced, which are essential for the qualification of school leaders. These could serve as recommendations or even as guidelines for the design and conception of future programs. They are listed here in a kind of catalogue that is not intended to be complete but tries to be open for supplementing. They are also meant to give an input that leads to new ideas and refinements. The following principles might also serve as standards which have to be considered by providers. Then, they could be the criteria for the accreditation of providers and programs or, in case of a certificate for quality assurance offered to the providers, serve as criteria for certification.

These recommendations include:

- Centralised guidelines for quality assurance combined with a decentralised implementation

- Suitable recruitment of teams of highly qualified trainers with appropriate backgrounds
- Selection of participants
- Clear and explicitly stated definition of aims, using the core purpose of school as a focus
- Alignment according to values and educational beliefs
- Development as a continuous process
- Importance of declarative and procedural knowledge
- Suitable balance between theory and practice
- Orientation towards the actual needs of the participants
- Active involvement of the participants
- Inspiring collegial learning and intensive collaboration
- Problem-based training in workshops
- Learning opportunities at the workplace
- Focus on the personal and professional development of the participants as well as on improving their schools
- Self-organised and reflective learning processes, supported by communication and information technology
- Academically grounded and authentic training material
- Presentations of learning results and self-evaluation of learning processes
- Certification of participants
- Conceptually established support for the actual transition

Centralised guidelines for quality assurance combined with a decentralised implementation

The responsibility for designing the programs and for assuring their effectiveness should be shared by the profession itself and the state. More centralised forms of quality assurance (for example, by determining guidelines and standards, the accreditation of providers, the certification of participants, etc.) in combination with a decentralised implementation of the programs (together with corresponding possibilities of collaboration with school authorities and schools, etc.) appear to be suitable. The intensive collaboration with universities should guarantee reasonable academic foundation and support.

Suitable recruitment of teams of highly qualified trainers with appropriate backgrounds

Special consideration needs to be given to the suitable selection and recruitment of the instructors, facilitators, trainers, mentors, etc. They are ultimately those who implement the program's concepts, are in immediate contact with the participants, and are responsible for the teaching and learning processes. Teams of trainers and instructors, which also include university faculty and representatives of the profession itself but also from the business world are especially suitable. They should not only design the implementation of the development program as a team, but develop the concept and plan together as well. They should be highly qualified and experienced in their field, but also have an understanding of other areas so that an interdisciplinary, integrative approach can be implemented.

Selection of participants

Careful selection of participants is needed to choose suitable candidates who meet the program's requirements and have a strong motivation to succeed as (aspiring) candidates for school leadership positions. Since the school officials or the ministry for education may pay the fees for the program either by offering scholarships for individual participants or by financing the entire program, these public entities should be given the opportunity to get actively involved with the selection of participants. Therefore, selection criteria have to be developed and agreed on collaboratively.

Clear and explicitly stated definition of aims, using the core purpose of school as a focus

Development programs should begin with an explicit statement and a clear definition of their aims. Goals and objectives should be clearly established and the programs' curriculum as well as macro-didactic and micro-didactic considerations should becomc an outgrowth of these aims. The goals should not be dominated by a set knowledge base that should be imparted to the participants, nor should that knowledge base be established as a result of external pressures. On the contrary, a concept of the purpose of school and schooling, and the function of educational leadership within that context should be clear, that is to lead communities of children and adults in a way that teaching and learning processes are promoted, supported, and genuine educational processes are realised. In the end, it is this goal of school and school leadership activities from which the goals and objectives for the development of school leaders should be derived.

Alignment according to values and educational beliefs

In a world of changing values and a broad range of different values, the development for educational leadership must not be subject to a positivistic management-oriented paradigm, but should be based on a value-centred paradigm. The participants should reflect upon their own values in general, and upon their educational values in particular. In the end, the individual should be able to develop rather than simply be made 'suitable' to fulfil a certain fixed school leadership role effectively. Besides, leadership must be madc legitimate in society and above all to those who are 'led'. Power must be handled carefully, and the balance between influence and confidence has to be maintained. The main principles of education in schools have to be respected: maturity has to be encouraged when dealing with pupils, teachers, and parents, acceptance of oneself and of others has to be practised, autonomy has to be supported, and cooperation has to be realised. Development programs should be aligned to these beliefs.

Development as a continuous process

The development of school leaders should be seen as a continuum, beginning with the initial teacher training, ongoing professional development for teachers, adequate orientation, preparation, and induction programs. For established school leaders continuous professional development should be provided, tailored to their individual needs and those of their schools. For experienced school leaders a reflective phase provides development opportunities through learning by teaching, supplemental train-the-trainer-programs, and the exchange with younger colleagues. This would lend itself to a multi-phase design.

Importance of declarative and procedural knowledge

In recent years of swift social, economic, cultural, technical, political, and educational-policy changes, along with experiencing increasing information overload, it would not be sufficient merely to enhance the quantity of (declarative) knowledge that aspiring school leaders should know. The development program would rather have to prepare aspiring leaders for something that they or others do not know. Consequently, there must be a paradigm shift away from programs that impart a fixed body of knowledge and towards the development of procedural and conditional knowledge. The acquisition of important knowledge should be accompanied by the creation of knowledge and the effective management of information. The participants shall be supported to further develop their ability to understand cognitive processes, and shall achieve what Giroux (1988) calls 'conceptual literacy'. Preparation programs must prepare aspiring school leaders to work in a complex, sometimes chaotic work environment.

Suitable balance between theory and practice

The time structure and scheduling should take into consideration that learning in terms of changing behaviour and thought patterns, is a process which should be supported by stimuli and information; a process, which, however, needs a lot of reflection and exchange, and which occurs over time. Therefore, development programs should have a good balance between theory and practice as far as both the content and the methods are concerned. Consequently, programs are designed which comprise a higher frequency of short events over a longer period of time.

Orientation towards the actual needs of the participants

The starting point of any program should be the participants, their experiences, needs, views, problems, and maybe their own prejudices or bias about their view of leadership. The programs should be needs-oriented. Here, the self assessed needs of the participant (What do I need? Where do I feel unsure?), those assessed by others (Where are her or his weaknesses?), and the demands of the school she or he comes from (internal or external: what does the school need, which competences are required?) may indeed be divergent. Different evaluations based on different levels of professional experiences, at different stages of the career cycle are essential. For a systematic evaluation of the qualification needed, a needs assessment might be helpful for the individual participant. Additionally, feedback should be given continuously to the participants about their individual performance. It is easier to realise a needs-oriented concept if programs are using a problem- and practice-oriented approach.

Active involvement of the participants

Previous knowledge and previous experiences should be collected thoroughly and used systematically throughout the program. The participants should be provided with the opportunity to contribute actively to the planning and the design of the program. Trainers and participants should collaborate and interact as much as possible. During the entire program, and throughout all its phases, the program should contain ample opportunities for trying new ideas and opportunities for collaborative reflection.

Inspiring collegial learning and intensive collaboration

The participants should be given the opportunity to collaborate with colleagues and to learn with and from colleagues in various contexts such as peer-assisted learning, learning tandems, or critical friends. Mutual participation in the tandem partner's school life can offer valuable insights. The formation of a professional network, which can outlive the duration of the development program, should be encouraged and supported. As a result collegial support and counselling for practising school leaders will become second nature.

Problem-based training in workshops

The teaching strategies and learning methods should be problem-based and foster both individual learning and collaborative teamwork and learning. This is important for adult learners. Problem-oriented learning environments that offer complex tasks can be most effective. A learning context like this has the best chances to be authentic and congruent with the working context that the participants will find in their everyday school life.

Learning opportunities at the workplace

In addition to the workshop, learning at the workplace is particularly useful. The practical applications of the development programs should be carefully considered. Internships supervised by mentors that also include opportunities to shadow practising school leaders, as well as active participation in leadership tasks, which cover as many aspects of school leadership activities throughout the term as possible, have proven particularly effective.

Focus on the personal and professional development of the participants as well as on improving their schools

The development of the individual participants in terms of 'individual development' should be linked to 'school development'. Modules that match the personal development needs and, additionally those, which include the present needs and demands of the individual school, should be encouraged. Moreover, the development should take place on site. Participants should have the opportunity to provide some input into the managing of the school and they should recognise the school itself as one of the essential places to learn school leadership skills. Therefore, it would be important to integrate other members of the school's staff into the program. This will result in the creation of school leadership teams in addition to the individual school leader. The respective school indeed has to benefit from this concept in order to attract its staff's support and to extent the leadership capacity of the school.

Self-organised and reflective learning processes, supported by communication and information technology

The participants have to be the designers of the learning processes. As mentioned above, they should be partners in the program and be actively involved. This also means that they plan their own learning processes according to their particular needs. Moreover, the training and development programs should also support individual reflection processes through writing a learning journal, assignments, etc. Here, communication and information technologies play an important role. New forms of self-designed and interactive learning can be applied throughout the program by using CD-ROMs, email platforms, and web-based learning environments.

Academically grounded and authentic training material
The materials used in development programs should be based on topics that focus on current topics in education. They should use authentic documents from current school leadership practice. Cooperation with university faculty seems suitable and necessary here as well. Carefully selected media will result in a broader variety of teaching strategies and learning methods.

Presentations of learning results and self-evaluation of learning processes
The participants should present the results of their work to the cohort. Self-evaluating the learning achievements stimulates one's own awareness of the teaching-learning experience. Objective, external feedback complements the evaluation processes and conveys a feeling of achievement. It should include both a summative and a formative feedback. The summative one gives a feedback about what has been achieved so far (i.e. is looking back), the formative one tells one what else con be done (i.e. is looking ahead).

Certification of participants
Upon successfully graduating from the program, participants should be awarded a certificate. This will give them the opportunity to provide information to others about their level of qualification as well as document their experience in the program. Additionally, the committees or boards who recruit school leaders may use the certificate as a selection criterion.

Conceptually established support for the actual transition
The participants should not be expected to assimilate and transfer what they have learned and experienced by themselves. Frictional losses due to transfer should be as small as possible. The sustainability of the development program should be both a goal of the program and one which is realised by follow-up events. This could be a week-end seminar after half a year to recapitulate learning outcomes and share experiences. Altogether, as many support opportunities far as possible should be provided. These could include establishing learning cohorts and networks of the participants, which remain in existence beyond the development program and of course, continue throughout ongoing opportunities for those who have successfully completed the programs.

4. What Has to be Done by Research?

Based on this explorative study, further research desiderates become apparent, which will be outlined briefly:

1. There is still some need of further pure research into educational leadership and the training and development of educational leaders.
2. In some countries, there is still some need of research into tasks of and demands on school leadership, as the results of existing job analyses cannot easily be transferred from one country's context to another. Among these should be surveys of school leadership estimated as 'good' by teachers, parents, and pupils.
3. Of particularly interest is the connection between school leadership and classroom development. Teaching and learning, or education and instruction are the core activities of schools. In terms of an organisational-educational approach (Rosenbusch, 1997b), it is this from which the core purpose of school leadership must be derived: what should school leadership activities be like in order to have the best possible effect on classroom instruction in a twofold sense, providing the best possible organisational conditions on one hand, and in terms of having an (immediate) effect on classroom instructions and classroom development on the other hand?
4. It is necessary to conduct analyses concerning the training and development needs of school leaders at different career steps and in different school contexts.
5. Specific research has to be carried out in the ways school leaders develop competences. How do they generate knowledge? How do they develop expertise? How can the transition of the knowledge acquired in the development program into practice be improved? How does this change across the various career stages? What is considered supportive?
6. Further international comparative studies, which include a more detailed investigation of some of the programs described in this study particularly concerning their effectiveness should be carried out. This will provide insights in the quality, and sustainability of these development programs. A standardised research design, and not only one for a meta-study would be desirable here.
7. There should also be educational-economic studies on the efficiency of training and development programs. These could provide information for educational-policy decisions concerning the overall efforts taken[5].

---

[5] The author assumes that the effects of training and development opportunities have been clearly underestimated. For this, studies on human resource development from the fields of andragogy, organisational psychology, and human resource management should be made use of. Calculations like these are certainly highly complex since many different parameters need to be taken into consideration.

8. Moreover, it would be interesting to investigate how the development of individual school leaders could be linked effectively to the development of individual schools in terms of qualifying school leadership teams and other change agents in the individual schools (including studies of schools that have realised alternative leadership conceptions such as shared leadership, etc.).
9. The importance of selecting and recruiting school leaders and its connection with school leader development should not be neglected. Additional research would be desirable here as well.

It certainly would be useful to link empirical research to planning and designing development programs in terms of research-based training conceptions. National and international experiences should be considered and integrated, and international cooperations should be promoted. As a basis for this, national and international networks should be developed further. In these networks, educationalists and practitioners should have a forum for exchange and cooperation. The ICSEI-Network 'School Leadership and Leadership Development' as well as the planned virtual service facility 'EduLead.Net'[6] are meant to be another step in that direction.

[6] The Educational Leadership Network (EduLead.Net) will provide a web-based learning community, which will offer counselling and information in the context of school leadership and leadership development at www.EduLead.net (or .com/.org). This web-based networking will also be linked to the International Congress for School Effectiveness and Improvement (ICSEI).

IV. Country Reports

In the following fourth part, 15 country reports provide an overview of current practices to develop school leadership competences. These reports have been arranged in geographical order. They are based on a consistent structure to make a problem-oriented reading of the single reports easier and to promote a more systematic comparison.

Firstly, in the 'Background' section, information about the educational and school system of the country is provided, recent educational changes are outlined, and consequences for the role of school leadership are described.

Secondly, in the section 'Development of School Leaders', information about the development provision of the country is given, partially including information about its development and historical course (when relevant), about the range of offerings, as well as about the general approach to training and development. After that, the qualification program of the respective country or – if several opportunities exist – exemplary programs are described in detail. Additionally, a table summarises the key features of each program.

Thirdly, a 'Summary' tries to pull all essential aspects together.

Fourthly, the section 'Recent Developments' was added in summer 2002. It deals with the modifications that have been made in the program since the time the description is based on.

Fifthly, a 'Personal Statement' by the internal expert is given, which may add some information or judge the overall country situation and the general approach to leadership training and development or the program(s) quality, or add what is relevant to the topic.

In spite of efforts to achieve a standardised structure and similar chapter lengths, the extent of the single sections differs in some cases to appropriately take into account the individual country-specific situation. For example, information about the historical development of the training and development situation is given especially in the country reports of England and the USA. For both countries, this topic is well-researched. Especially in the USA there is a distinct tradition of scientific documentation and analysis of the evolution of training and development programs which is worth reading.

Sometimes the descriptions of the programs also deviate from the systematic order to some small extent. This, however, enables a more coherent, idiographic approach. We aimed at using and keeping to a standardised terminology to improve comprehension, clarity, and comparability in the country reports. If it seemed necessary to use the country-specific terminology, it was included in the text or added in brackets.

Europe

Sweden: Split Responsibility between State and Municipalities

Stephan Huber and Olof Johansson

1. Background

The Swedish school system has a long centralist tradition, based on the principle of guaranteeing the highest possible level of education for all citizens. The intensive work towards equal opportunities goes back to the introduction of the nine-year comprehensive school 'grundskola' in 1962 and also the uniting of all secondary schools under the roof of 'gymnasieskola' in 1970. Career training paths were also integrated step by step with the aim of creating equality among general education and career training courses. Now over 95% of 'grundskola' graduates take one of the 16 training routes at the now standard three-year 'gymnasieskola'.

Since the end of the 1980s, further school regulation reforms have been introduced. There were various motives and intentions behind this. Firstly, in terms of economics, limited budgets led to rationalisation and the more efficient organisation of state administration, including schools. Secondly, centralised control was the subject of much debate, it seemed to be outdated and had become complicated and disorganised. Decentralization measures promised more freedom for independent initiatives and therefore greater efficiency. Parliamentary decisions at the end of the 1980s and start of the 1990s specified that Swedish schools move away from direct control to goal-oriented control (see Eikenbusch, 1995). Since the start of the 1990s there has been a clear tendency towards further decentralization. The county school authorities[7] were abolished and their responsibilities, which were few in any case, were handed down so that a complete administration level was removed (see Fuchs, 1998). Transferring the authority to appoint teaching staff from the state to municipalities in 1991 was the most effective step. Since then, teachers and school leaders are no longer state civil servants, but are independently selected and appointed by the municipalities (taking the statutory qualification requirements into account). The financial duties of schools are split between municipalities and the state. Since 1991, administration of this budget has mainly been transferred to the local authorities, breaking away from the governmental accounting tradition. The Swedish state has since assigned a lump budget sum to the respective local authorities, which then allocate this to their institutions themselves. The movement towards a local authority-managed school system allows local authorities to decide in their school committees, 'skolstyrelse' (school board) or 'barn och utbildnings-

[7] Sweden is divided into 24 counties, called 'län', each with a centrally appointed governor.

nämnd' (children and education committee), for example, whether they want to form school associations or whether they want to combine the school authorities with other Departments, for example, the Youth Department (local school supervision is then no longer always performed by a qualified teacher, but by another official, for example, the Head of the Youth Department). They appoint teachers and can plan their continuous professional development in such a way that they become part of the development strategy for local schools.

The state is still responsible for specifying nation-wide goals in the form of guidelines which ensure the equality of education, a framework for the responsibility and involvement of pupils and parents, guidelines for assessment and grading, structures for the different training courses, specifications for school management responsibilities, etc. The Ministry of Education itself is relatively small because the actual administration work is dealt with by the Skolverket (National Agency for Education, NAE) and its eight regional offices. Its tasks include supervision of schools, the evaluation of school systems (by means of formal inspections of individual schools), the further development of curricula, the coordination of school research as well as the organization of teacher training and the state qualifications for school leaders (see INES-database/DIPF, 2000d; Eurydice, 2000; Fuchs, 1998). The Skolverket will be divided into two agencies on the 1st of March 2003. One will be responsible for control and statistics and the other for support, school improvement and development activities. The second new agency will also be responsible for the National Headteachers Training Program.

Today most decisions in Sweden are made on-site by the schools themselves. The headteachers[8], known as 'rektor' in Swedish, together with the teaching staff (and preferably in collaboration with the pupils) draw up a medium-term school program on the basis of the state framework and the possibilities for the municipality. In the annually compiled work plans, the school program proposals are put forward so that they can be evaluated. School leaders tend to be seen as being on the other side and instruments of the education and municipality policy by teachers since they have the job of applying the new regulations at schools together with representatives from the municipality. Due to their function, they are also more or less employers and superiors for the staff working at the school. Theoretically, decentralization should reinforce their role as driving forces for school development and lesson development. However, according to the experiences of many school leaders, in fact it increases their share of administrative work to the detriment of their teaching work and their educational work at the school. They are often seen just as administrators and not as educationalists (see Eikenbusch, 1995). The extent to which the job of the 'rektor' has shifted from educationalist to administrator can be seen in the changes to the law from 1991; school leaders are no longer required to have completed a teacher training program. In most cases, however, they actually still have.

The tasks of the 'rektorer' include preparing the annual budget, specific assignment of the funds and educational measures, the organisation of lessons, the coordination of the subjects on offer, ensuring appropriate school career advice for pupils (above all at 'gymnasieskola'), adaptation of the lesson organisation to the

[8] Notably the number of female school management staff has increased greatly over recent years: from 10% in 1985 to 59% in 1998 (Nygren & Johansson, 2000).

requirements of pupils, the provision of suitable training for staff, the development of structures for cooperation between the school and parents, the school and workplaces, the school and universities, etc. Furthermore, the school leader is obliged to evaluate the results obtained at his or her school on the basis of the state specifications as well as with regard to the targets set in the medium-term school program by the municipal council and in the annual school work plan. As over a third of Sweden's 'grundskoler' are relatively small with less than 100 pupils, several schools are often combined into a single administrative area (rektors områden) and are all managed by one 'rektor'[9]. He or she mostly works with one or more 'biträdande rektorer' (assistant headmaster/headmistress) or 'studierektorer' (dean of studies), who are mainly responsible for educational tasks.

2. Development of School Leaders

Towards the end of the 1960s, model projects for the training of school leaders were started in Sweden in the form of short courses with educational and administrative content. During the first half of the 1970s, the need for basic training was emphasised repeatedly by the state on the basis of the work of the Commission on the Internal Work of Schools (SIA). In 1976, the Swedish parliament (Riksdag) then decided that all school leaders at state schools should follow a two-year training program in tandem with their work. After a nine-year trial period and evaluation, the parliament decided to introduce a modified version of this program on a permanent basis in 1986. Both the state and the municipalities should be responsible for different areas of basic training in the form of a cooperation project. The state-run training program is intended to focus the national goals of education, the planning and implementing of school development as well as evaluation and quality assurance. The municipalities contribute to the project by offering management training to the new school leaders and other leaders from the intermediate and higher levels of administration; this training is specifically tailored to suit the local circumstances and needs. This model is still applied today. Four steps of training were introduced:

1. A recruitment training program for persons who wanted to become principals. Training should give a broad view of different school leadership functions but retain a focus on the national goals for education. The Riksdag's intention in introducing this program also had three other elements. Parliamentarians wanted more women as school leaders, more recruitment from other municipalities and more people with other educational backgrounds.
2. An introduction training program was introduced to help new principals during their first years in office. The main part of the education would focus on the practical and administrative tasks of the principal, but it was also made very clear that the principal should be introduced to pedagogical leadership.

[9] In 1990, the 4 700 schools in the primary and secondary level I areas were divided into 1 300 of these 'Rektor' areas (see Fuchs, 1998, p. 24).

3. A National Headteacher Training Program was to be followed by all principals after about two years in office. This program lasts two years and comprises around 30 seminar days. The purpose of the training is to deepen the principals' knowledge and increase their understanding of the national school system, the national goals for the school and the role of the school in society and the local community.
4. Continuing development program in form of university courses for school leaders.

The two first programs were to be run by the municipalities. There is a great variation between different municipalities when it comes to how well the different school boards worked with these two types of programs. The third program – the National Headteacher Training Program – has functioned very well. The reason for this is that the state, through the National Agency for Education, organised the headteachers' training and provided the resources needed to run the program. The fourth type of program, academic courses, has been offered at different universities. Unfortunately, university courses have not been able to attract large number of principals for continued school leader education (Johansson, 2001).

The State Basic Training Program

The tendency towards decentralization since the start of the 1990s has highlighted the importance of the suitable qualification of school leaders even more because of the additional tasks. A new basic training course was therefore introduced following a parliamentary decision in 1992, which dealt with these additional areas of responsibility. Since July 1, 1993, the Swedish state has been offering this basic state qualification for 'rektor' through the Skolverket.

The fact that, alongside the state, the municipalities themselves should also bear responsibility for the qualification of their school management staff, is being stressed. The state sees itself as obliged to guarantee a basic training course preparing participants for the key role of school leadership in order to ensure equal schooling throughout the country at a high performance level (see Skolverket, 1999). The qualification can be taken at four universities, Umea, Uppsala, Linköping and Göteborg, as well as at the Teacher College in Malmö and the Skolverket in Stockholm.

The target group for the program are all newly appointed school leaders at schools. There is one exception: in a cooperation project between the universities of Uppsala, Örebro and Dalarna with the Stockholm Teacher Training Institute, students in the second half of their education who are interested in school management or graduates upon completion of their degrees can take courses at these universities as part of extended studies and conclude their studies with a corresponding Master's degree or doctorate. That academic qualification program is optional and is not linked at all to the National Headteachers Program. So far there are no plans to require completion of the course before appointment. Instead, school leaders should contact one of the Swedish universities where the National Head-Teachers training Program is offered after one year in office. Agreement from the respective municipality as a service authority is required for the acceptance.

According to the Skolverket, the starting point for their basic training concept is a standard image of the school as an organizational unit which has close ties with its environment and is characterised by specific internal and external conditions. Conveying the skill of reflection, analysing information critically and solving problems should be at the forefront of the course. The starting point should be the Head's own range of work experience. Ideas, terms and models from theory and practice should help increase understanding for their experiences and the work at the school. Particular importance is attached to a cooperative work atmosphere, openness, reflection and learning.

The goal specified by the Skolverket is deepening the school leaders' and Heads' understanding and knowledge of the Swedish school system, the national educational and training aims of the school as well as the role of the school in society and in the respective municipality. The knowledge of the function of leadership in a goal-oriented school system should be increased, as well as the ability to plan, implement, evaluate and further develop school activities. School leaders should obtain the skills to analyse the results of school activities, draw conclusions and present them to the respective committees. They also should increase their ability to cooperate with members of the school and community and represent the school to the outside world.

From these main factors, more specific education goals can be determined, which are assigned to the four areas of school management (see Skolverket, 1999).

The areas under which the training aims are organised are:

1. Aims of school,
2. School management,
3. Pedagogical development of school,
4. Evaluation.

Aims of school

The national education goals of the Swedish school system and the role of individual schools in communities are examined. This also includes the function of schools in the preservation and further development of a democratic state. The connections between the general Swedish national education goals, the medium-term school program established by the municipal council and the specific annual work plan of the school should be analysed and made transparent just like the interaction between the individual decision levels of the school system and school leader or all others working at the individual schools. The school leader is responsible for actually directing the work at his school towards the aims. Additionally, the rights of the pupils and parents to be involved in the decision-making processes are preserved or implemented. The qualification should support school leaders in their work to increase cooperation between the school and the parents and pupils and to take responsibility both for the content of lessons and the working conditions at the school.

School management

The course deals with the basics of national and local administration and coordination of schools. Different control systems are looked at from political, legal and financial viewpoints. The relationship between the influence of the different political levels of the school system and the role of the individual school leader in the management of the school should be clarified. Through the qualification, she or he should realise the significance of the tradition and culture surrounding the school for the work at this particular school against the background of its aims and the continuous change processes of society. He or she should be able to see future developments from different perspectives and should be able to analyse and assess development tendencies within the school and society.

Pedagogical development of the school

School leaders are responsible both for daily business and for long-term planning at their schools. The qualification thus includes methods for analysing the immediate environment of the school, the continuing processes which change society, and strategies for school development. The school leader should learn to create the necessary conditions so that everyone associated with the school learns to take responsibility, develop understanding and be able to feel a sense of involvement with the development of the school as a whole. The qualification should convey a deeper understanding for learning processes and the development potential of organizations, and it should determine the effect that the behaviour of the teaching staff and the school administration has on people and on learning processes and on the work of the school in general. This covers theoretical and practical knowledge of how an active supporting role in pupils' and teachers' work can boost school development processes and keep them alive.

Evaluation

The qualification should make participants aware of the meaning of an efficient auditing and evaluation system, both from a national and a local perspective. They should gain knowledge of how suitable evaluation measures can be applied to the respective school. The close ties between evaluation and school development should be highlighted.

Taught in groups of 25, the qualification program is led by two trainers respectively, one is an educationalist from the university and the other a school leader working part-time in school leadership training. This collaboration should link theory and practice as closely as possible by guaranteeing an academically solid basis, which is then discussed in relation to application and can also be compared with practical experience.

Forms of work used in the Basic State Training Program are seminars (held at the respective colleges) with presentations, impulse lectures, debates, etc. as well as reading in private study and written reflection – individually and in small groups. Coaching at the participant's own school also plays an important part. This on-site support – for school development projects and their documentation, for example – encourages increased reflection on and further development of the participants' own management practices. In addition to scientific publications, the material used will include written group work projects produced by the participants themselves, case studies and experimental games on the basis of organisation development theories.

The course runs over a period of two and a half years and comprises a total of 30 training days. Seminars lasting two to three days are held six times a year. As a rule, two courses are offered each year. It is hoped that the course groups will also go on study trips abroad to find out about the tasks and functions of school management in other cultural contexts.

The training costs of 60 000 Swedish crowns, or approx. 6 700 euros per participant, are paid by the state, while the municipalities cover travel costs, the costs for substitute teachers at the school and of the literature needed.

Table 11: Outline of the State Basic Training Program in Sweden

Training and Development of School Leaders in Sweden The State Basic Training Program	
Provider	Skolverket (National Agency for Education) through four universities: Umea, Uppsala, Linköping and Göteborg, as well as at the Teacher Training College in Malmö and the Skolverket in Stockholm.
Target Group	Newly appointed school leaders in their first year in post
Aims	Deeper understanding and knowledge of the Swedish school system, the national educational and training aims of the school, the role of school in society and in the respective municipality, development of skills in the areas of school management; pedagogical development of the school and evaluation
Contents	The aims of school (the role of school in society, state, community); school management; pedagogical development of the school; evaluation
Methods	Lectures; spontaneous presentations; debates; case studies; experimental gaming; private study; written reflection; coaching at the participant's school
Pattern	30 course days (timetabling: several two to three-day seminars per year) plus literature studies and coaching dates within 2 to 3 years
Status	Optional; appreciated by municipalities appointing the school leader
Costs	Ca. 6 700 euro (60 000 Swedish crowns) per participant; state-financed; additional costs for travel, substitutes and reading material are covered by the municipality
Certification	Certificate of participation; from 2002 restructuring with possibility of Master's degree

3. Summary

The Swedish school system has a long centralist tradition with a marked social aim. Since the end of the 1980s, however, there has been a development towards increased decentralization; the school system has been placed under the control of local authorities to a great extent, above all in terms of budget administration and staffing. The 'rektorer', the school leaders, have thus been assigned a larger share of administrative tasks.

The qualification of school management staff, however, has moved towards a central system since a parliamentary decision in the mid-1970s. The state sees itself as obliged to guarantee a basic training course for all school leaders to prepare them for their key task and to equal schooling at a high performance level across a country with a varied population density. The State Basic Program was developed centrally by the Skolverket, the National Agency for Education. It is run decentrally at universities and education colleges. The main focus is on the aims of schools in a democratic state; questions on school management, the pedagogical development of schools and the evaluation of school development processes are emphasised as well.

It is interesting, however, that the municipalities have to make a substantial contribution beyond this basic program. The Swedish qualification model is based on dividing the responsibility for the qualification of school leaders. The municipalities are responsible for regional qualification offerings, for example, in the form of management courses, in which the 'rektorer' together with other management staff from the municipality are trained especially for local conditions. This is intended to simplify and improve cooperation between management staff within this district later on.

4. Recent Developments

The program has been under review and a new program will be in place soon. There will be substantial changes in the content but the basic organisational structure will probably remain.

In the new National Head Teachers Training Program the vision is to create a school leader who is a democratic, learning and communicative school-leader (DLC-leader). Democratic means that the leader him or herself is leading the school in accordance with the national curriculum and the democratic ideas expressed in those documents. The democratic reflective school leader understands that it is not sufficient that education imparts knowledge of fundamental democratic values. Education must also be carried out using democratic working methods and preparing pupils for active participation in civic life. By participating in planning and evaluation of their daily education, and exercising choices over courses, subjects, themes and activities, pupils will develop their ability to exercise influence and take responsibility. The school leader also understands that school democracy is for all those who are working in the school. And this knowledge governs his or her actions.

The DLC-leader is a learning leader and understands that all leadership is about constant learning and that this is especially the case for school leaders. To be able to lead such a group, the leader him or herself must be a learner. The school leader needs to be a learner in relation to the goals of the curriculum. If a school leader can live up to the demands in the national curriculum, she or he must be a learner as well as a person who understands that governing power is not power over money, buildings and personnel, it is an authority based on discursive power. If she or he is able to live up to the very high demands placed on a democratic environment in the school, where everybody can feel that they are seen and appreciated for whom they are, the school leader must be the change she or he wants to see (see Johansson, 2003).

A school leader who is both democratic and reflective in these matters must be a learner, a person who creates and merges school cultures and school structures by re-thinking and leading by the power of dialogues and discussions. But they are also persons who are aware that the learning process and the control of related emotions and anxiety have an impact on educational leadership. The democratic reflective school leader, whose task is to support and promote interactive professionalism, is essential. Therefore the main tasks of the democratic, learning and communicative school leaders are:

1. Leading the teaching. As the leader with overall responsibility for the quality of the teaching offered to pupils, she or he must have insights into teachers' planning, implementation, and evaluation. The leader must be able to guide and supervise teachers in these matters.
2. Leading the learning of teachers. Teachers are the most important persons in challenging and facilitating pupils to learn. They must be constantly learning, reflecting and collaborating on didactical matters. The leader is responsible for promoting and challenging teachers' learning processes.
3. Developing language, values and culture. In order to facilitate the collaboration and work of pupils and teachers, the school must discuss the values underlying the actions and decisions. The leader is responsible for the school meeting the challenge of this discussion.
4. Leading relations to the outside world. The stakeholders of the schools demand more information and justifications. At the same time the legitimacy of the school within the local environment is questioned constantly. The leader is in charge of pro-active relations with the outside world.
5. Reading, thinking, reflecting, learning and acting. A very important basis for legitimacy within and outside the school is the power of knowledge with which the leader can capture the power to lead (see Johansson et al., 2000).

The democratic dimension of the power struggle is about how the agents handle their relations and positions. In all educational systems there are different approaches to this leadership role. Some leaders do not want to engage with the power struggle; they deny it. Some leaders take over all power, not giving teachers any latitude for influence or dialogue. And some take on the responsibility to initiate a continuous professional discourse about developing and ensuring quality in schools. A school leader can – and should because of the democratic objectives of the institution – act democratically. He or she can be the one agent responsible for having the struggle made public and visible so that everybody can examine what is going on and can have a say in decisions. Accordingly, the leader should be responsible for creating the agenda for professional dialogue in the school and for the associated democratic educational practices. The democratic reflective school leader's task as a supporter and promoter of interactive professionalism is essential and, therefore, training of communication skills is of great importance in the new Swedish program (see Johansson, 2000).

'Authentic leadership' (see Begley, 2003) implies a genuine kind of leadership as a hopeful, open-ended, visionary and creative response to social circumstances, as opposed to the more short-sighted, precedent-focused and context-constrained practices typical of management (see Begley and Johansson, 2000). This is a values-informed leadership, a sophisticated, knowledge-based, and skilful approach to leadership. It is also a form of leadership that acknowledges and accommodates in an integrative way the legitimate needs of individuals, groups, organisations, communities and cultures – not just the organisational perspectives that are the usual preoccupation of much of the leadership literature (see Johansson, 2003). Finally, authentic leadership is a form of leadership which is shared, distributed, and democratic, even when the leader has to make the limits clear. The qualities of the authentic leadership style is therefore a pre-requisite for a successful democratic, learning and communicative leadership.

5. Statement by Olof Johansson

The new Swedish program started in January 2002 and the present preliminary regulation document issued by the National Agency for Education contains the aims, goals and guidelines for carrying out The National Head Teachers Training Program.

The national training program for headteachers should be offered, in the first hand, to newly appointed headteachers. Headteacher refers not only to school leaders but also holders of positions with leadership responsibility within the public school system, nursery education, after-school child care and recognised independent schools that are eligible to receive grants and are subject to national supervision (according to chapter 9 of the School Act 1985: 1100). School refers to the institutions embraced by the previously mentioned activities.

Admission to the training will take place in consultation with the school board for the school to which the headteacher belongs. The Board will also be responsible for each participant being offered reasonable time for his or her own studies and adequate financial support. The training should include at least 30 days of lectures and seminars and take at least 2 and at most 3 years.

After completing the training, the headteacher should, on the basis of democratic principles and with regard to individuals' integrity and equal value, be able to lead and develop the school as well as asserting the rights of children and pupils to the education guaranteed in the government's legislation and regulations. The Head-teacher should also have the ability to direct the organisation's learning towards better goal achievement and thereby bring into focus children's, pupils', co-workers' as well as his or her own learning.

Understanding the school as a learning organisation, his or her role as chief and leader in a politically controlled activity, and having insights into how control within the national and local government sectors affect the school's activities are other areas covered by the program. Together, these embrace: understanding the school's role in society, being able to work in accordance with public education's task, being able to conduct him or herself in a professional manner with respect to the national and local government goals for the school, and developing his or her own goal direction and behavioural strategy. Additional skills examined and fostered during the course include: being able to explain and argue in support of the school's national and local government goals, leading his or her co-workers' efforts to interpret the goals, analysing the consequences the goals have for the activities in his or her own school and municipality, and being able, verbally, in writing and in dialogue, to achieve better goal fulfilment in the school.

Another important training goal is to prepare the headteacher so she or he can together with co-workers be able to use different methods for following up and evaluating the quality of the service and its need for development. In a pedagogical way, too, the headteacher's goal is to make explicit the school's conditions and assert these needs not only with the school board but also among children, pupils, parents, co-workers and other members of society.

Universities responsible for arranging the training should be able to set the curriculum for the course. It is their responsibility to continually follow up and evaluate the training as well as presenting an annual report to the National Agency for Schools[10]. Every participant who has successfully completed the course will receive a certificate recording the length and the content of the course.

The implementation of the program will probably not be exactly the same for the eight universities who run the program, contracted by the National Agency for Education. Nevertheless, key features of the program at the Centre for Principal Development, Umeå University are described below.

The program will run for three years and participants will be together at seminars for at least 30 days. There will be two main instructors working on the program in each course and expert instructors can be called in for special parts of the course. All participants will be active principals and will have at least a ten-per-cent reduction in their workload.

The learning activities will include interplay between practices, theory and research, and one training goal is to create new knowledge in the inter-section between the three arenas. Practical work will be based on examples from the different schools of the participating principals. This case method may also include activities and/or discussion at the school of one of the participants.

Two types of supervision will be used, individually in each participant's school, and in small groups with participants that work on similar tasks. Participants will create their own course portfolios, which will be the basis for course evaluation in conjunction with the instructor. During seminars and lecture presentations, the participants will practise different forms of communication skills. The course group of principals will also be an arena for practising different leadership skills and analysing participants' performance.

Comparative approaches and reading of international literature are built into the pedagogical model of the program. The program's learning philosophy is process learning with fellow participants. All participants are requested to be active learners and, in that role, also take part in the group learning processes. Together with the instructors, participants will seek ways to inform their superintendents about their progress in the program.

The focus of the program will be on democratic, learning and communicative leadership and that focus will be the screen through which the program goals – presented above – will be analysed.

---

[10] As stated earlier, the Skolverket will be divided into two agencies in March 2003, one of which will be responsible for control and statistics and the other for support, school improvement and development activities.

Denmark: No Need for Regulations and Standards?

Stephan Huber and Lejf Moos

1. Background

The Danish school system has always been very strongly decentralised. The Danish Parliament (Folketing) and The Ministry of Education provide only the broad goals and objectives, which are then open to interpretation by local education authorities in the municipalities and later on by schools and teachers. The Ministry is responsible, for example, for school tests.

In the Danish school system, primary and lower secondary schools (kindergarten class and grade 1-9) are referred to as the 'folkeskole'. Two governing bodies, the municipal council and the school board, are active on behalf of this sector of the school system. The municipal council is responsible for opening or closing down schools, appointing and dismissing teachers and school leaders, administering budgets and developing visions and principles for this municipal school system. There are 275 municipalities in Denmark, each with a municipal council of up to 25 members. The school board of each school advises the municipal council and governs the school. In relation to the municipal council, the school board draws up the work plan for the school, and makes suggestions for the curriculum and budget. In relation to the school itself, the school board decides on principles for the school life and education and for the budget. The school boards are comprised of five to seven parent representatives, two teacher and two pupil representatives. Representatives from the municipal council can be members of the school boards, too. This is a matter on which the council itself decides. The school leader is secretary to the school board. The day-to-day leadership and management of the 'folkeskole' are left to the school leader in collaboration with teachers.

The senior high school (grade 10-12) in Denmark is referred to as the 'gymnasium'. This is governed by a special division of the Ministry of Education in so far as it is this division that makes up the curriculum and the exams. The counties of which there are 12 in Denmark have the local authority. The day-to-day management of the 'gymnasium' is left to the principal in cooperation with the following two authorities: firstly, the school board, composed by council representatives, parent representatives, and teacher and pupil representatives, and secondly the Educational Committee, which is composed of the principal and the teachers. This Committee is advisory to the principal and considers areas such as school year plans, subject range, educational and organisational development, project days, formation of class groups and of project groups. If pupils want it, a Pupil Council must be formed. The relations between all of those committees and councils create a system of control and balance which is less bureaucratic than it would appear because of the small-scale society and of the Danish tendency to stay informal (see Pedersen, 1995).

All parts of the Danish educational system have experienced a restructuring- and transformation process since the mid-1980s. One example is the action plan from 1993, "Education for All". This was provided by the Ministry of Education, and focused on improving the educational situation for young people and lowering the drop-out-rate, the 10% of young people who leave the educational system after having completed the obligatory nine school years. It is felt that the system provides a broad range of opportunities for education for these young people, but that each individual should be treated more as an individual and should be motivated to stay on in the system. A suggested approach to this should be an increase in project organised education and new educational offers that would further pupil's creativity and personality.

Aspects of this transformational process in the Danish educational system are:

- management by objectives instead of management by rules and regulations,
- increased autonomy of schools and other educational institutions,
- more market orientation with consumer-influence by forming governing bodies (see the committees described above),
- increased autonomy for parents who can choose the school for their child,
- more information flow from one level in the educational system to another.

The latest (1993) Act on the 'folkeskole', the comprehensive school, underlines the educational leadership of schools. The Act defines explicitly that the duty of the school leader is to lead the school organisation in developing collective values and a shared work profile in each school. Decisions on curriculum and the administration and management of budget, and school development plans are devolved in individual schools and are therefore the responsibility of the school leaders. The responsibility for those transformations is accompanied by a new responsibility for quality assurance[11], evaluation and accountability that is very much the responsibility of each school and each local authority. In 1988 the Ministry of Education established the Danish project for subject- and quality development called Quality in Schools, which included all schools. In this context, it was underlined by the Ministry of Education that quality in education couldn't be defined or assessed in the same way as quality is understood in industrial production. In the field of education there would be the risk of placing too much emphasis on the indicators that were easily measurable and using too many quantitative tests. Schools were

[11] Quality assurance is distributed among several levels in the Danish educational system: The test system in the Gymnasium is located in the Ministry of Education, who make up the examination paper and nominate the external examiners. Every Gymnasium is visited once every year by an inspector from the ministry. The Ministry publishes a statistical report on the results of each school (quoting the number of passed and failed students). The Ministry commits schools to publish a report of the annual work and a plan for next year which mentions what the school intends to do about the weaknesses pointed out by school consumers. All Danish schools must carry out self-evaluations, which include assessments by students, parents and representatives of the local community. Finally the assurance of quality in the Gymnasium is left to the market forces because there is free choice of schools and the numbers of students determine the State funding for each Gymnasium (the management of the budget is left with the individual school).

explicitly asked to involve other and more complicated factors. The impression of a clear cut request for accountability should be avoided as it could be counter-productive because the educational staff of schools could be de-professionalised and a climate could be created which would deter the most suitable from becoming teachers (see Pedersen, 1995).

Parallel to the public school system there are private schools attended by 12 to 15% of the children. It is relatively easy to found a private school, and the state funds 85% of the overall budget.

The distribution of responsibility to individual schools was a part of decentralising decision making from state level to municipal level. On average there are 10 to 15 schools in a municipality, which is the school-owner and carries most of the finances (the state distributes its part of the finances in big block grants to the municipalities). The municipalities have formed an association, The National Association of Municipalities, which has gained a strong influence on educational matters.

School leaders in Denmark (both school leaders in the folkeskole and principals in the Gymnasium) are in the middle of several contradicting expectations. Some interest groups would like them to be public middle managers with unbroken loyalty to the political hierarchy. Others would like to see them as responsible to their own school only. On one hand they are responsible to the Parliament and Government for ensuring that their school complies with the national objectives for the schools. Here the first and foremost interest is in the overarching objective of contributing to the education (the 'Bildung') of pupils so they come of age as citizens in a democratic society. On the other hand the local authorities would expect school leaders to be more cost effective because funding is decreasing. In addition to this, Danish teachers are very competent and self-confident and they want more autonomy and less management or leadership. They don't easily accept decisions from outside the school on educational or financial matters. The school leader is clearly responsible to the local authorities on administrative and educational matters, she or he must take care of the day-to-day administration of the school, the maintenance of buildings, the administration of finances and the quality assurance and evaluation. He or she must make a work plan and make statements on maximum number of pupils in classes and how to spend funding. This must be negotiated with the Educational Committee. He or she must advise the municipal council on appointment and dismissal of teachers and other staff. In educational matters she or he has great latitude and must see to the timetable, class composition, distribution of work, planning of optional subjects etc. School leaders' teaching duties are calculated according to school size, but most Danish school leaders do not teach on a regular basis.

2. Development of School Leaders

There are very few, if any, systematically gained experiences or research on how to qualify school leaders in Denmark. There are no formal requirements for pre-leadership training (except for a teacher training certificate), nor is there any formal requirement for continued professional development for school leaders. It is left with the municipalities to decide whether they want to offer or demand development opportunities. For some years now a number of programs have been available, but none of them are compulsory.

The National Association of Municipalities has its own training institute, The Local Government Training Centre, which has developed a four-week course. Newly appointed school leaders can apply for this course. It is organised as a four-week full time course concentrating on pragmatic and management aspects. One aspect of this course is the formation of leadership networks where school leaders consult and cooperate on many matters, even after the course has finished.

There are no national standards on content or organisation for these training opportunities. Nor are there any formal standards for the qualifications to be gained. There are no initiatives from the Ministry to establish any kind of either national leadership education or standards or certifications of any kind. The municipalities reject any initiatives of that kind. One might notice that the local level seem to be more interested in management than leadership and more interested in the financial qualifications of their leaders, while the national level is more interested in educational competences in order to achieve educational goals.

Example: The Diploma in Educational Leadership of Schools and other Educational Institutions at the Danish University of Education and at Teacher Training Colleges

For about five years The Royal Danish School of Educational Studies (RDSES) – which is the predecessor of The Danish University of Education (DUE) – and a number of Teacher Training Colleges have offered a part-time two year long diploma in Educational Leadership of Schools and other Educational Institutions. More branches of the RDSES offered this training and from 2002 on it will be offered by a number of regional training centres (Centre for Videregående Uddannelse: CVU's).

This training is not compulsory either. The target group for it is acting school leaders and teachers from the 'folkeskole' and other educational institutions. Until now the group primarily utilising this training have been middle managers from basic schools (Deputies and Heads of Department). There have been only a few acting school leaders and a group of teachers who have taken advantage of this course. This training seems not to have included principals or teachers from the Gymnasium. They are now being offered studies of leadership as part of a Master's course at The South Danish University in Odense (Syddansk Universitet).

The vision of a democratic, reflecting school leader is guiding the training, which builds on the objectives of the 'folkeskole' to promote democratic education/'Bildung'. So the understanding is that school leaders should lead according to the objectives of their organisation.

Themes are:

- educational leadership,
- staff leadership and staff development,
- team leadership,
- power, conflicts and micro politics,
- values, ethics and legitimising leadership,

- school development,
- theories on organisations, organisational culture, learning in organisations,
- school evaluation,
- leading the relations to the community,
- administration and strategy.

Participants must deal with all themes, there are no options. The teaching is a blend of instruction, group work and individual studies. In the lectures, theories, models and concepts are introduced and discussed. In group work and individual studies the participants apply these theories, models and concepts to describing and analysing cases from their work practice based on personal experience. The whole program is spread over 13 days each year. In each of the 4 modules there are 46 lessons, which are distributed over two whole workdays, and eight afternoons/nights.

Table 12: Outline of the Diploma in Educational Leadership of Schools and other Educational Institutions at the Danish University of Education in Denmark

Training and Development of School Leaders in Denmark Example: The Diploma in Educational Leadership of Schools and other Educational Institutions at the Danish University of Education	
Provider	Danmarks Laererhojskole, now Danmarks Paedagogiske Universitet, and CVU's
Target Group	Teachers aspiring to a leadership position and established school leaders of the folkeskole (primary and lower secondary school) and other educational institutions
Aims	Development of leadership competences in a very decentralised school system; a vision of a democratic and reflecting school leader guides this training; the understanding is that the school leader should lead according to the objectives of their organisation, which is democratic education
Contents	Educational leadership; staff leadership and staff development; team leadership; power, conflicts and micro politics; values, ethics and legitimising leadership; school development; theories on organisation, organisational culture, learning in organisations; school evaluation; leading the relations to the community; administration and strategy
Methods	Combination of seminars with instruction and discussions and with group and individual projects
Pattern	Ca. 26 course days (timetabling: 4 modules of 46 lessons each within one semester (half year) divided into 2 full-days and 8 part-days) plus group and individual studies within 2 years
Status	Optional; welcomed by the employing community
Costs	Ca. 3 380 euro (24 000 Danish crowns) per participant
Certification	Diploma issued by the DPU

The state and the municipalities pay most of the cost, but a number of participants have paid the registration fee themselves. There has been no formal evaluation of this course, but in informal evaluations we have learned that participants appreciate being introduced to theories, models and concepts that can be practically utilised for getting a broader and deeper view of the duties and problems of school leadership. The theories have been most valuable to them when they could be put into use to analyse practical problems and to form new plans and actions.

For a few years branches of the RDSES have offered a special course for teachers who are interested in becoming school leaders. This course is called From Teacher to School Leader. There is even more stress on the situated learning aspect of this training in the sense that every participant is shadowed and counselled by his or her own school leader as a mentor. This course is equivalent to Module One in the Diploma in Educational Leadership.

Example: Leadership Development by Praxis

The private consultancy firm Praxis has developed a continuing professional course Leadership Development (LeaD). The concept of this course is that schools should be learning organisations that develop and transform. School development means to Praxis:

- an individual and organisational capacity to find a dynamic balance between the demands and expectations from the outside and the inner goals and conditions,
- a capacity to learn on various levels that include a rising capacity to learn and solve problems and develop new goals and visions. This includes developing methods suitable for transforming objectives and for analysing and evaluating events in a critical and constructive way.

The target group is school leaders, teams of school leaders from different schools and educational institutions, and employees at local education authorities and consultancy centres.

It is an intention of this training that participants gain insight into the structures and organisational processes of their own school and the opportunity to critically reflect upon aspects of their own thinking and attitudes that would impede the development of the schools. It should be emphasised that it is not the intention to communicate ready-made answers or guidelines for action. It is rather the intention that participants should develop their own analysis- and cognition processes. Therefore it might be necessary to de-learn rigid views and thinking structures. The close connection between theory and practice is stressed.

The themes of the seminars are:

- School development: Organisational learning and school development. The balance between demands from the outside and inner goals and conditions. How to handle barriers to development. The continuous process of clarifying values and objectives. Developing a professional school culture.

- Leadership in times of change: What are we leading? What does effective leadership mean? How are we leading?
- The learning organisation: The learning organisation in theory and practice. Systemic thinking. Development strategies: From knowledge to action.
- Learning by doing: Development of and in practice. Examples of models and methods for development. Double-loop learning. The biggest challenge: Implementation and institutionalisation.
- Analysis and evaluation in schools: Quality control versus quality development. Analysis and evaluation in theory and practice. Action research.
- Leadership and management in practice: Job development versus staff development. Participation and devolution of responsibilities. The school leader as coach. Supervision and counselling. Counselling competence and techniques.
- Handling of problems and conflicts: Creative conflict handling techniques. Theory of conflict. Strategy for conflict handling.
- What's worth fighting for? Leadership in theory and practice. The way from theory to practice. Development or restructuring, or both? Vision for the future.

The course takes between a year and a half and two years to complete and is organised in eight two- or three-day seminars, two per semester, and projects in schools. Participants from up to four schools form a network group that collaborates closely and meets at least once in between the seminars. The knowledge that has been communicated at the seminars is discussed and reflected upon at these meetings. School leaders employ the methods and procedures that they get to know at the seminars and in the school development projects in their own schools, and therefore the network meetings can function as a forum where critical yet supportive peers give confirmative and adjusting feedback.

The participants get a certificate for having participated. This certificate allows transfer of credits to the Diploma in Educational Leadership at the RDSES. For those not in leadership posts, participation in this training increases the likelihood of being appointed to a leadership post in the future. Many municipalities, among them Copenhagen, combine in this way the need for individual school leaders' professional development, the support for school development and the building of working networks within the educational system.

The costs for this training are approximately 1 200 Danish crowns (160 euro) per seminar. On top of that there is accommodation and consumption. The total for the whole course is 9 600 Danish crowns (1 300 euro) plus accommodation and consumption. Some schools pay these fees themselves, although in some cases it is the Local Authority that pays.

Table 13: Outline of Leadership in Development by Praxis in Denmark

Training and Development of School Leaders in Denmark Example: Leadership in Development by Praxis	
Provider	The private consultancy firm Praxis
Target Group	School leaders, leadership teams of different schools, and staff of local school authorities and consultancy centres
Aims	Insights into the structure and organisational processes of their school and into their own thinking and attitudes (analysis and recognition competences)
Contents	School development; leadership in times of change; the learning organisation; learning by doing; analysis and evaluation in school; leadership and management in practice; handling of problems and conflicts; What's worth fighting for?
Methods	Seminars; project works; critical friends; case studies
Pattern	16 course days (12 course days in the more compact form) (timetabling: 8 two- to three-day seminars) plus project works in schools plus meeting in network groups and individual studies within 1½ to 2 years
Status	Optional; appreciated by the kommuns, accepted as a part of the Diploma-qualification in Educational Leadership
Costs	Ca. 1 300 euro (9 600 Danish crowns) per participant
Certification	Certificate of participation

3. Summary

In the strongly decentralised Danish school system, the state is restricted to providing some goals, aims and standards to schooling, particularly in the primary school sector. Hence the responsibility lies at the level of the community, the municipalities. The school leaders are faced with having to deal with often contradictory expectations, those of the state, those of the municipality, those of the parents, and those of self-confident teaching staff. In Denmark, there are no formal qualification requirements, nor are any standards provided as to content and other aspects of school leader development, all opportunities are optional. The municipalities, who are the employers and the school owners, decide what to expect of future applicants. Their association, the National Association of Municipalities (kommunernes landsforening), rejects too much state influence and at the same time offers their own program.

For this to be done, they offer courses for acting school leaders via their own Educational Institute, where the emphasis is put on pragmatic issues relevant on the individual school level, e.g., budgeting. The profession itself organises interactive networks and creates knowledge pools. The Danish University of Education (until 2002 and from then the regional training centres) provides an optional development program, administered in regional training centres, which is designed for acting school leaders as well as teachers who want to become school leaders of the 'folkeskole'. The emphasis of this program is on leadership and school development and the participants are awarded a diploma. The municipalities favour other programs, one of them is a program that is provided by the institute Praxis and called LeaD. This program offers a combination of development opportunities for individual school leaders and support for school development processes in their respective schools as well as a regional network for cooperation, which should continue after the program ends.

4. Recent Developments

The Danish University of Education is applying to the Ministry of Education for approval of a plan to offer a Master's study in Educational Leadership. It is difficult to get this approval, among other things because The National Association of Municipalities is opposed to the idea. This association wants all municipal leaders to have the same training at master level, irrespective of what kind of institutions they lead.

The new, liberal Minister of Education (appointed in November 2001) has stated that a postgraduate diploma course for school leaders should be prepared. There are no details of that proposal yet.

In the spring of 2002 a committee under the Ministry of Education (The Marius Ipsen Committee) has produced a proposal, suggesting that school leaders – like all other public middle managers – should be given an opportunity to be educated through a postgraduate diploma course in leadership/management. The neo-liberal Danish tradition to make all kinds of leadership the same seems to continue.

5. Statement by Lejf Moos

The examples mentioned, the diploma course at the DPU, the Teacher to Leader course at the DPU and the Leadership in Development course by Praxis, share many similarities as professional development courses. They are building on the experiences and the day-to-day practice of schools involved, and on communicating theories, models and concepts for participants to reflect on and employ in their daily leadership practice. It would seem that most leadership training in Denmark is to a certain extent building on the situated learning concept, combining leadership education and school development. It seems (from unsystematic evaluations) that leaders who have participated are satisfied with this training.

But the fact that there is still not any formal requirement for teachers applying for school leadership posts is a very short-sighted policy because it increases recruitment problems and it adds to the conflicts, problems and stress for newly appointed school leaders and their schools.

The fact that leadership training is left to the municipal level (and therefore to The National Association of Municipalities) is worrying, because they are more interested in the New Public Management style of leadership (i.e. school leaders are to be municipal middle managers, who are first and foremost loyal to the municipal level, in control of finances and not different from leaders of other municipal institutions) and less in educational leadership.

England: Moving Quickly towards a Coherent Provision

Stephan Huber and Mel West

1. Background

Nationally, in England, the responsibility for education policy lies principally with the Department for Education and Skills (DfES)[12], though responsibility for implementation and monitoring is shared with Local Education Authorities (LEAs). Within this responsibility falls the development, interpretation, implementation, and control of the national educational policy through a framework of Education Acts passed by Parliament. The Department oversees the National Curriculum, monitoring both the content and the quality of teaching in schools, and also, through the Teacher Training Agency, has established a framework for the initial training of teachers. Most recently, it has extended its interest to the continuous professional development needs of teachers and headteachers. The Department's role also includes devising formulae for the allocation of budgets to Local Education Authorities, and, since 1988, also directly to individual schools. The trend has been towards increasing financial and managerial autonomy at school level, with an accompanying diminution of LEA influence.

The Local Education Authorities do remain responsible for the performance of publicly financed schools in their respective districts, and their tasks include ensuring that there are sufficient school places and school buildings suitable for the education of children living in the district. The regional differences which shape the school system in England can be accounted for by the freedom with which the LEAs can establish schools. However, their capacities to determine the distribution of funds to schools, to develop curriculum locally, to appoint teaching staff, to inspect schools, have all been eroded during the past dozen years, as the national policy has moved towards a partnership built around strong government and strong schools that has squeezed the LEA's powers. Increasingly, the LEA seems marginalised by policy direction and legislation, and there is now considerable debate about whether LEAs will survive – or, if they do, whether they still have a meaningful role to play.

Within the individual school, the school governing body is in charge of the delegated budget, and of the management of the school. Members of this committee include the school leader, elected representatives of the parents, representatives of the teaching and the non-teaching staff and of the LEA – the latter being representatives of the local political community. Since the 1988 Education Act, School Governing Bodies have had considerably increased powers, which extend to the selection of teaching staff, the establishment of salary and promotion policies

[12] Scotland and Northern Ireland have different far-reaching ranges of freedom of decision in education policy and therefore differ from what is described here for England.

and, significantly, the appointment and suspension of the teachers and of the headteacher. Generally, responsibility for the day-to-day management of the school is delegated by the Governing Body to the headteacher, who consequently needs to have a close relationship with and the confidence of the school's governors.

However, changes in the education system that had consequences for the role and practice of school leaders can be traced back as far as the 1960s and 1970s, when the reorganisation of the selective system into Comprehensive Schools created many large schools. The headteachers of these schools were perhaps the first to feel the burden of school management alongside the professional leadership role, as the sheer size of these schools brought new problems of structure and control. For school leadership, this meant a much stronger management orientation within the job, more complex organizational structures and more complex patterns of decision-making and delegation. The pace of change has accelerated within the last decade as devolution and decentralisation has continued to be pursued by successive governments committed to local management. The range of reforms introduced during this period is unprecedented, and includes:

- a National Curriculum,
- national, standardised testing for all pupils at the ages of seven, eleven, fourteen, and sixteen,
- the nation-wide publication of the individual school's results in the centrally administered school leaving exams for the 16-and 18-year-olds in national newspapers in school ranking lists, the so-called League Tables,
- increased parental choice of (and so competition between) schools (Parents' Choice),
- significantly increased powers for the Governing Body for each school, by which the influence of the parents was institutionalised,
- annual reports on the school's progress from the headteacher to the Governing Body,
- annual reports from the members of the Governing Body, the School Governors, to the parents, the community, the ministry or the school authorities and the general public,
- Local Management of Schools (LMS), a formula under which school funding levels are determined by pupil numbers,
- a nation-wide accepted assessment procedure for teachers and school leaders,
- regular school inspections (originally at four-year intervals) against national standards of the quality of teaching, learning and management in each school,
- the publication of the results of these inspections,
- the obligation to draw up a school development programme taking into account the recommendations from the report of the inspection,
- the possibility for schools to leave their LEA and to become directly funded (grant maintained), that is to receive their budgets centrally from London.

Out of these reforms a number of new responsibilities and additional duties have emerged. The role of headteachers in England has become much more demanding, as the longitudinal study by the National Foundation for Educational Research (NFER) reveals[13].

In 1993, 90% of the headteachers interviewed by NFER claimed that their role had changed dramatically within the previous five years. They regarded themselves as much as managers of their schools as headteachers. The Local Management of Schools was, nevertheless, welcomed by a majority of them, albeit with a certain ambivalence. To headteachers it means new freedoms and self-government, for example, in relation to decisions regarding the administration of the human and financial resources of the school. But there has been conflict perceived between decentralisation of management and the simultaneous centralizing of what is taught and how it will be assessed (for example via the National Curriculum and Testing, and School Inspection). Many interviewees felt much exposed to the central preconditions made by the government in London, which could have a paralysing effect on initiatives at the individual school level, which are also needed. The central preconditions are considered by many as interference, increases in bureaucratisation, and exerting too strong an influence on internal decision-making processes of the individual school. Statements from the headteachers suggest that they feel their role has been shifted from essentially an educational one, in the context of external management policies and arrangements, to essentially a management one, in the context of externally determined curriculum policies and arrangements.

Alongside the introduction of Local Management of Schools under the Education Reform Act 1988, the influence of the local authority was reduced enormously. Originally, schools were given the chance to become grant-maintained (GM schools), and to leave their LEA in favour of direct funding from the government. This meant that they no longer fell under the jurisdiction of the regional education authority, but under the Department for Education directly, and, as a consequence, received their budget directly from London. The amount of the budget, which still depended on the number of pupils, was greater this way, as no deduction to support local educational services was applied. On the other hand, the services of the LEA were no longer available, except on a paid for basis. The schools could buy services on the open market, in which the LEA is one, but only one, of the providers. Schools opting out in this way affected all schools in the district, as the government did deduct the amounts paid to such schools from LEA budgets, making it more difficult for LEAs to maintain services for schools that remained within the local system. There was a general expectation that this strand of government policy would be reversed when the Labour Party came to power in 1997- however, it was not. Indeed, the replacement of grant maintained status with foundation schools

[13] A cohort of 188 headteachers (of Secondary Schools) in England was accompanied during ten years, from 1982 on, when they took over their first headship. The second survey was done in 1988, after the Education Reform Act. In the third stage of the study, in 1993, data from 100 headteachers were collected, who had meanwhile been in headship positions for 10 or 11 years (Weindling & Earley, 1987; Earley et al., 1990; Weindling et al., 1994-1995). The findings of this third survey of 1993 offer particularly rich insights into the role-change that has taken place.

effectively extended the arrangements to more schools, making the role of the LEA even more difficult to sustain.

The accountability of schools towards the parents, and to the community in general has also increased and become sharper. It is now one of the central areas of focus for school leaders. Preparing for one of the regular school inspections held by the Office for Standards in Education (OFSTED), for example, means a lot of additional work and creates considerable strain for headteachers and staff alike. During inspection, a team of inspectors can seem to turn the school upside down for a week, and after their findings have been published, the headteacher is responsible for setting up a school programme within a given time and with a clear timescale, which takes remedial action against any deficiencies stated. In England, inspection reports are made public, that is they are available to parents and extracts from them are frequently published in local newspapers. (This procedure has been described as a Name and Shame Policy.) This additional pressure created is meant to stimulate the school's improvement efforts.

Recent education legislation has transferred a great deal of authority to the School Governing Body. The headteacher has to cooperate with this committee continuously in all decision-making processes. Many of the head teachers interviewed by the NFER, however, questioned the competences and even the availability of Governors. Many schools have found it extremely difficult to recruit persons of appropriate experience and commitment But, as they are directly accountable to this committee, the headteachers reported that a high number of meetings were necessary, which required a lot of time. Political skills that helped to manage the governors were also seen as a necessary part of the headteacher's repertoire.

Due to the new market orientation, there is sometimes an intense competition among schools. A good reputation for the individual school is an important aim, so that gifted and high-achieving pupils come, or at least their parents are encouraged to opt for the school under local selection procedures. This is decisive because the amount of the budget allocated to the school is dependent on the number of pupils, and because the number of pupils trying to get into the school is heavily influenced by reputation – determined substantially by average school leaving results published in the League Tables. This ranking has, therefore, become extremely important to establish and protect. Consequently, schools and their school leaders are very much interested in the performance and image of the school, on which they are dependent for their income. Teachers and headteachers are necessarily ever aware of the placement of their school in this ranking. Many educationalists consider these ranking lists a crude and simplistic measure of school performance that have had an unfortunate influence on public perceptions. Certainly, the consequences that these rankings and their interpretation can have for the individual school, as well as for the individual pupil[14], are often negative, and it is clear that the construction of the tables favours schools that are already advantaged. It can easily be inferred which

[14] Less successful schools have to fight against the following vicious circle: bad reputation, worsening school atmosphere, decreasing identification of the pupils with their school, decreasing number of pupils, reduction of resources, decreasing job satisfaction and motivation among staff, lack of applications of well-qualified teachers for this school, worse quality of lessons, decreasing pupil achievement, worse results in the League Tables.

conflicts the school leaders have to sustain in this context. In the NFER study of 1993, the majority of headteachers unsurprisingly disapproved of the great competitive pressure open enrolment and league tables have produced, and consider the strong market orientation as educationally misconceived, even harmful.

In England, even in the largest schools, traditionally the headteacher has retained some teaching commitment. The NFER study shows that in 1988 two thirds of the interviewees still carried some teaching. By 1993, however, only half of them were teaching. Many headteachers regretted being less involved in the key activity of school, that is working together with pupils, but felt that administrative tasks had become the priority. Headteachers wanted to take a part in what they saw as the core activity of the school – teaching, for a variety of reasons: for "they can give some support where needed, they know about what is expected, know what the pressures are and gain understanding, and they get 'street credibility'" (see Huber, 1997b, p. 30). Sadly, finding time for such activities had become more difficult.

Another central concern for British headteachers is the development of the school around a set of values (see Weindling, 1998; Southworth, 1998). Feeling most often closely linked to their school, identifying strongly with it, and seeing the school's goals and achievements as indicators of personal influence, the borderline between professional and personal identity becomes permeable. The development of the school is regarded as closely connected with the development of oneself and one's own ideas, so that the role brings with it an important personal dimension. In spite of their personal commitment to the improvement of their schools, many headteachers in England feel frustrated by hindrances and regulations when seeking to implement this vision. They complain of lack of time, and of being overworked due to administrative pressures. Southworth (1998) suggests that it is an irony of the recent changes in educational policy that, right now, when school development is so high on the agenda and could be tackled with so much commitment from headteachers, they are literally prevented from doing this due to the additional administrative tasks that have been transferred to them in the course of decentralisation.

2. Development of School Leaders

As early as in the beginning of the 1960s in England, the systematic training and development of headteachers was called for. In 1967, the first significant step was taken by the Department of Education and Science (DES), when it established the Committee on the Organization, Staffing and Management of Schools (COSMOS). Their strategy was to set up a series of short courses, to collect or create the materials, and to devise the methods through which the skills of senior school staff might be developed. During the 1970s, increasing numbers of school management courses were offered by a range of providers. Nationally, the DES and, regionally, higher education institutes (HEIs) and local education authorities (LEAs) offered training, though there was little coordination between these courses as the providers were acting quite independently. The range of provision loosely termed 'educational management' was quite wide, and such courses proved to be popular with teachers and headteachers. But, all too often these courses focused narrowly on specific tasks or activities such as timetabling or selecting staff, and were lacking any coherent conceptual foundation in 'management' as a discipline. The situation can be described as a rich diversity of provision, which was somewhat chaotic, and others concluded that because of the variety and the confusion amongst what was being provided, a well-resourced national initiative was desirable to bring order to the field.

In 1979, the University of Birmingham was awarded a research grant "to obtain a clearer picture of provision [...] and thereby to have a firmer basis for policy making by central and local government in relation to future provision" (see Hughes et al., 1981). The report of this research concluded that government should tackle the problem through the creation of an executive agency, which should function – independent from the Ministry for Education – as a promoter of school management training and establish a knowledge pool that would help to unify and support regional initiatives, conduct independent research and establish connections outside of England. Two important initiatives followed: first, the DES Circular 3/83, which identified educational management training as one of four areas of training for teachers which were deemed to be priorities. This circular provided for the setting up of one-term training opportunities (OTTO) for headteachers, and of basic courses (with a minimum of twenty days study) for headteachers and deputy-headteachers, which were offered by some twenty DES approved institutions. Second, the DES funded the establishment of the National Development Centre for School Management Training (NDC), located at the School of Education of the University of Bristol, to stimulate management training and research, and to disseminate practice. Alongside these national initiatives, in the mid 1980s, LEAs too started to construct management development strategies and programmes, though again these varied in content, detail and approach. Unfortunately these measures were seen to make little impact, and by the end of the decade the NDC had been wound up.

In 1989, seeking a fresh perspective on the problems and issues, the Government set up the School Management Task Force (SMTF) to examine the development and training of headteachers and other school staff with school management roles. The SMTF recommended an approach that placed emphasis on management development as a school-based activity. This meant that management training should be seen as an integral part of the management role, rather than as something that primarily takes place on courses away from the job. Another important recommendation was that LEAs could carry out their training role more effectively if they acted together, pooling expertise and resources and thus providing their course participants with a more comprehensive range of opportunities and support.

In 1994, the Teacher Training Agency (TTA), which is central for England, was set up, with the overall purposes of improving the quality of teaching, raising the standards of teacher education and training, and promoting teaching as a profession, in order to improve the levels of pupil achievement and the quality of their learning. As part of this work, it carried out a fundamental review of in-service training needs, and it continues to influence perceptions of what training is needed by those in management roles in schools and to promote activities aimed at improving the availability of sound professional qualifications. The TTA stated in their initial advice to government, in July 1995, that leadership and managing schools should be targeted as one of eight national priority areas for in-service training, and it launched the headteachers' Leadership and Management Programme (HEADLAMP) in September 1995.

This centrally funded, national training scheme is specifically designed to support newly appointed headteachers in the first two years of headship, through support to develop their leadership and management skills and abilities. For this purpose, grants of 2500 pounds are available to individual headteachers through the TTA, which can be spent individually with TTA approved training providers, which include LEAs, HEIs, professional associations, management consultants, and a range of private sector organisations. The training should build on and extend the previous experience of participants, be appropriate for the size, the age range, and the overall aims of the particular school and relate to the priorities set out within the school's development plan. Additionally, the training must meet the criteria identified by the TTA in relation to the HEADLAMP leadership and management tasks and abilities. Activities might include attending courses and workshops – for example, organised by higher education providers that offer the HEADLAMP programme as a series of modules – but also could cover attendance at lectures or seminars, undertaking distance learning, going on placements, engaging in mentoring activities and so on. These activities can take place on local, regional or national levels, though "In all cases, particular care should be given to considering the relationship between the content of the event, the expertise which will be acquired as a result, and the effect that expertise will have on practice within the school." (TTA, 1995a, p. 12)

The initial advice from the TTA also voted for an extension of HEADLAMP and stated that to target leadership and management of schools as priority implies extending training opportunities to headteachers in post for a number of years. For this target group, the TTA has subsequently developed the Leadership Programme for Serving Headteachers (LPSH), which has been offered nation-wide since 2000. Besides the opportunity of intense professional exchange of experiences in management and leadership of their schools among colleagues, the initiative Business in the community gives the participants the chance to cooperate with an experienced partner who holds a leading position in industry or commerce. The aim is to enable a fruitful exchange of experiences among leaders within and without education, with corresponding benefits to both parties. This notion of exchange between educational managers and their counterparts in other sectors is not new – indeed this is one of the activities that the OTTO programmes sought to promote fifteen years earlier, but the scale of the operation and its organisation are more substantial than anything seen previously.

However, the main thrust of the TTA's programme for school leaders focuses on those who have yet to become headteachers. It is here that the largest and most tightly designed programme is targeted.

The National Professional Qualification for Headship

With HEADLAMP and the LPSH, headteachers are offered two development programmes that are meant to respond to the need for support and training after taking over leadership positions in schools. But, at the invitation of the government of the day, the TTA in 1995 started to develop a new, national qualification programme for aspiring headteachers, a programme completed before taking up a headship post. Combining a detailed training programme with rigorous assessment of the candidates, this qualification seeks to become established as the common route into headship, guaranteeing that future headteachers have met the standards required to embark upon headship, and have developed the professional abilities and skills necessary to do the job well. In the preliminary consultations about the content and structure of the programme, ideas and suggestions from interested parties, including the profession itself, were collected. Then, the programme was drawn up, a programme specified in great detail. The conception of this qualification, of the assessment and appraisal procedures, of the key principles and working assumptions underpinning it, was generally well received, as were the National Standards for Headteachers (that define the knowledge, understanding, skills and attributes required for headship) on which the content of the programme is based. Subsequent to its analysis and the corresponding modification of the concept, the first trials followed from autumn 1996 till spring 1997. After some modifications of the programme, the first cycle of the new National Professional Qualification for Headship was launched in autumn 1997[15].

The target group of the NPQH are teachers aspiring to a leadership position in a school. They will usually have some experience of leadership at subject or department level, and may well have experience as members of the senior management team or the school's leadership group (especially if they are drawn from the secondary sector). It is hoped that the opportunity to go through the NPQH before applying for headship will demonstrate which candidates are most likely to prove successful headteachers *before* putting them in post. Up to now, the qualification has been voluntary; but from 2004 on, it will become mandatory for all new headteachers to hold the NPQH, as the government sees this a major strategy for professionalising leadership roles and improving leadership performance.

The National Professional Qualification for Headship is based on five key principles (properly, these should perhaps be regarded as assumptions):

1. The qualification is rooted in school improvement and draws on the best leadership and management practice inside and outside education.
2. It is based on the National Standards for Headteachers (NSfH), which set out a range of school leadership and management tasks for each of the five key areas of the standards, in addition to the professional knowledge, understanding, skills and attributes that are required from headteachers for their headship.
3. The qualification signals readiness for headship, but does not replace the selection process.
4. It is rigorous enough to ensure that those ready for headship gain the qualification, while being sufficiently flexible to take account of candidates' previous achievements and proven skills and the range of contexts in which they have been applied.

[15] In the meantime, responsibility for the NPQH as well as the other two programs (HEADLAMP and the LPSH) lies with the DfEE.

5. It provides a baseline from which newly appointed headteachers can subsequently, in the context of their schools, continue to develop their leadership and management abilities.

Figure 4: The procedure of the National Professional Qualification for Headship in England

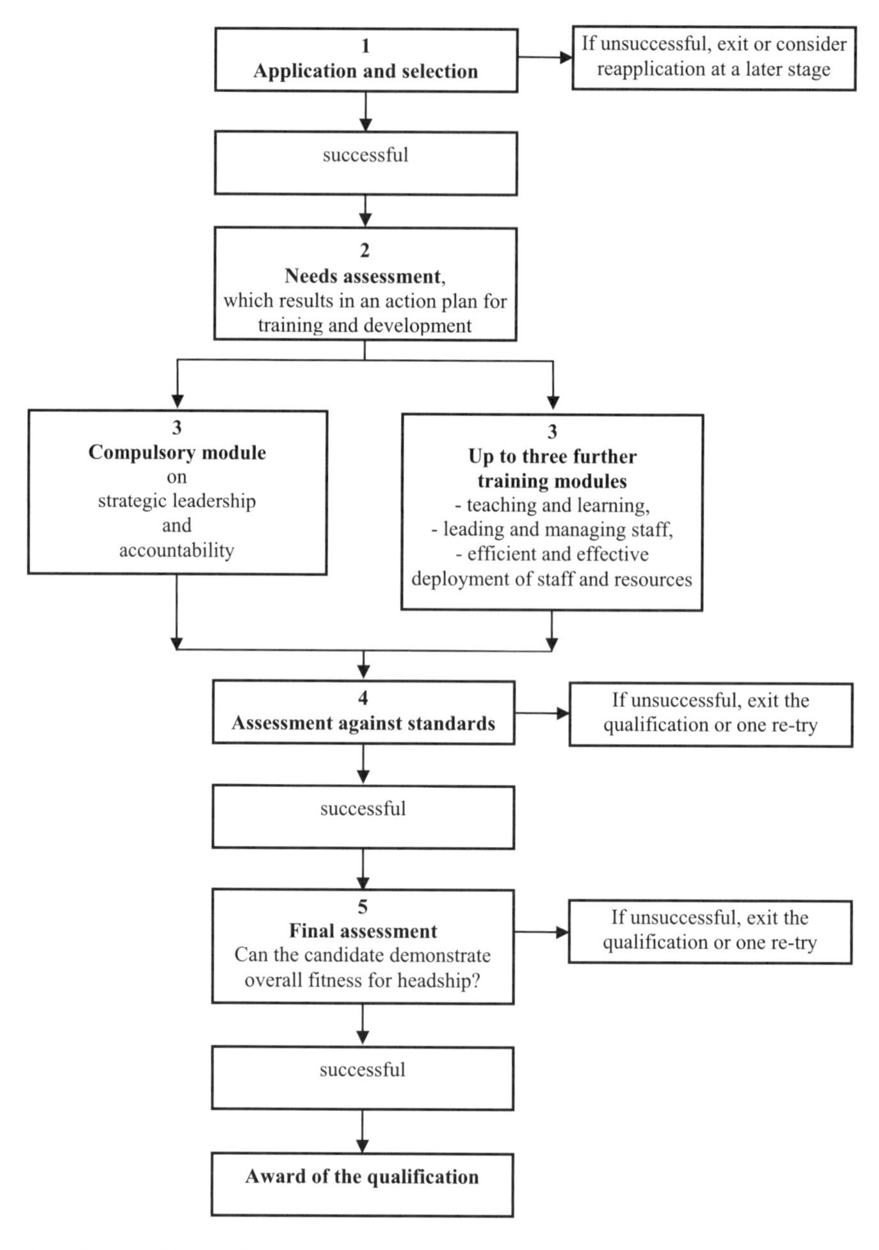

(from: Huber, 1997b, p. 20)

As indicated above, there are five stages in the qualification process: In the first stage, the application and selection process, applications from suitable, experienced teachers are invited by means of a nation-wide advertisement campaign.

> They should already have substantial, successful experience and expertise in leadership and management in schools and elsewhere, combined with relevant professional knowledge and understanding. In short, they should already be demonstrating the potential to become a headteacher. (TTA, 1997a, p. 1)

Candidates have to provide references from their headteacher or education authority that endorse their suitability. Then, in a selection process, appropriate candidates are identified. If judged not suitable, the candidates are entitled to be given reasons for rejection and another chance to apply again in the following year.

In the second stage, the "rigorous, supportive and confidential procedure" (TTA, 1997a, p. 3) of needs assessment should help candidates to identify their particular training and development needs in relation to the National Standards for Headship, which underpin the qualification programme. The goal is to generate an individual action plan that determines which modules the candidates should participate in – there is only one compulsory module, though most candidates opt to follow the whole programme. The number of modules attended then defines the total duration of the training, which varies from one to three years.

The third stage embraces the actual training and development modules. These modules cover the key areas of headship that are outlined in the National Standards for Headship, which underpin the qualification programme. All participants must complete the compulsory module, focusing on strategic leadership and accountability[16]. The module is made up of four units that are seen as central to all headteacher roles:

Unit 1: Developing a strategic educational vision committed to raising pupil achievement. This unit is intended to impart skills and abilities that enable the course participant to develop a suitable vision or a model for his or her school; this model should state the overall direction of development for the school and inspire improvement processes. The participants themselves develop this vision during the course. This is not considered a static act but rather a process, which means that their drafts are constantly being revised, questioned and newly designed throughout the programme.

Unit 2: Translating the vision into practice in order to secure high-quality teaching and learning. The participants should be enabled to concretise the developed vision of the goals of school step-by-step, and translate these into practical activities. It is considered an essential criterion for leadership to have the capacity to develop appropriate goals through collaboration with colleagues, and then to help implement these and translate them into actions.

Unit 3: Monitoring, evaluating and reviewing the effectiveness of a school. This unit is meant to address the often neglected processes of monitoring and evaluating

[16] Ranking these themes as the only compulsory module mirrors the British education policy of favouring Local Management of Schools with a far-reaching financial autonomy with all the chances and disadvantages linked to it.

school development activities. Furthermore, it intends to provide the future headteachers with theoretical knowledge and applicable strategies concerning evaluation processes in the school, and assessing the impact of school improvement processes on pupil learning.

Unit 4: Being accountable for the efficiency and effectiveness of a school to governors, staff, parents and pupils. The increased accountability at school level and, therefore, of headteachers is the main topic of this unit. In addition to the formal requirements of stewardship, the dilemmas which result from the fact that headteachers have to mediate between the demands of their political masters and the public on the one hand and teachers and pupils on the other side, are also covered.

This first module involves a total of 180 hours of study, of which 60 hours are contact time, spent face-to-face in training sessions, and 120 hours are spent on school based projects, individual study and preparing the assessment tasks. In addition, depending on the individual trainee's action plan, NPQH candidates may take further modules, as follows:

Teaching and learning:

This module focuses on the identification of possible actions to enhance pupils' learning outcomes. "It will include training in establishing and implementing policies and practices which promote high-quality education and in setting challenging and realistic targets for improvement." (see TTA, 1997a, p. 4)

Leading and managing staff:

This module focuses on how best to motivate staff and develop their professional capabilities. It also covers establishing and implementing policies, appraisal, planning, allocating, supporting and evaluating work undertaken by teams and individuals.

Efficient and effective deployment of staff and resources:

This module includes training relating to the recruitment and selection of staff, legal aspects of employment, and financial and resource management issues.

These additional modules cover 90 hours each (30 hours contact time and a minimum of 60 hours on school based projects and assignments). Training methods are varied, and include taught sessions, seminars, workshops, case-studies, simulation exercises, group reviews and presentations. The materials used include inspection reports, research findings, video materials, and, increasingly, computer based activities. Offering materials close to actual practice is the main objective here, but these are not only drawn from education and school management, but also from management practice in other sectors. "Training will draw on authentic, practical examples of management from schools, (but also from) business, commerce and industry." (see TTA, 1997a, p. 5)

The fourth stage involves the assessment (based on the National Standards) that module objectives have been met. All candidates will be expected to meet these objectives, whether they have undertaken the specific training units or not. At the fifth stage, where candidates present their portfolio of work compiled during the programme, "the final assessment will show whether candidates can demonstrate their fitness for headship" (see TTA, 1997a, p. 5) through interviews, group activities and presentations etc. Here again, they must demonstrate the potential to operate across all areas of the National Standards. Candidates who have been successful will be awarded the NPQH.

The approximate costs of the various stages are for the needs assessment 350 pounds, for the compulsory module 900 pounds, for each further module about 450 pounds, for the assessment of areas which were not assessed through training 150 pounds, and for the final assessment 300 pounds. Altogether, the costs will be between 2 000 and 2 900 English pounds (3 200 and 4 700 euro), according to how many modules have been taken. There is funding for the programme available through LEAs (for candidates from LEA-maintained schools) or central funding (for those from directly funded schools), but self-funding or alternative sources of finance are also possible.

National Standards for Headship

As mentioned above, the NPQH is based around the National Standards for Headship as far as the training content and the assessment criteria are concerned. This catalogue of requirements relevant for the qualification and for assessing candidates consists of five sections: 'Core purpose of headship', 'Key outcomes of headship', 'Professional knowledge and understanding', 'Skills and attributes', and 'Key areas of headship'. Under these headings the various tasks and competences required are spelt out.

For the TTA, the "Core purpose of headship" is "to provide professional leadership for a school which secures its success and improvement, ensuring high quality education for all its pupils and improved standards of achievement" (see TTA, 1997b, p. 1). This is approached through a focus on several whole-school tasks for which the head is responsible.

The section on "Key outcomes of headship" hypothesises that "effective headship results in effective schools" (p. 2). It focuses on the nature of this relationship and then gives a picture of an effective head in an effective school. It emphasises the need to establish a positive ethos, and helpful working conditions based on an effective cooperation with teachers, pupils, parents and governors, and an "efficient and effective use of [...] staff, accommodation and resources" (p. 2).

The competences needed by heads are split up by the TTA into two areas. In setting out the key areas of 'Professional knowledge and understanding' needed, the TTA draws upon school effectiveness research, and emphasises the insights to be gained from this source. But it also lays emphasis on management (e.g. budgeting and recruitment of staff), legal and statutory knowledge (e.g. school law, industrial law) and knowledge and understanding of leadership styles and learning and teaching methods. Supplementing the above-mentioned domains, 'Skills and attributes' comprise leadership skills and attributes, professional competence, decision-making skills, communication skills, and self-management skills. Above all, the capacity to work productively with colleagues seems to be emphasised here.

The fifth section describes the 'Key areas of headship', in relation to which school leadership and management tasks are set out. These are the 'Strategic direction and development of school', 'Teaching and learning', 'Leading and managing staff', 'Efficient and effective deployment of staff and resources' and

'Accountability'. Inevitably, some have found too much overlap in these National Standards, while others have pointed at gaps, but generally the standards have proved useful to all those interested in headship, and not just those studying for the National Qualification.

Table 14: Outline of the National Professional Qualification for Headship in England

Training and Development of School Leaders in England, Great Britain The National Professional Qualification for Headship	
Provider	The Education Ministry, formerly the Teacher Training Agency (TTA), now the National College for School Leadership (NCSL) through approved 12 approved Training and Development Centres and 11 Assessment Centres
Target Group	Teachers aspiring to a school leadership position, i.e. before application
Aims	Providing the participants with leadership and management competences in order to prepare them for headship following the five key principals and the National Standards for Headship
Contents	Mandatory module: strategic leadership and accountability (developing a strategic educational vision committed to raising achievements; translating the vision into practice in order to secure high-quality teaching and learning; monitoring, evaluating and reviewing the effectiveness of a school; being accountable for the efficiency and effectiveness of a school to governors, staff, parents and pupils) additional modules: teaching and learning; leading and managing staff; efficient and effective deployment of staff and resources
Methods	Self-assessment; taught sessions; seminars; workshops; case studies; simulation exercises; group reviews and presentations; materials used include inspection reports, research findings, video materials etc.
Pattern	10-25 course days* (according to the number of modules) (timetabling: mandatory module: 180 hours (60 hours contact time and 120 hours for school-based projects, individual study and preparing for assignments); 3 further modules: 90 hours each (30 hours contact time and 60 hours for school-based-projects and assignments)) plus school-based projects, individual study and preparation of assignments within 1-3 years
Status	Optional; appreciated by the employing committees at the individual schools; from 2004 on mandatory
Costs	Ca. 3 200 euro to 4 700 euro (2 000 to 3 000 English pounds) per participant depending on the number of modules taken; different sources of funding are offered, but self-funding is possible as well
Certification	Award of the qualification after successful completion of the final assessment

* *If there is no specification by the provider as far as the number of days is concerned, we converted the contact time in hours into the unit course day, taking 6 hours training as one day.*

3. Summary

In the UK, efforts to provide headteachers with management training date back to the 1960s. Early efforts were, however, provided on a rather piecemeal basis, with little coordination of provision. This fragmented approach came under scrutiny in the early 1980s, when the first systematic training programmes were offered nationally to serving headteachers. Though these training opportunities were generally well regarded by participants, the numbers able to secure a one-term release from school to undertake the programme were very small, and the programmes withered away. Driven by the Teacher Training Agency (set up in 1994), in recent years there has been increasing formalisation of the preparation of headteachers in England. While there have been training programmes available for some time, the piecemeal local arrangements have now been replaced by national schemes that target three groups: aspiring headteachers, newly appointed headteachers, and established headteachers.

The largest of these programmes, the National Professional Qualification for Headship (NPQH), is intended as a preparation for headship. It is available to senior staff in schools that are nearing that point where an application for a headteacher post is possible. There is a general expectation that programme members will take one year to three years to attend the programme (depending on their experience and the assessment of their strengths and needs).

The scheme for newly appointed headteachers, the Headteachers Leadership and Management Programme (HEADLAMP), offers a range of modules based in approved centres and covering management topics to those in the first year of headship. The programme for established headteachers, the Leadership Programme for Serving Heads (LPSH), was introduced in 2000 as a further piece in the jigsaw, ensuring that updating on management issues is available to all those who seek it.

These programmes are widely accessible, due to a deliberate strategy to provide regionally based training opportunities, and the relatively generous funding available to pay for training, administered through local education authorities but originating from central government. The government has made it clear that it expects NPQH to become a mandatory requirement in due course, so it is likely, once all headteachers have been through this route that the HEADLAMP and LPSH programmes will alter to reflect this prior training.

4. Recent Developments

Three years after the initial implementation, the NPQH has been revised. The modified scheme is arranged in three parts and thus will be shorter: The Needs Assessment will be cancelled. Instead, the candidates themselves will decide – with the help of a self-evaluation survey – (in addition to the evaluation carried out by the NPQH-Centre based on the candidate's application), whether they participate in an Access Stage (which lasts up to one year) or if they skip this stage and directly step into the next, the Development Phase. This is only possible if the candidates already have some experience with leadership tasks. The Development Phase is concluded with an assessment held at the participant's school and is followed by the Final Stage. After another evaluation, the certificate is finally awarded to the participant.

There is a general expectation that programme members will take one year to work through stage one and another to work through stages two and three.

Currently, the programme is widely accessible, due to a deliberate strategy to provide regionally based training opportunities, and the relatively generous funding available to pay for training, administered through Local Education Authorities but originating from central government.

Up to now, the qualification has been voluntary; but from 2004 on, it will become mandatory for all new headteacher to hold the NPQH. It is likely, once all headteachers have been through this route that the HEADLAMP and LPSH programmes will alter to reflect this prior training.

In 2000, the UK government established a National College for School Leadership (NCSL), and the responsibility for coordinating and further developing headteacher preparation programmes has transferred to the College. Under the College umbrella, the revised NPQH scheme has been implemented across the country, and the LPSH programme has been piloted. The Headlamp programme is currently being revised and, elements of this new support programme for headteachers in the first year of appointment are also being piloted, under a programme called 'New Visions'. Most recently, the College has commissioned the design and piloting of a further strand in school management development, 'Leading from the Middle'. This programme, in some ways the most ambitious of all, since it targets more than 20 000 teachers who hold middle management positions (typically subject leaders in the primary school, departmental heads in secondary schools). The intention is to ensure that systematic training, again underpinned by the National Standards for School Leaders, is available earlier in the potential headteacher's career, and also to increase leadership density within all schools – underlining the notion that schools need leadership at all levels and that leadership is most effective when it is distributed.

In 2001 the NCSL produced its 'Leadership Development Framework'. This sets out the five 'key stages' around which school leader development activities will be targeted in the coming years, as current provision is revised and sharpened. These are:

1. Emergent leadership: when a teacher is beginning to take on management and leadership responsibilities and perhaps forms an aspiration to become a headteacher.
2. Established leadership: comprising heads of faculty, assistant deputy heads who are experienced leaders but who do not intend to pursue headship.
3. Entry to headship: including a teacher's preparation for and induction into the senior post in a school.
4. Advanced leadership: the stage at which school leaders mature in their role, look to widen their experience, to refresh themselves and to up-date their skills.
5. Consultant leadership: when an able and experienced leader is ready to put something back into the profession by taking on training, mentoring, inspection or other responsibilities.

The College's purpose here is to create, for the first time in the UK, a coordinated and structured approach to leadership progression. "There is, of course, already a solid foundation of national programmes and initiatives to support school leaders.

NCSL is proposing changes to some of the existing national programmes and introducing proposals for subject and specialist leadership and consultant leadership. It is also considering programmes to meet major skill needs of headteachers, such as ICT and inspection skills." (see NCSL, 2001)

The setting up of the NCSL has provided a mechanism for organising and coordinating headteacher preparation on a national basis. The funding of development programmes from central and local budgets means that opportunities for training are widely available. England is, then, taking significant steps towards a coherent programme of school leader development, and the next generation of headteachers will certainly be better trained and more carefully prepared than their predecessors.

5. Statement by Mel West

Undoubtedly, the preparation and training of headteachers has become a much more systematic process in England in recent years. The present series of opportunities stretches across several key points in the headteacher's development, reflects current thinking about and research into school leadership, and, through the tight central control of content and methods, guarantees these regardless of where in the country the programmes are followed. There is also no doubt that the training provided- especially the NPQH- is becoming well established in the minds of participants and those who appoint headteachers as a credible and worthwhile preparation. The importance attached to appropriate training and the prominence of the training programmes themselves have been significantly enhanced in the past two years by the activities of the National College for School Leadership. The College has focused attention on both needs and provision, and has developed a range of activities that complement existing activities. These include the stimulation of research, particularly practitioner research, the building of networks of headteachers around the country who have begun to share experiences and practices, the setting up of on-line information sources and forums for discussion and the (to a degree) re-shaping of established programmes in light of participant feedback and changing need.

The programme for middle mangers currently under development brings some exciting new opportunities, such as collaboration in learning activities among staff from the same school, and the identification (and, of course, training) of school based coaches to help participants find their way through the materials and the activities. A virtual school package is also being developed within this programme, which will create new e-learning opportunities and offers the possibility of simulation and review of decision-making skills.

Nevertheless, there have been some questions raised about whether the programme has got everything right. For example, there is some feeling that if any component is to be compulsory, this should be the Headlamp programme, since we can be sure that all heads will be new at some point and all new heads have needs. There has been some criticism of the control of curriculum content in NPQH. While it is an important quality control mechanism, it may also mean that potential headteachers do not always have their thinking challenged as it might be in a more diverse programme that drew in more contributions from outside the relatively narrow band of approved providers and trainers. Some would like to see a clearer link between the new professional qualifications and conventional masters' programmes- with the facility to cash in the one against the other. It has also been suggested that the programmes do little specifically to increase the representation of women- a continuing concern as men still achieve a disproportionate number of headships- and of members of ethnic minority communities in the most senior roles in schools.

However, despite these criticisms, it is indisputable that in England the most complete and focused preparation for headteachers that has ever been available there is now in place. There is greater coherence in provision, scope for differential progress through the various elements according to need and readiness, and the different levels are beginning to function as targets or stepping-stones on the way to headship. Of course, there is scope for further improvement, but in England a very solid platform has been built and the development of future school leaders is being carried forward with some confidence.

The Netherlands: Diversity and Choice

Stephan Huber and Jeroen Imants

1. Background

The basic legal principle in the Dutch educational system is the 'freedom of education', as determined in article 23 of the Constitution. This article gives parents and citizens in general the right to establish schools based on diverging religious, philosophy of life, or pedagogical principles, as long as these schools come up to certain general regulations. As a result, the Dutch educational landscape is characterised by about one third of public schools and about 65% of private schools: Catholic and Protestant schools, and schools based on specific pedagogical principles (for example Montessori or Pedersen), besides, small amounts of Islamic, Jewish, and Hindu schools have been founded. The financial resources of these schools and school boards are guaranteed by the national government for the full 100%. Depending on the resources of their associations and foundations, private schools gain additional resources from private funds.

As compared with other countries, the Dutch school system is rather decentralised. Although a mix of centralization and decentralization characterises the actual situation, the dominant trend is that the national government and the Ministry of Education focus on the determination of general guidelines and the creation of conditions for education.

Each school is under the formal governance of a school board. School boards differ considerably in size (1 school to 100 schools) and membership, related to the identity, history and demographic background of the school. The Board of the public schools in a community is the democratically elected community council. Members of school boards can be representatives of the founders, representatives from business, churches, and politics, and professionals with an educational background. Especially the smaller Boards of private schools (Boards with one or a few schools, mostly in rural areas) consist, for example, of some parents with a strong commitment to the school, the vicar, and individuals with a special position in the local community. An important responsibility of the school boards is the selection of personnel, school leaders as well as teachers.

Local school boards have a large amount of autonomy to start to develop their educational policy from their philosophy of life or religious identity. In this respect the educational landscape is quite diverse. Because teachers and school leaders are in the service of these local school boards, school boards have certain discretion in personnel policy within the limits of national regulations. Moreover, within the limits of general regulations and a national curriculum, schools can take decisions about the use of financial resources, the elaboration of the school curriculum, the methods of teaching, and the development of the school organization. Many Boards have mandated decision making responsibilities towards the school leaders. At the same time many school boards, with more than one school, develop a new management level between the Board and the schools, the super school management, and they mandate decision-making about educational, personnel and financial matters to this super school management structure.

Each school is obliged to maintain a participation council (Medezeggenschaps-raad), in which parents and teachers (school personnel) are represented. This participation council has limited formal decision making power, more specifically the right of approval, about such topics as school identity, use of financial resources, planning of holidays, and changes in staff. Councils for parents, personnel and pupils (in secondary education areas) undertake more or less activities at most schools.

Complementary to this school autonomy is that schools are accountable to the national government and to the public for the school curriculum as compared to the national curriculum, and for the pupil learning results related to national standards. Within a regular timeframe each school is evaluated by the National Inspector on a fixed set of standards. Although the Inspectorate is formally independent from the Ministry of Education, it functions within the framework of regulations by the Minister of Education. The policy of the National Inspector is to transform the traditional control relationship with schools, based on formal power, into a support relationship, based on influence. However, schools still experience a strong external pressure from the authority of the Inspector to express an opinion about their functioning and their results.

As a result of decentralization policy and the strategic responses of school boards towards this policy, the complexity of the role of school leaders and their strategic responsibilities have grown very fast during the last ten to fifteen years. Their autonomy has increased, but the same holds for their responsibility and the chance to get involved in conflicts. School leaders are expected to develop the school at a continuing base and in close connection to the local community, to implement and evaluate school policy and innovations, and to be accountable for school results to the school board, the parents, the Inspector, and the public. They are responsible for the development of the school as a learning community for pupils and teachers. Moreover, they have many organizational and managerial duties among others in the field of personnel policy. The focus in their work is changing from working with pupils to working with adults, and from working inside the school to working outside the school in local networks and policy councils. More and more, an 'entrepreneurial' and market-oriented attitude is expected from school leaders (see Imants & De Jong, 1999). This market-oriented and entrepreneurial attitude is illustrated by a quote from a Dutch school leader: "It's the number of pupils which brings the money in for our schools". These developments in the function of school leaders can be summarised in four roles (see van Agten, 1997; Verbiest, 1998):

- organization, administration,
- educational leadership,
- support and consultation of adults,
- 'entrepreneurship'.

As a result, school leaders work in a field that is filled up with complexities and paradoxes. Many school leaders interpret these complexities and paradoxes as the tension between the educational leadership role and the general management role. However, close examination shows that at least three paradoxes are apparent in the school leaders' role at the same time (see Imants, 1996):

1. educational leadership and general management,
2. internal and external orientation and responsibility,
3. being the coach of colleagues and the employer of employees.

2. Development of School Leaders

About twenty years ago the Dutch Minister of Education published a policy document, in which the changes in the role of the school management were identified. For the first time in the history of Dutch education the training of school management was stressed as a priority in educational policy (see Minister van Onderwijs en Wetenschappen, 1984). Essentially, the priority of training school management was based on two pillars. One pillar was the growing insight in the role of school leaders, especially their educational leadership role, in school effectiveness and school improvement. The second pillar was the decentralization policy by the national government, which was accompanied by a stronger focus on output control, market relationships in the field of education, and the role of pupils and parents as critical consumers. In the early years, the demand in secondary education for school management training was rapidly growing. After about five years, the field of primary education followed as an exploding market for school management training.

Although the importance of professional school management is stressed, formal education and qualification to acquire a position as a school leader is not required in the Netherlands. The only formal requirement is that candidates for a school leader position should be experienced teachers (a certain amount of years of experience as a teacher is required). However, school boards (the employers of school leaders) and the national government expect that school leaders who are in post and teachers who aspire to become school leaders do take care for their professional preparation, and do participate in training sessions and courses. Because national regulations are absent, it is up to the school boards to decide about content and application of formal requirements and qualifications for their school leaders.

Since the early 1980s three generations of programs for the preparation of school leaders can be distinguished in the Netherlands:

1. Short and practically oriented management training sessions from a diversity of providers; these training sessions contained isolated and fashionable themes and practical skills, the contents of these sessions were fragmented, the training was in-service and off the job.
2. Two-year part time school leader courses that cover the field of school leader competences; besides their practical orientation, these courses have stronger conceptual roots in educational, organizational, and economic sciences. Providers in the field of secondary education operate in a free market, with universities, schools of education, and commercial institutes. The field of primary education can be characterised as quasi-decentralisation. By means of a protective policy of the national government five selected programs of traditional schools of education receive a hidden

monopoly. This hidden monopoly is based on two measures by the national government: formal recognition of the certificates of these programs and financial compensation to the schools of school leaders who participate in these five courses. In secondary education the content and didactics of courses are more diverse as compared to primary education, both in primary and secondary education the fees for the courses are paid by the school leaders, but it is up to the schools (school boards) to return these fees to the school leaders from the professional development budget.

3. A point of discussion is whether integration between separate themes of these secondary generation courses has been established. From a critical point of view, the fragmented study of separate subjects is continued in these courses. The practical projects of these programs can be regarded as the beginnings of a more integrated approach of the school leader's role in the professional development programs. These programs are courses off the job, but the practical projects are closely connected to real life school contexts.
4. A recent development is the application of insights in the cognitive aspects of leadership in school leader development. In this initiative towards a third generation of programs a central aim is the development of integral leadership in schools. The starting point is that instructional leadership and general management should be treated as interdependent fields of effective school leadership, and that a prerequisite for effective school leaders is an integral view of their school. Characteristic is a systemic approach of the school leader's job in schools and a theoretical or conceptual closeness. In the first initiatives the hidden monopoly of selected providers is passed by because this program is realised by a university institute and a school consultancy institute apart from the schools of education which traditionally do all the training of school leaders.

In the next two sections, two Dutch programs are described. In the field of secondary education, a program from the second generation is selected: the program Management- en Organisatieopleidingen by the Nederlandse School voor Onderwijsmanagement. This program has a long tradition and its reputation is very good. In the field of primary education a program from the third generation is selected: the program Meesters in Leidinggeven by the Gemeenschappelijk Centrum voor Onderwijsbegeleiding yn Fryslân (GCO fryslân)[2] in cooperation with the Instituut voor Leraar en School (ILS) at the University of Nijmegen[17].

[17] Other examples of school leader preparation programs are the four programs provided by the institutes that are selected by the Ministry of Education: Hogeschool van Amsterdam (Esan), Hogeschool Gelderland (Interstudie), Hogeschool Katholieke Leergangen Tilburg (Magistrum), Christelijke Hogeschool Zwolle (Windesheim) (see van Kessel et al., 1999).

Example in the Field of Primary Education: Meesters in Leidinggeven of the GCO fryslân in Cooperation with the Katholieke Universiteit Nijmegen

In the Netherlands a network of local and regional school consultation services has developed in the last four decades. Schools can consult these institutes with regard to individual pupils (pupil assessment for example) and school development (improvements in instruction and capacity building). These institutes can also provide staff development programs to schools. The biggest school consultation service is GCO fryslân. In cooperation with the University of Nijmegen, GCO developed an in-service program for experienced school leaders in primary schools. In January 2000 the program started for the first time with a group of 16 participating school leaders. Some of the reasons that GCO began the cooperation with a university were the combination of specific know-how at the practical and theoretical level, academically based quality control, and the status of the certificate.

The program Meesters in Leidinggeven is developed for school leaders of single schools (locatiedirecteuren) with several years of experience. These school leaders are expected to be looking for further personal growth in their role and for an integration of experiences and insights. An important aim is the development of a 'helicopter view' on their school, besides the development of practical and reflective skills at a higher level of conceptual integration. Because participation in the course is not externally subsidised, the participants share a strong intrinsic motivation. A potential career perspective for participants is that the certificate will help them in gaining a position as Director of a cluster of schools. One of the effects of decentralization policy in Dutch education was that small school boards strengthened their position by merger rewarded by the Ministry of Education. Large school boards have developed the new function of Super School Manager or 'bovenschoolse directeur': a team of directors of a cluster of schools (varying from less then 10 schools to 40 schools or more) supervising the local school leaders, the 'locatie directeuren'.

The program Meesters in Leidinggeven builds on the concept of integral leadership. Integral leadership is defined as an approach of school management in which educational leadership and general management are looked upon as entities which necessarily complement each other (see Imants, 1996; Imants & De Jong, 1999). The concept of integral leadership starts from a definition of leadership in which the steering of the behaviour of others is central. The meaning of educational leadership is thereby, being able to steer the instructional behaviour of teachers and the learning process of pupils. The idea behind integral leadership is that to be effective as a school leader the steering of educational processes and the coordination and supervision of the school should be effectuated from an integrated perspective. This integrated perspective puts a lot of pressure on the shoulders of school leaders of single schools in primary education. Most of the 'locatie-directeuren' (school leaders of single schools) are only partially exempted from teaching duties.

The central question for school leaders is how to affect the instructional behaviour of teachers and the learning process of pupils in a positive direction. The school leader disposes of a wide array of instruments or linkages for steering, but none of these instruments is very powerful. Moreover, the conditions in terms of time and resources are rather restricted. This implies that school leaders should put into action smart combinations of available resources and opportunities as efficiently as possible. Moreover, the opportunities to put these smart combinations of instruments into action for steering should be identified by the school leader in her or his school, and therefore these school leaders should be able to construct at a mental level a cognitive map of their own school. The merit of the Meesters in Leidinggeven program is the insight school leaders develop in their own school and its environment. The clearer the image of one's own organization, the more adequate one can act and take steering initiatives. This school leader will interpret diverging kinds of tasks, educational leadership tasks as well as general management tasks, in a coherent perspective, having in mind the realization of the educational goals of the school.

Participating school leaders construct the cognitive map of their own school in close cooperation with a critical friend, who is the school leader of another school. During the program these pairs of school leaders interview each other and ask each other critical questions about the other's school, and during this process the school leaders are writing down images of their schools. The writing of these documents, the discussion with the critical friend and the feedback from a university professor on these written documents, altogether constitute a process of reflection that is much deeper than the usual reflection based on verbal discussions. During each monthly plenary session a pair of school leaders presents the results of their inquiries of the preceding month. The discussion of these results helps all the participants to lift their descriptions of their schools towards more concrete, complete and coherent images of their schools. As a result the relationship is strengthened between the elaboration of the cognitive map of the school and the ongoing processes in daily reality of school life during the program.

The themes of this part of the program are organised and described in a textbook, called 'De School Meester' (Master your School) (Imants, 1999). Themes are:

- introduction of the concept of integral leadership and the construction of a cognitive map of your own school,
- the school as an open system,
- the organization of instruction and pupil guidance in the school,
- school structure,
- organizational behaviour by teachers and school leaders in the school (motivation, climate and culture, leadership),
- decision making in the school, including micro politics in the school,
- innovation and school development,
- school autonomy and quality control,
- the construction and analysis of the cognitive map.

Each chapter in the book starts with an introduction of central concepts and the discussion of some underlying theoretical insights and practical examples. These concepts should be elaborated in the description and analysis of the school leaders' cognitive maps. Starting from these concepts questions are presented that should help the school leaders during the construction of their cognitive maps. At the end of each chapter examples are given and discussed of school descriptions by students who worked with these questions during university courses.

At the end of the program participants combine their descriptions organised around the themes into one coherent description of their school. In the next step this description underlies a SWOT analysis of the school (strengths, weaknesses, opportunities, threats), in which internal weaknesses and strengths of the school are combined with environmental opportunities and threats. This analysis is the starting point for developing a strategy for the school. This strategy for school development should help the school leaders in selecting concrete priorities and in putting these priorities into a plan with operational points of action. This operational plan is accompanied by a financial and personnel budget. The estimation of this budget should help the school leaders in focusing the selection of specific priorities for their schools and in developing realistic and efficient planning documents. During the last plenary sessions these school strategies and operational plans are presented in a role play in which the participants are playing the school leader's faculty. Finally the school leaders receive feedback on their plans from the participants and the university professor.

In addition to the process in which participants develop a cognitive map of their schools, the school leaders participate in concrete training sessions around three themes: inspiring faculty members, coaching, and team building. These training sessions are supervised by trainers of GCO fryslân, and they take place on the afternoons of the monthly plenary sessions.

The program Meesters in Leidinggeven takes 14 months. Each month the school leaders participate in a one-day plenary session together with two supervisors of GCO fryslân and a professor of the University of Nijmegen. The participants also receive individual consultation sessions with a GCO supervisor. The pairs of school leaders (critical friends) meet between the plenary sessions according to their own agendas. The idea behind this time frame is that the participants need time for literature study, writing and reflection in the pairs, and application of methods from the training part (inspiration, coaching, team building) between the plenary sessions. Moreover, the absence of one day in a month for professional development purposes is close to the maximum of what school leaders of small primary schools and their schools can afford.

Central in the didactics of this program is the principle of the pairs of critical friends, which is related to the peer-assisted learning method. These critical friends should provide practical and intellectual assistance to each other, stimulate data based reflection, and give feedback on a continual basis. During the learning process the participants are expected to develop insight into their own school together with the ability to reflect at a professional level and to give and receive feedback. Moreover, professional development of school leaders is closely linked to the development of the school.

The fee for one school leader is about 2.400 euro, this fee is paid by the school leaders or their school boards. Occasionally the fee is paid from the school's professional development budget. The size of the group of participants (16) reflects an optimum between the costs of the course and the didactical principles of the course (monthly plenary sessions, the central role of pairs of participants, and individual consultation).

Requirements for successful completion of the course are a written cognitive map of one's the school, including a school strategy, an operational plan and a budget, written proofs of the application of practical exercises from the ICT training part, and sufficient participation. The participants receive a certificate of the

University of Nijmegen and GCO fryslân. Although this certificate lacks a formal status, it is expected that it will help the participants when they are in the running for a career step, for example, a position as Super School Director.

Table 15: Outline of Meesters in Leidinggeven of the GCO fryslân in cooperation with the Katholieke Universiteit Nijmegen in the Netherlands

Training and Development of School Leaders in the Netherlands Example in the Field of Primary Education: Meesters in Leidinggeven of the GCO fryslân in Cooperation with the Katholieke Universiteit Nijmegen	
Provider	GCO fryslân in cooperation with the Katholieke Universiteit Nijmegen
Target Group	Established school leaders in primary education
Aims	The development of competences for leading a school in a more systemic way (integrating the steering of educational processes and administrative activities) and the development of a cognitive map of one's own school as a starting point for school development plans
Contents	Integral leadership and cognitive mapping; the school as an open system; the organisation of instruction and pupil guidance in primary schools; school structure; organisational behaviour of teachers and school leaders; decision making in schools; innovation and school development; school autonomy and quality control, the construction and analysis of the cognitive map of the school; concepts and practical exercises regarding inspiring and motivating staff, coaching staff members, and team building
Methods	Lectures and invited speakers; training sessions and practical exercises; consultation sessions; role plays and simulations; peer assisted learning; literature study; writing the school analysis and school development plan; presentations
Pattern	14 course days (timetabling: 1 full day per month for the plenary sessions) plus study meetings (pairs of school leaders); coaching sessions; literature study; homework; preparation of school analysis within 14 months
Status	Optional; appreciated by employing school boards; the certificate can serve as a stepping stone for school leaders in obtaining a position as 'bovenschoolse directeur' (a Super School Leader overseeing several schools)
Costs	Ca. 2 400 euro per participant; financed by the participants themselves, sometimes funded by the school budget
Certification	Certificate of the University of Nijmegen and GCO fryslân

Example in the Field of Secondary Education: Management- en Organisatieopleidingen of the Nederlandse School voor Onderwijsmanagement

The Nederlandse School voor Onderwijsmanagement (NSO) is an institute, in which the Universities of Amsterdam, Nijmegen, Utrecht, Leiden, and the Free University of Amsterdam cooperate in the professional development of established school leaders, middle managers, and teachers who aspire to a school leader position in the near future. The focus of the two-year part time program Management- en Organisatieopleidingen is in the field of secondary education, but the program also has participants from the fields of vocational and adult education. The entry requirement is at least five years of professional experience in education. This experience may vary from primary and secondary education to vocational and adult education. Some experience with coordinating activities in schools is desirable. Expectations are very explicit about the available time and motivation of students. After a first selection that is based on information from the subscription form, all potential participants have an interview with a committee, with the final selection being after this interview. One criterion for the composition of the final group of participants is an equal share of female and male participants.

The vision underlying the NSO program is that school leadership is a profession on its own, and that a profound preparation for future school leaders is a necessary prerequisite for this job. Starting from this point of view the program focuses on established and aspiring school leaders. The aim of the program is to create a context for participants in which they are supposed to take charge of the educational process in (part of) a school, and by doing so, to create opportunities to learn relevant knowledge, skills, and attitudes. Some of these skills are problem-solving skills, conflict resolution skills, and the ability to handle diverse perspectives and to reflect on personal behaviours, feelings, and thoughts. Learning as well as 'unlearning' is supposed to be important in this process, because existing beliefs and cognitive schemata have to be subject to critical reflection. As leaders of their unit the participants are expected to be a source of inspiration and motivation, and to show an open attitude. They develop skills in fields like cooperation, intervention, risk taking, contextual thinking, and network development. At the end of the course the successful students receive a formal grade (Master of Educational Management). This grade should be helpful to the students in taking a position as school leader.

Five didactical principles underlie the management course Management- en organisatieopleidingen of NSO:

- a maximum group composition of 20 participants and heterogeneous group composition to create a context in which learners learn from each other,
- the work in the plenary group and in groups of five participants to be switched at a regular base,
- a high level of autonomy and motivation for homework and an active role in group work,
- a close connection from course themes to examples from practice plays an important role in the application of theory,
- an intensive learning process based on an interplay of action and reflection, learning and working in school practice.

The program Management- en Organisatieopleidingen has a competence-oriented approach. It is based on the results of interview studies among school leaders on the central competences of school leaders (see NSO, 1998 and 1999). Five general competences are regarded as very relevant:

1. Pedagogical, educational, and instructional competences
   A key competence of school leaders is to initiate, stimulate, and review learning processes. School leaders should be focused on the improvement and effectiveness of these learning processes, which means that they should focus on such fields of action as:

- the relationship between general school management and the aims of the school as an educational institute,
- the relationship between school management and pupil guidance,
- the role of school management in educational change processes,
- the effectiveness of the school and quality control,
- the implementation of the national curriculum at the school level and curriculum development.

2. Controller competences
   Schools function within the boundaries of financial conditions. Starting from the need for school leaders to handle their budget with care, important fields of action are:

- the school as an economic unit, which has to deal with scarce resources,
- year planning and long term planning,
- a solid financial administration,
- personnel management,
- legal issues and issues from industrial law,
- utilisation of computer resources in school information systems and administration.

3. Leadership competences
   People are the most important resources in schools. Frequently recurring issues with regard to the human resources in the organization are collaboration, communication, leading meetings, effective decision making, and conflict resolution. Fields of action are:

- decision making, collaboration, feedback, and dealing with defensive behaviours,
- dealing with strong and weak characteristics as a leader,
- feeling for control and balance in change processes,
- communication and behavioural change,
- strengthening strategic thinking in meetings,
- conflict resolution,
- leadership style,
- time management and stress management.

4. Organizational competences
   Issues with regard to the organization are the development of school policy, organizational effectiveness, external relations, planning and evaluation. Fields of action are:

- school rules and other formal conditions,
- strategic school development,
- dealing with people inside and outside of school,

- public relations, acquiring sponsors, nourishing the public image of the school,
- development of an organizational vision and productive organizational processes,
- initiating quality circles in decision making processes,
- development of the identity of the school,
- school self evaluation and quality control,
- utilisation of external conditions for internal developments.

5. Consultation competences
   School leaders are expected to motivate people, to recognise and develop their qualities, and to find a balance between their aspirations and the school vision. Fields of action are:

- human resources development,
- performance evaluation meetings with teachers,
- professional development, merit pay, and career development,
- supervision,
- understanding and using socio-psychological processes.

The general starting point is that these competences should be treated in relationship to each other. More specifically, the relationship among these competences is emphasised in the practical school project.

Five didactical methods can be identified in the program:

- seminars (Wednesday afternoon and evening),
- training sessions (Friday and Saturday),
- consultation meetings with discussion,
- keeping up a log,
- practical projects in schools.

The seminars focus on the learning of knowledge and the development of a vision with regard to the previously mentioned competences with the participants expected to be actively engaged. Discussion of real life examples, homework, and exchange of personal experiences strive for an interaction between theory and practice. A leader of the learning group supervises each seminar. His role is to meet the learning goals of the participants, to supervise the teachers, and to coordinate the learning process and make corrections if necessary. In most cases a seminar covers two days, which are separated by a period for homework. The themes of the seminars are closely related to the fields of action. These themes are mentioned above in the discussion of the core competences for school leaders.

Training sessions consist of five parts of three hours each. As with the seminars the parts of the training sessions are separated by a period for homework. An exemption is the start session, which consists of eight parts. Themes of training sessions are communication, leadership, meetings and negotiation, conflict management, and career consultation.

The consultation meetings and the writing of the log on a regular base should be helpful in identifying strong and weak points as a school leader, and in keeping the participants motivated during the two-year course. In the first year the number of meetings varies from six to eight. Each group consists of five participants (male/

female) and a supervisor. Aims of the group work are feedback, reflection on alternatives for action, and the development of insight in alternative courses of action. As the course progresses, the participants are expected to gain insight in their own learning goals and in their learning progress. The log is expected to play an important role in this respect.

Participants start their practical projects in the second part of the second year. Projects are selected by representatives of school leader associations, and are executed outside the participants' own schools. In these practical projects the goals of the program are elaborated in an integrated way. Dyads of participants are engaged in a problem-solving process, which is closely connected to the work of school leaders, and in which they assess their own competences. Examples of themes of practical projects are the implementation of a new system for grading pupils or a new framework for school finances, or the development of a plan for staff development at the faculty level. The teachers and supervisors of learning groups are at the disposal of participants to play a coaching role. For each project a written report is demanded, and each project is presented to the other participants.

Didactical methods are regular lectures and lectures by guests (speeches), seminars, training sessions, consultations, role plays and simulations, case study, peer counselling, readings, writing a study journal (documenting one's own learning process)/ reflective writing, and school projects/internship.

The team of teachers and supervisors is composed of three groups: school leaders, organization consultants, and scholars in the field of education. Theoretical knowledge and practical insights and experience (also from business) are combined in each module.

The workload for students is 350 hours per half year. About 175 hours are spent at training sessions, 20 hours at seminars and 15 to 20 hours at consultation meetings. For the first three half years approximately 140 hours for literature study, homework, and practical projects have to be added; in the fourth half year again approximately 140 hours are spent on conducting the final practical project and writing the report about it.

The price of one half year or semester is 1 800 euro. The costs of the total course for one participant are about 7 200 euro. Costs for literature and other learning materials are included. Not included are costs for hosting and dining (140 euro per seminar).

Each semester the participants receive a certificate based on a written examination and three homework projects. In the fourth semester a written report of the practical project is demanded. The grade of Master in Educational Management requires a master thesis and a final examination.

Table 16: Outline of Management- en Organisatieopleidingen of the Nederlandse School voor Onderwijsmanagement in the Netherlands

Training and Development of School Leaders in the Netherlands Example in the Field of Secondary Education: Management- en Organisatieopleidingen of the Nederlandse School voor Onderwijsmanagement	
Provider	Nederlandse School voor Onderwijsmanagement (NSO) a cooperation of five universities
Target Group	Teachers aspiring to a school leadership position and established school leaders (and deputies), particularly at secondary level
Aims	Development of competences for leading schools and other institutions in the educational sector; improving the chances of the participants to be employed in a leadership position due to a formal qualification (i.e. by holding a certificate)
Contents	Context and strategic management; organisation management; operational management; theories of management and organisation; models of educational organisations; organisational diagnosis; decision-making; school management and school boards; marketing and public relations; contract activities; control of the school culture; leadership styles; personnel management; recruitment, selection and guidance of new staff; job evaluation interviews; guidance of sitting staff; labour-relations and collective bargaining; instructional leadership; curriculum and instruction; modularization; productivity and quality care; implementation of innovations; internal and external guidance; management information systems; management of information technology; facility management; financing and budgeting; selected problems of school management; selected practices of the school leader
Methods	Lectures; speeches; seminars; training sessions; consultations; role plays and simulations; case study; peer counselling; readings; writing a study journal (documenting one's own learning process)/ reflective writing; school projects/internship
Pattern	Ca. 144 course days* (timetabling: 4 semesters with 215 hours contact time each divided in seminars: 20 hours per semester, every Wednesday (afternoon/evening); training sessions: 175 hours per semester, Friday and Saturday; supervision: 15-20 hours per semester; school project within the internship: 140 hours per semester) plus time for preparing and implementing the school project within the internship (4 semesters with 140 hours each), and time for literature research and readings, and for the assignments within 2 years
Status	Optional; appreciated by the employing school body as the NSO is well renowned
Costs	Ca. 7 200 euro per participant; financed by the participants themselves, sometimes funded by the school budget
Certification	Master in Educational Management

** If there is no specification by the provider as far as the number of days is concerned, we converted the contact time in hours into the unit course day, taking 6 hours training as one day.*

3. Summary

The Dutch school system is strongly decentralised. The state provides guidelines and creates a basic framework for schooling in the form of a national curriculum and national standards. The state holds schools accountable for the quality of their education according to the guidelines and standards. The school is administered locally and possesses a high degree of autonomy.

As to the development of school leaders, the state does not interfere directly. It is up to the individual school council employing the respective school leader to determine their expectations of the candidate's expertise. The provision of school leader training and development is stimulated by the national government. The actual supply is driven by the market, which is characterised by diversity and choice, especially at the secondary and higher levels of education. There is a wide range of providers and programs, which differ in content and methods as well as quality; they differ in that sometimes they are preparatory programs and at other times they are programs for experienced school leaders. However, at the primary school level, the Ministry of Education is indirectly involved in that they choose to finance the candidates participating in certain programs offered by traditional teacher training institutes.

What is remarkable in the Netherlands is that in certain programs there is an innovative approach to integral school leader development. For example, the program Meesters in leidingsseven uses mainly peer-assisted learning, in which the participants build pairs who act as critical friends, to support each other intellectually and in their daily tasks and challenges. Based on the concept of 'integral leadership' (see Imants & de Jong, 1999), they are encouraged to develop cognitive maps of their schools. This process of cognitive mapping will then form the basis for concrete school development planning. The development of individual school leaders is integrated with the development of the individual schools.

A different and especially extensive and renowned program for both experienced and aspiring school leaders is offered by the Nederlandse School voor Onderwijs-management (NSO). The NSO, which is a joint institution of five universities, uses a competence-based approach over a long program of 176 study days with extensive time for internships and school-based projects, which leads to the academic degree of Master in Educational Management. The program seeks to develop a broad range of competences, starting with pedagogical and educational competences, followed by counselling competences and ending with controlling and organisational competences.

4. Recent Developments

A recent trend in Dutch teacher and school leader education is the development towards competence-based learning. For example, in the NSO program principles of competence-based learning can be identified. In competence-based learning the integration is effectuated at the level of mastery of competences in daily school practice. Potentially, competence-based learning can contribute to an integral approach of school leadership in school leader education. In this respect, the second generation programs can develop towards a more integral approach of school leadership by implementing principles of competence-based learning into their programs.

In the Meesters in leidinggeven program integration is defined at the level of cognitive control. In the following section some reflections are made on the first experiences of the Meesters in leidinggeven program, and some ideas are developed about what the meaning of cognitive control in the context of school leader education might be.

5. Statement by Jeroen Imants

In this personal reflection I would like to focus on the need for integral leadership in schools and the development of cognitive control.

In 2000-2001 the third generation program that was described in this chapter was executed for the first time. During this period a further specification of the development of integral leadership and cognitive control of school leaders over their school occurred. The development of strategic thinking by the school leaders became a central theme in this course. When the course started, the participating school leader felt very dependent on the Ministry of Education, the Inspector, the Board, parents, etc. During the course they learned to look at their school as an autonomous entity and to interpret their own role of school leader as a leading strategic actor in a complex political field and educational context. Questions asked by these school leaders changed during the course from 'how can I defend my pupils and teachers against these negative external pressures?' towards 'how can I make use of the opportunities that are offered by the complexity of the field for the actualisation of the aims of our school?'. Before the course, these school leaders were not used to making strategic choices and to specifying priorities. Their routines were to treat all expectations from parents, policy makers, local networks, and inspectors as equally important, and to draw up unrealistic action plans which combined all these expectations. During the course they learned to make an integrated analysis of these external pressures and the internal characteristics and processes of their schools, and to develop a school strategy based on this analysis. A very difficult step for most of the participants was to specify their strategic choices in concrete priorities for the schools' action plans. It turned out that most action plans still were overloaded by diverging initiatives. However, the discussions that followed up the presentations of the action plans showed that the participants were able to discuss their own plans from an integrated strategic perspective. It can be concluded that this third generation program with a focus on integral leadership and cognitive control contributed to the empowerment and emancipation of the participating school leaders.

A challenge for GCO fryslân is to pass by the monopoly of selected leadership development programs, and to gain a position in the training market in uneven competition with protected programs of schools of education. After the first experiences with the third generation program, and more specifically with bringing in participants in the program, the next step is to supply tailor-made programs with core elements of the prototype that was implemented in the first year. These tailor-made programs should be based on specific demands of customers (school boards, participants, etc.).

In the field of secondary education many of the established school leaders had already developed some skills in strategic thinking about their school as an autonomous entity. Some important topics in secondary education schools are how to deal with the tension between an external and an internal orientation of school leaders and how leadership functions can be distributed throughout the school. Another important topic is about how schools can be developed as attractive places to work and learn for teachers, in order to deal with the shortage of qualified teachers on the national labor market. These topics combine in the effort to make the school an attractive employer for beginning teachers and a stimulating context for teacher learning. For these efforts to be successful, initiatives in the field of personnel policy should be taken by schools and school leaders. However, many aspects of personnel policy and human resources management are completely new for schools, and personnel policy is not the first priority of school leaders. It can be concluded that the shortage of teachers and the initiatives to train aspiring teachers in schools are creating new opportunities for school leaders to bring external pressures and internal conditions into balance, to develop new leadership functions in the school, and to involve staff in these leadership functions. This example shows that the development of integral leadership and the cognitive approach of leadership development can be regarded as a priority in the field of primary and secondary education.

France: Recruitment and Extensive Training in State Responsibility

Stephan Huber and Denis Meuret

1. Background

Due to state sovereignty, the education system in France is organised and implemented in a centralised way. Schools are answerable to the central Ministry for Education in Paris, the Ministére de l'Education Nationale (MEN). This centralisation of educational administration is particularly apparent with regard to the curriculum; the contents of lessons are laid down in centralised curricula, which are obligatory for the entire country. As well as being responsible for the development and drafting of curricula, the Ministry for Education is also directly responsible for the structuring of the schools, for the training, leadership and payment of staff, for the advertising of vacant positions, for the setting-up and closure of classes and for the didactic material. This does not mean that all is decided in Paris, since some decisions are taken at the regional level. The Minister for Education is represented at this level by the 'recteurs d'académie', who are appointed by the President of the Republic at the recommendation of the Minister. The 'recteur' has extensive powers, she or he takes care of the implementation of all legal and administrative regulations and distributes the budget allocated by the state among the individual schools in his or her district.

Since the Decentralisation Act of March 2, 1982, some of the responsibilities have been transferred to local bodies. They are now responsible for the construction, maintenance and renovation of public school-buildings, for the organisation of school-transport, for the maintenance of school-canteens and for the assessment of the specific educational requirements of the region. Suggestions at local level are submitted to the regional council, the 'conseil régional'.

The community is responsible for pre-primary and primary schools, the 'départements' look after the 'collèges', and the 'regions' take care of the 'lycées'. In each individual school there is a 'conseil d'administration', an elected schoolbody made up of representatives of the teaching staff, the parents, the pupils and the appropriate authorities concerned. This administrative council is responsible for the organisation of lessons and school business (see Eurydice; INES/DIPF).

The Chairperson of the conseil d'administration is the school leader, the 'principal'[18] or 'proviseur'. School leaders fill the role of state representatives and are

[18] A school leader in a state-run secondary school (the "collège", with students aged 11-15) bears the title "Principal"; in a "lycée" (with students aged 15-18), his/her title is "proviseur". The state-run primary school system is only presented here in some aspects, the government funded private sector, which enrols about 20% of the students, not at all.

regarded as a type of general manager of a public institution, who have been given responsibility and authority by the state. As representatives of the state, their most important function is the role they play in school administration. They bear the responsibility for the smooth running of school affairs, as well as for the choice of subjects offered in the school. They are responsible for order and safety in their schools and deal with the implementation of disciplinary measures for pupils. Another of their functions is dealing with officials and people of authority in public life. At secondary school level, this means above all the sharing of responsibility for school funding and school planning between the local community and the state. Unlike the pre-primary and primary schools, which have no funds of their own, the 'collèges' and 'lycées' enjoy the legal status of being public institutions with the corresponding legal recognition and financial autonomy. The 'conseil d'administration' gives its consent to the school budget, which from then on is controlled by the respective school authorities: the 'département' in the case of the 'collèges', the 'region' in the case of the 'lycées', and the 'inspection académique' in the case of other types of schools. In secondary schools, the school leader is responsible for the budget (see Eurydice, 2000).

Within the school, the French school leader's responsibilities are traditionally focused on primarily administrative tasks[19]. As regards teaching a subject, this is something which is done solely by the teachers themselves (usually one subject per teacher[20]). The instructions for teaching are given in the form of centrally administered curricula and final examination exercises, which teachers receive directly from the 'inspection générale' and not from the school leader. Nevertheless, individual primary schools and especially secondary schools have taken over a certain amount of pedagogical responsibility in recent years, even if the curricula and the basic pedagogical guidelines are laid down by the Ministry for Education. This is particularly reflected in the various school programs, which are worked out by the 'conseil d'école' for primary schools and by the 'conseil d'administration' for secondary schools. Within these programs, the particular school's specific conditions for adapting the national educational aims and curricula are defined, and the resources of its socio-cultural and economic surroundings are taken into consideration. The school leader plays an important role in the adaptation of these programs. Moreover, she or he guarantees for the information and consultation of the pupils, and the development of a plan for the continuous professional training for her or his teaching staff. The school leader is supported by one or more assistants (adjoints). Some claim that the tasks of 'principaux' and 'adjoints' are not clearly distinguished, which creates some problems (see Thélot & Joutard, 1999). In French schools there is no middle management level, e.g. in terms of subject advisors or Heads of Department, who undertake organisational tasks (see OFSTED, 1995).

Since administrative tasks have become more decentralised in the past few years, the position of the school leader in France has been strengthened. She or he is faced with new tasks, which require him/her to adapt her or his working style, which

---

[19] Secondary school leaders have no teaching duties; primary school leaders have no teaching duties in Paris, but outside of Paris most of them have to teach.

[20] Everything that takes place outside of lessons (e.g. attendance checks, dealing with discipline and penalties, counselling, etc.) is not the responsibility of the teacher. Specific staff is employed for these tasks.

should now be based on a more individual interpretation of legal responsibilities. Therefore the school leader is expected to show her or his own initiative in order to meet the demands of the state, while at the same time fulfilling the requirements of the public. School leaders must be good motivators, strategists and coordinators in order to plan the school program. They must secure quality standards within the school and act as suitable representatives of their schools in public.

They are particularly responsible that:

- the school's performance meets the acceptable national standard, as prescribed by the state. (The 1989 law of orientation of the education system states that 80 per cent of pupils of a certain age must reach the preparatory year of the 'baccalaureat'.),
- the school opens up to the community (and forms connections with industry, with the community in the region around it and further a field in Europe). Some school leaders of vocational or technical schools make a strong use of this opportunity.

With the way that responsibilities are shared, as is common in French secondary schools, it is not easy for the school leader to motivate the teaching staff to take on tasks which are above and beyond the call of duty.

2. Development of School Leaders

The qualification process for school leaders in France differs greatly, depending on whether they are from primary or from secondary schools. The more complex procedure for the secondary school will be explained here. The qualification procedure for the position of 'directeur d'école' in primary schools will be outlined briefly afterwards.

National Program for School Leaders of Secondary Schools

In France, the qualification procedure for 'principaux' or 'proviseurs' of secondary schools begins before one takes office, and can therefore be seen as a type of training for a job of great importance. Since there are no middle management functions in schools, the training must ensure that a series of leadership and management qualities are developed, and that the necessary knowledge pertaining to these subjects is imparted.

The selection, recruitment and qualification of school leaders in France is carried out by a single body. Before 1988 the selection was carried out on an interview basis, after which candidates who were deemed suitable, received training for their new position. In practice, a large proportion of these school leader candidates had already occupied a school leader position on a full-time basis, or if not, then at least part-time. Since 1988 the selection of candidates for this post (or at least for the training for this post) has been based on a competitive examination procedure – a competition. Since 1988 there have also been candidates who only have experience in educational guidance and support[21].

---

[21] See footnote above.

This procedure, the 'concours', consists of two parts (see Lafond, 1993). The first part consists of:

- a four-hour written exam, involving an essay about a topical theme relating to education and the institutional, economic and social environment of the school. In this way, the methodical and analytical skills of the candidate can be examined. (Recently it has been found that this written exam is not evidence enough of the competence of the candidates and it has therefore been replaced by a dossier in which the candidate presents him/herself and should prove her or his motivation on the basis of her or his career experience thus far (see Kunert & Kunert, 1998; Thélot & Joutard, 1999),
- a second facultative written examination about the area of data processing and information technology, and their applications in school.

Candidates who are successful in the first part must then face an oral examination in the second part. This in turn consists of:

- a fifteen-minute presentation. The candidate has two hours to prepare and present a type of exposé on a text drawn at random,
- a forty-five-minute interview. During the interview, a commission made up of an 'inspecteur général', a school leader and a third person (either a representative of a school, a local politician or a representative of a locally situated firm) should examine the candidate's knowledge of important philosophical, political, scholastic and economic issues. However, their most important objective is to get an impression of the candidate's personality and motivation, and to see how these may be suitable for a job as leader of an educational institution. Needless to say, the candidate's interpersonal skills are also evaluated.

This selection procedure takes place annually from September (written part) to December (oral part). In 1992 there were approximately 2,800 applications for about 670 places (see Lafond, 1993). There was a similar ratio in 1993 (see OFSTED, 1995). Following the first part of the examination, i.e. the written part, the number of hopefuls who survive to compete for the oral part is halved. Kunert and Kunert (1998) state that there are almost 1 200 examinees annually and estimate that by the end of the 1990s between 600 and 700 of these would be admitted. The results from both parts of the examination are added together, although it is noteworthy that the oral exam has twice the weighting of the written[22] exam. As a rule, whoever graduates from the course, usually gets a post as a school leader.

Because the school structure system leads to a lack of experience at middle management level, the entire selection process is a means of redressing the balance, and of highlighting the qualities necessary for a school leader. In order to prepare candidates for this examination, the Centre National d'Enseignement à Distance publishes past exam papers and offers a preparatory correspondence course. The Ministry itself does not offer any assistance for preparation for the exams.

It is the Ministry for Education that is solely responsible for the qualification of French school leaders. The qualification process is administrated by the Centre Condorcet in Paris (see OFSTED, 1995). The implementation of the qualification programs is the responsibility of the 28 'académies', which are scattered throughout

[22] The optional part has less weighting compared with the mandatory four-hour written examination, it accounts for 1/7 of the overall results of the written examination.

the country, each academy is run by a 'recteur d'académie'. The Centre Condorcet has the task of ensuring that the prescribed legal guidelines, the 'charte de la formation des personnels de direction', are followed and that the training offered by the academies conforms to these parameters, in order to minimise differences between the various training and development programs. Additionally, the Centre advises and supports the academies with regard to the setting up of programs. The organisation, methodology and content of the individual programs are left up to the academies. Each academy must, however, submit an annual report to the Centre, giving information about the programs on offer, as well as their budget for the year.

The principles that training should follow according to the national 'charte' include:

- an effort to reconcile the preparatory theoretical training with the real life situations in schools and practical aspects of a future school leadership position,
- a similar effort to allow a certain amount of individuality during the course of training, which above all will take the candidate's previous experience into consideration,
- developing the candidates' ability to prove themselves in certain situations and teaching them how to reflect on the way they deal with said situations,
- an effort to create a balance between imparted theory and internship experiences.

The explicit goal of the French school leader training and development program is to help future school leaders, most of whom have no experience whatsoever in running a school, to rise above their roles as teachers in a classroom or as mediators of education, and to develop them into leaders of state institutions.

This goal is elaborated in three explicit aims (see Centre Condorcet, OFSTED, 1995, p.5):

1. The candidates should be able to gain insight and experience with regard to the role of a school leader.
2. They should have the opportunity to develop a series of qualifications, skills and knowledge, which should help them to act effectively as state representatives.
3. Their cultural awareness, their pedagogical abilities and their understanding of leadership should be furthered in order to prepare them for taking up a position as a pedagogical leader in a school.

The actual training consists of three phases: the 'Formation initiale' or 'Formation au Premier Emploi' (the basic training for the first position), the 'Formation d'Accompagnement' (the further training course which is run simultaneously with entry to the first position) and the third phase, the 'Formation Continue' (continuous professional development throughout one's entire working life).

Formation au premier emploi

On successful completion of the 'concours', the candidate immediately begins this qualification phase, i.e. in January. It lasts for the following six months (until June), during which time the aspiring school leader gets release time from her or his school and any other duties. This phase consists of:

- a total of four to six weeks of seminars at an academy,

interspersed with

- a total of twelve weeks internship in schools,
- a four to six week period of internship at a company,
- a two week period of internship at public authorities.

At the academy modularised seminars are offered about:

- administration,
- budgeting,
- school law,
- management techniques,
- teacher evaluation,
- interpersonal and communication skills,
- leading conferences and staff groups,
- assessment in practice,
- youth psychology etc.

A total period of four to six weeks is designated for these lessons, whereby one-week seminars are alternated with two-week periods of internship in schools, since the school is regarded as the most important environment for training (see OFSTED, 1995)

During these periods of practical training, the future school leaders should have the opportunity to gain practical experience and to lead a project in conjunction with the respective school leader. The schools are selected on the basis of the leadership qualities of the school leader there. Having already completed a three-day preparatory course through the academy, she or he acts as a mentor for the up-and-coming school leader. She or he therefore plays a key role in the training of the aspiring school leaders who are assigned to her or his school. A maximum of three future leaders can be appointed to her or his care, and she or he should:

- emphasise the role of a school leader, without favouring a particular style of leadership or forcing the newcomers to adopt her or his style.
- give them the opportunity to observe and analyse different types of leadership behaviour.
- devise a program which is tailored to suit the needs and future goals of the future school leaders.

There is a training team at each academy, whose members all have first hand school leadership experience. The team plans the training program, organises the seminars, invites experts to give presentations, selects and trains the school leader mentors, supervises the internship, provides chances for reflecting on experiences and evaluates the academy's program.

In addition to the school internships, the future school leaders have a four to six week long period of training in companies, as well as two weeks of training at a local authority or a regional administration body. The aim is to allow the participants to gain insight into the day-to-day running of an organisation which is not a school, in order to put them in the position to draw useful analogies between the situation in a different organisation and in a school. This is particularly important as regards management (e.g. decision-making processes, communication processes etc.). The selection of the company, the establishing of necessary contacts and the collaboration with the employer about a study project are left to the initiative of the school leader. However, the company selected must satisfy the criteria set out by the academy's training team.

Table 17: Outline of the National Program for School Leaders of Secondary Schools in France

Training and Development of School Leaders in France The National Program for School Leaders of Secondary Schools	
Provider	Centre Condorcet in Paris through 28 regional state academies
Target Group	Aspiring school leaders at secondary level, who have successfully come through the selection process and passed a written exam (now a dossier) as well as an oral exam
Aims	Imparting of leadership and management skills in preparation for the task of leading a secondary school
Contents	Administration; budgeting; school law; management techniques; teacher evaluation; interpersonal and communication skills; leading conferences and staff groups; assessment in practice; youth psychology; etc.
Methods	Modularised seminars at state academies interspersed with practical training in schools (with the school leader as mentor) as well as in companies and public authorities in the second phase: courses
Pattern	Ca. 41-51 course days (or 141 days altogether) Phase 1: Formation au Premier Emploi: 6 months full-time = 24 weeks = ca. 120 days (20-30 course days) directly after successfully passing the selection process, from January to June (timetabling: 4-6 weeks at an academy; regularly interspersed by a total of approx 12 weeks internship in schools, 4-6 weeks practical training in a company, and 2 weeks practical training at a local authority) Phase 2: Formation d'Accompagnement: 21 days 1 or 2 day courses immediately after taking over as a (deputy) school leader during the two year probationary period both phases within 2½ years
Status	Mandatory; selection and training are interdependent: training cannot begin without first having been selected; both are preconditions for taking over a position as a school leader
Costs	Unknown; state-financed; participants get release time from school for the duration of the first phase, full salary will be paid throughout this time
Certification	Employment as a (deputy) school leader on probation after successful completion of the Formation au Premier Emploi

The Centre encourages the academies to draw up a list of the skills and knowledge necessary for leading a school. This information should be of assistance to the mentors and should give the future school leaders an idea of what is being asked of them so that eventually after the training they can evaluate their own performance. However, according to various sources, this list is rarely used in practice, and in fact most participants find out what is expected of them by speaking directly to their mentors. This makes way for flexibility and the freedom to adjust to one's own training and development plan.

Formation d'accompagnement
On completion of the preparatory phase of their training, the graduates receive a full-time post for a two-year trial period from the beginning of the next school year. However, this post is sometimes only that of a deputy school leader. Some candidates are also directly appointed to the position of school leader, nevertheless they have a trial period as well. However, should school leaders come to the decision, within these two years, that they do not want to carry out their new functions, they may return to their old job, which would only have been filled on an interim basis. During this trial period, the Formation d'Accompagnement takes place. This is a comprehensive 21 day further training program in the form of one or two day courses. The contents of these courses are especially tailored to suit the needs of the participants. After the trial period, the persons concerned are then appointed as school leaders at the recommendation of the 'recteur d'académie'

Formation continue
During their service, school leaders have the possibility to take part in continuous training. Internships are also offered on a regular basis.

The various phases of qualification (particularly the first with the 'concours') are very expensive. The seminars, the administration and the release time of the participants from school all have to be funded. This financial expense is borne solely by the state.

Qualification of Primary School Leaders

There are no comparable large scale preparations in the area of primary schools. A reason for this is probably that the hierarchical structure of the administration does not allow for such extensive responsibility since the primary school leaders, the 'directeurs d'école', are supervised by the 'inspecteurs d'education nationale', who in turn are supervised by an 'inspecteur d'academie'. Their common objective is essentially to ensure compliance with the prevailing legal and administrative regulations for the smooth running of school business.

Applicants for school leader positions at primary schools must have at least five years teaching experience and be at least thirty years old. The 'inspecteur d'education nationale' evaluates the applicant on the basis of her or his knowledge and work as a teacher. The final decision about the appointment then lies with the 'inspecteur d'academie'. The academies do not offer any formal qualification programs, although there are courses (e.g. in Academie de Créteil a one-week course between appointment and entry into office), where questions about administration, the organisation of school projects, the settlement of conflicts or computer usage are addressed. A further two-week course is attended during the new school leader's first school year. The themes discussed here include time-management, personnel management, leading conferences and school projects. There is a kind of mobile reserve, who stands in for the school leader at her or his school during these weeks.

Further training is regularly offered by the 'inspecteur d'education nationale' for school leaders within her or his jurisdiction. In addition, there are monthly meetings of school leaders of the area to exchange experiences and to advise each other.

3. Summary

The educational system in France is centralised. The school leader, particularly the 'principal' or 'proviseur' of the secondary school sector, is seen as the director of a public institution and a representative of the state. The work emphasis is on administrative tasks. In the last few years the scope of site-based educational responsibility has been enlarged, bringing with it new tasks for the school leaders.

In France, recruitment and preparation of school leaders, particularly those of secondary schools, are strongly centralised. Preparation is designed by the Centre Condorcet, according to government-provided standards and guidelines. The program is carried out in a decentralised way by state-run academies, with slight differences depending on the region. Recruitment and training are interlinked. First the applicants have to go through a selection procedure driven by competition, 'le concours', which is also the employment assessment. This decides whether a candidate will become a school leader or not. Having passed it successfully, the preparation training program, Formation au Premier Emploi begins. This full-time program is comprised not only of seminars and traditional taught courses, but also of an intensive internship, for which to take part in the candidates receive leave with full salary. After the overall program they take over school leadership positions, most often as a deputy. During the two-year probation they undergo another support training, the Formation d'Accompagnement. Having successfully finished the probation phase, there are another range of training and continuous development opportunities, the Formation Continue.

Hence, among the countries in this comparative study, France has got a unique combination of recruitment and qualification, based on a great financial involvement of the state. This comprehensive and expensive qualification program is needed, because the responsibilities of the school leaders have been increased to some extent and because there is no middle management level in the French secondary school system, and hence teachers aspiring school leadership positions could not have gained any experience in school leadership.

4. Recent Developments

In 1998, the Minister of Education mandated a report on what should be the function of the school leader. This report (see Blanchet et al., 1999) made some propositions, which were discussed between the Ministry and the Trade Unions. They reached an agreement (see Bulletin Officiel de l'Education nationale spécial no 1 du 3 janvier 2002), of which the main points are the following:

1. A 'conseil pédagogique', a committee of representatives of school teachers, has to lead the educational policy of the school. This 'conseil' is chaired by the school leader.
2. The school leader makes a diagnosis of the school. From this diagnosis she or he deduces a mission statement, which defines some objectives for the school. Each year, the school leader writes a report on the steps taken, and every three years she or he makes a formal evaluation of the school, involving all stakeholders and assessing if the aims have been reached. This evaluation of the school is used for the evaluation of the school leader him/herself, which also occurs every three years.

At the time of writing (July 2002), it is too soon to know if these procedures will be endorsed by the newly elected government and will truly be implemented.

5. Statement by Denis Meuret

The role of the school leader has increased in relevance in France, in a system where the teaching of each subject is traditionally governed by the 'inspection' and not by the schools, for several recent evolutions, of which the main are:

1. a growing importance of the non-academic aspects of school life (e.g. dealing with violence) and also of the school's desired outcomes (civic attitudes, health awareness, etc.),
2. a growing importance of the academic aspects which go beyond each subject,
3. a growing importance of some pedagogical patterns of which the precise content has to be decided at the school level.

The past professional activities of the future school leaders, most of whom are teachers, does not give them any experience of the tasks they will have to deal with as school leaders. This is, with the absence of middle management in the French schools, which Huber rightly insists on, a reason why the training of school leaders is somewhat more developed in France than in comparable countries.

Two other reasons are :

1. That the weakness of the mechanisms of accountability in France (see Meuret et al., 2001) leaves the school leader with a very difficult task. French regulation is not a regulation by outcome, but a regulation by the process, in which the school leaders have to persuade teachers to adopt new

pedagogical habits or patterns which are favoured at the superior administrative levels and which the schools are under pressure to adopt. Since neither a formal evaluation or accountability process, nor an inspection at the school level, tells if the school is moving in the right direction or not, much depends on the power of persuasion of the school leader.

2. That, as in every 'Professional bureaucracy' (Mintzberg), where incentives have a weak role, training is seen as the main way to obtain the desired behaviour from the agents.

Given the situation, the management of the recruitment and of the training of the school leaders probably goes the right way. Some misconceptions in the recruitment process have been corrected. This process has been enlarged to non-teaching staff. High priority is given to on the job training and to the ability of the agents to reflect upon their experience.

If they are needed, improvements in school leadership quality will come not from training, but from a more rewarding and formative content of the job and a stronger process of evaluation of their action. As for the latter, the system seems to go in that direction: when entering a school, school leaders will be given a 'lettre de mission', a mission statement, which will indicate some objectives for their action in this school, on which they will be evaluated.

Germany: Courses at the State-Run Teacher Training Institutes

Stephan Huber and Heinz Rosenbusch

1. Background

The German school system is under federal control. At a national level, independence in matters of education and culture lies with each state due to the federal principle. This means that each of the 16 federal states (the German 'Länder') has an individual school system ensured by jurisdictional and administrational laws. Each education administration is organised in a more or less centralised way encompassing, school structure, school types, curricula, etc. These administrations also encompass educational-political goals, different education and administration traditions, and unique regional characteristics. The organisation of the education administration in the individual states is the same as the organisation of the general administration. Even now, it has not lost its bureaucratic character, which it received in the first half of the 19th century when schools were integrated into the general administration. The Minister or Senator is head, with a succession of subordinate institutions, at the end of which the schools function as the lowest unit. In the large states, like Bavaria, North Rhine-Westphalia, and Baden-Württemberg, there is a four-level administrative organisation including the ministry, the regional administration, the school offices on the level of counties or county-independent cities, and finally the school leadership at school level. In city-states (like Bremen, Hamburg, Berlin) the organisation obviously is less complex. In Hamburg, for example, only two levels of administration exist.

The school types and school careers differ among the individual states. They show, however, a relatively similar structure: a common compulsory elementary school until 4th or 6th grade; secondary schools are differentiated into compulsory technical or vocational schools, secondary modern schools, grammar schools, etc. There is a relatively low number of comprehensive schools (ca. 5%).

In order to unify the variations that exist in each of the states, the Conference of Ministers of Education (Kultusministerkonferenz, KMK) was established. The influence of the KMK is sometimes very high and sometimes very low as it depends on the subject of debate and its political dimensions.

Compared to other school leaders internationally, school leaders in Germany have limited authority due to the bureaucratic traditions. They are basically not responsible for staff employment and dismissal; they have hardly any influence on the schools' curricula and have only very limited financial resources. Even though there have been recent attempts to change this situation and to shift responsibilities, their authority is still restricted. Notwithstanding, school leaders are responsible for enforcing valid regulations and for the daily management of school life and lessons. They have to take over above all administration tasks. Furthermore, they are responsible for representing the school, which may entail keeping in contact with parents, neighbouring schools and institutions, as well as the community. Generally, however, their main task is to keep the school going. However, further school-based responsibilities are emerging as more and more states try to decentralise decision-making to the school level. School-based management has only been implemented to a small degree in a view states yet. Additionally, school leaders' influence is restricted as teachers are quite free to make didactical and methodical decisions in their own right. This is called the 'Institut der pädagogischen Freiheit'. Moreover, the true decision-making body in school is the staff conference (or the school community conference which consists of teachers and parents). The school leader is not in charge of decision making here, whereas decisions made in the staff conference are obligatory for him or her to implement and follow. In certain kinds of schools (for example in grammar schools and secondary modern schools in Bavaria) the school leader conducts the regular official assessment of teachers; in elementary, vocational and supporting schools this is taken over by the school inspectors. (In Baden-Württemberg, however, a teacher evaluation by the school leader is also conducted in elementary, vocational and special needs schools.) In general, however, the regular assessment of teachers (each 4 or 5 years) has largely been discontinued in the other states.

The school leader's teaching obligation depends on the kind of school, the number of classes and the number of pupils in her or his school. In a grammar school with over 1 000 pupils, the teaching obligation of a school leader is at least two lessons per week (the maximum is at 11 hours a week in certain states); teachers at grammar schools teach – depending on the state – 23 to 27 hours a week. School leaders in elementary and vocational schools have considerably more lessons to teach. In Bavaria, for example, 50% of elementary school leaders teach 18 hours or more.

School leaders are supported by vice school leaders and by other staff (for example, in the senior management team) who take over specific tasks, such as devising lesson plans, school career counselling, extra-curricular tutorship, the school library, etc.

Vacant school leadership positions are announced publicly. Applicants' backgrounds are checked including an assessment of their past achievements and their teaching skills. A basic precondition for being appointed as a school leader is teacher training for, and teaching experience in, the respective school type. Moreover, additional qualifications are an advantage, like experiences as deputy school leader, in leading teams, working as an instructor, who was in charge of the induction phase of teacher training, etc. Mostly, however, the state examinations after teacher training are decisive as well as the regular official assessments by superiors. The candidates who are evaluated as most suitable are appointed school leader for life.

2. Development of School Leaders

All 16 German states provide qualification programs for school leaders. However, the programs vary from state to state (see Huber, 1999, 2000; Rosenbusch & Huber, 2001). The only thing that all the states have in common is that in each state there is only one provider, which is the state-run teacher training institute of the respective state.

When analysed in detail, the differences become apparent, especially regarding the goals, the contents, and the time pattern. The competences necessary for taking over tasks as a school leader are obviously evaluated differently in the German states. As the process of recruitment shows, too, the assumption that a good teacher will automatically be a good school leader is still in existence. However, the significance of an adequate qualification is becoming more important and is being more and more acknowledged by politicians and educators. This is especially true in those states that further decentralise decision-making and give more power and self-responsibility to individual schools. Thus, school-based management is leading to increased responsibility for school leaders. Therefore, school leadership training and development becomes important. It will be on the agenda of educational-policy-makers soon, and, particularly, as international comparisons of different kinds will show the lessons which have to be learnt.

As far as the present states' programs are concerned, they differ in a variety of ways.

Aims

It is informative to investigate first of all the aims and central ideas of the qualification. In so doing, a very heterogeneous image with different foci appears. This is astounding regarding the above-mentioned far-reaching identity of school leadership tasks in the individual states. Certain states do not name any goals at all, they simply offer a description of school leadership tasks, and frequently the idea of 'professionalisation' can be found. In Baden-Württemberg the program states that the main focus is a new model for school leadership or professionalisation that includes competence in the capability of acting, leadership, communication, cooperation and conflict management. In Hesse the 'improvement of the school's quality' is referred to. Mecklenburg-West Pomerania names three main foci: leading a school should be based on evolutionary management; it requires communication that is able to deal with contradictory positions; there is no single ideal leadership behaviour, instead, there is the need for 'situational leadership'. In Saxony-Anhalt, the program states that the participants are supposed to feel, respectively experience encouragement and confirmation. The goal setting in Rhineland-Palatinate is referred to as professionalisation of leadership for quality improvement in the system of school.

Contents

A comparison of the content of the qualification offered reveals that a change of paradigms has taken place. The school leader is not regarded as an expert of school law in the first place any more, but on the contrary takes into consideration the importance of communicative and cooperative competences. This "communicative shift" (see Rosenbusch & Huber, 2001) shows the effort to deal with the new awareness of the role of school leader tasks, and the new models of school-based management. In general, there is a noticeable development in the training contents, from the image of the school leader as an expert in administration, to the notion of the school leader as an expert in communication. However, it may still be observed, especially in some of the new German states, that 'school law' is a favourite topic among many participants, possibly as a result of some uncertainties in this area since the political changes of 1989.

Methods

As can be seen from the analysis, practice- and action-oriented teaching strategies and learning methods are favoured. Similarly, application-oriented learning should be made possible. Among the methods mentioned are input seminar papers, partner and group work, question/topic catalogues, mind mapping, role-playing, video training, moderation and visualisation techniques. Lecturing has lost its former pre-eminence. Furthermore, very few states (for example Rhineland-Palatinate) organise hospitations in either model schools or mutually in the participants' schools or elsewhere in the education, but also non-education sector. The training material used consists of curricula, compilations of school law and regulations, seminar textbooks or manuals of the Ministry for Education, recent research-based literature and work sheets composed by the training institutions or the trainers.

Timing

In almost all German states, school leadership personnel are qualified only after being appointed and usually only after taking over a leadership position. A different approach in this field is applied by the city-state of Bremen, where applicants are selected from a recruitment pool, which consists of participants from the preparational short course. To what extent this is obligatory or if the pool can be avoided, is, however, not clear. Some more states (for example, Bavaria) offer optional orientation courses for interested teachers. In nine states there are, moreover, considerations about introducing an obligatory pre-qualification, that is a program before applying for a leadership position. In Baden-Württemberg, on the other hand, the pre-qualification has no longer been conducted since 1994 for jurisdictional and economic reasons. In six states, namely Berlin, Hesse, North Rhine-Westphalia, Rhineland-Palatinate, Saarland and Saxony-Anhalt, the participation in a qualification program is optional, while in the other states mandatory programs can be found. Brandenburg represents a special case, however, as officially the participation in the program is mandatory there. Due to a lack of places in the courses, however, not all school leaders can take part. In North Rhine-Westphalia, development programs will be made obligatory soon. In Lower Saxony, there were plans to establish obligatory qualification courses as a prerequisite for participating in an assessment-centre-procedure, which should help to select the appropriate school leaders. This procedure, however, has never been truly realised. Currently there are some considerations of whether an assessment before beginning training programs would make sense in order to select even the participants of the training program.

Number of training days and duration of the programs

Even more obvious differences can be seen when looking at the duration of the qualification programs. This varies between seven days and 31 to 34 days. This is a little bit simplified because in the case of Hamburg, for example, there are further opportunities for training and development. However, considering these aspects, in some states the discrepancy between the prevalent assumption of school leadership as a new profession in its own right, given the importance of school leadership, and the respective preparation for this role becomes clear (see table below).

Figure 5: Number of course days of the state-run programs in the German states

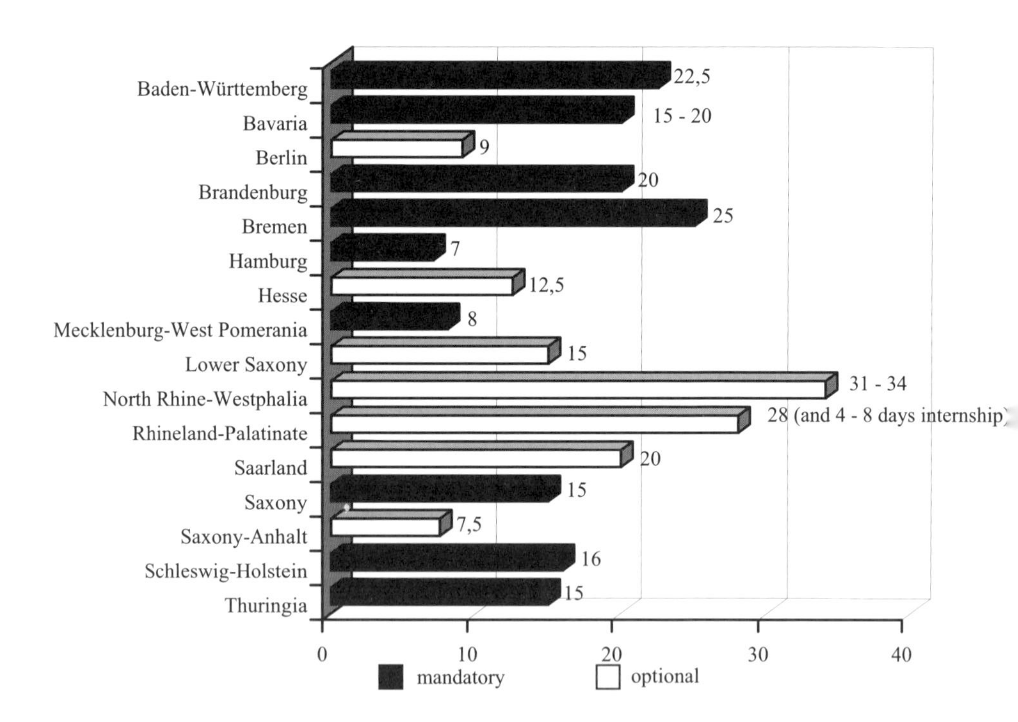

The overall duration which the qualification covers ranges from eight weeks to up to two years. These variations in the overall duration can be explained, naturally, by the number of qualification days on the one hand, and by macro-didactical

considerations on the other hand. In those five states, in which the qualification offer consists of 20 or more days, the overall duration is generally planned for the first two years in post, so that the school leaders can be supported in their new tasks. In order to accomplish this, some states offer their qualification in week-long seminars. Others, however, put their main focus on frequently conducted shorter sessions, which are sequential so that their contents can be linked to actual and relevant school leading activities at the point in time. Often, it is emphasised that during this time 'cooperative partnerships' may develop, which might last even longer than the qualification programs, and which could be useful for the school leader, as a network of mutual support.

Additional providers

Even though the school leadership qualification in all German states is in the first place the responsibility of the respective state's teacher training institute, there are still additional opportunities for training and development by other providers. For example, in some states, the offers are supplemented by subordinate state offices, or state or county governmental departments, or the local education authority. In Bavaria there are also institutions from foreign countries functioning as cooperation partners. In Bremen, Rhineland-Palatinate, and Thuringia, universities are involved. In Saxony there is a partnership with the training institute of Volkswagen. In Saxony-Anhalt systematic training and development programs are offered for school leadership personnel or teachers interested in school leadership at Halle University in cooperation with the Ministry for Education. At the University of Bamberg courses were established in 1986 within the 'Diplom für Pädagogik'. Moreover, the school leaders' associations offer programs. The Open University Hagen has recently begun to offer courses (since 1999), as has Kaiserslautern University (since 2000). Moreover, the Kiel Institute for Advanced Studies and a great number of private providers (among them some publishing houses) offer development courses. The Bamberg School Leadership Symposia should be mentioned here as well, since the first of them in 1988 was the first conference for and about school leadership in Germany (which shows how late the educational sector looked at school leadership). The symposia have taken place at two year intervals and offer a survey of recent research in the field of school leadership.

Three Examples:
The Programs of Bavaria, North Rhine-Westphalia, and Rhineland-Palatinate

In order to indicate the heterogeneity of the 'school leader training and development landscape' in Germany, the development opportunities in three German federal states will be outlined in the following sections[23]. These are the programs of Bavaria, North Rhine-Westphalia, and Rhineland-Palatinate.

[23] The authors thank Giesela Seidel (Rhineland-Palatinate), Herbert Buchen (North Rhine-Westphalia), and Gerhard Schmidt (Bavaria) for additionally validating this chapter.

Example: Basic Courses and Additional Courses in Bavaria

In Bavaria the Akademie für Lehrerfortbildung und Personalführung in Dillingen offers a centrally organised qualification for all school leaders. The working assumption, on which this training and development concept is based, is formulated as a description of successful school leadership. It comprises:

- relying on the resources and development capacities of each staff member,
- leading by developing goals and giving feedback,
- leadership and management as a service for staff, pupils and parents,
- motivation by providing development possibilities and positive reinforcement.

The school leader is regarded as central for initiating educational development processes and for focusing all activities at school on education and instruction for the pupils. At the same time, the school leader's role is seen in a centralised school system. She or he is in charge of implementing top-down reforms designed at the State Ministry for Education (Staatsministerium für Unterricht und Kultus, STMUK), the State Institute for School Pedagogy and Education Research Munich (Staatsinstitut für Schulpädagogik und Bildungsforschung, ISB), or the Akademie für Lehrerfortbildung und Personalführung itself.

The Akademie für Lehrerfortbildung und Personalführung offers a qualification program, which is obligatory for all newly appointed school leaders in Bavaria. Besides this, however, there is the possibility for interested teachers to attend orientation courses even before applying. These orientation courses are intended to assist candidates in making a decision for applying for a leadership positions. School leaders and local education authorities are required to support course applications from interested teachers who appear to be suitable as potential future leaders. Participation in the orientation courses, however, does not result in any legal entitlement to a leadership position. The program's contents and methods explicitly aim at supporting the participants in developing high-level competences. It enables a shift of perspective from a teacher to a school leader, which is beneficial even if they do not exercise leadership afterwards. The courses are conducted as one-week seminars during the Easter, Whitsun or autumn holidays.

The courses consist of modules about leadership topics with different foci, which may be attended in varying order. Themes are:

- role and role conceptions of school leadership (responsibility, self-perception and perception by others, trust and cooperation),
- educational leadership (theory and practice of leadership in economy, staff development, school development and quality management),
- administration and law (communication and counselling, moderation of conferences and team meetings, opening of schools to their environment).

Since 1978 a combination of basic and additional courses on the topic of leadership and administration in school has been offered. The three training sequences, which each take one week, are designed for participants from different school types and are supposed to assist them in the first year of their new function. The first course takes place between the appointment and the beginning of the actual work (usually during the summer break). The two following courses usually take place in November or December, May or June of the ongoing school term, so that the whole program is to be completed within 12 months.

The qualification usually includes 15 days; in the field of elementary, compulsory secondary, and special schools 20 days, because a fourth week is organised regionally by the respective school department of the county government or by the Bavarian School Leader Association.

The contents of the qualification offered for school leaders in the field of elementary, compulsory secondary, and special schools include the following:

Course I 'Basic Package'

- Self-reflection phase and understanding of roles (self-image, the role as school leader, gain and give trust, critical faculty),
- School organisation/school administration (class organisation, timetable, teacher employment, office organisation, budgetary planning),
- School law (legal basis, basic knowledge, problematic issues).

Course II (taking place, if possible, 5 to 8 weeks later):

- Moderating conferences and discussions (moderation and presentation techniques, leading conferences and setting up agendas),
- Leading staff (basic issues of leadership functions, guidelines and styles, management techniques),
- Communication training (basic issues, ways of communication, negotiation and discussion strategies),
- Additional contents: team work, school concepts of an alternative school.

Course III (after approximately 8 to 12 months in office):

- Leading staff (conflict management and coping),
- School development and school quality (school vision, school profile, creating a vision, corporate-identity models, TQM-techniques),
- Eco-friendly school.

Topical information is provided by the State Institute for School Pedagogy and Education Research Munich and by the Bavarian State Ministry for Education, as well as by members of the education authority and by practitioners.

The final event in Dillingen consists of a reflection phase about the role of the school leaders, who will then have been in office for about one year. Such reflection is also continuously encouraged and stimulated during the program.

The fourth course for school leaders of elementary, compulsory secondary, and special needs schools, which is organised regionally by the school department of the county government or by the Bavarian School Leader Association, comprises:

Course IV:

- Public relations work
- Parents' consultation
- Deployment of stand-in teachers
- Education of foreign pupils

The contents are taught with various methods of adult learning. These include seminar papers, short lectures, group moderation, role-playing, simulations, and possibilities of 'learning by doing'. There are no costs for the participants during this state-run program. Within the continuous professional development programs targeted at experienced school leaders, excursions to innovative schools and school districts in other countries are offered to those interested.

Table 18: Outline of the Basic Courses and Additional Courses in Bavaria, Germany

Training and Development of School Leaders in Bavaria, Germany The Basic Courses and Additional Courses	
Provider	Akademie für Lehrerfortbildung und Personalführung in Dillingen (for the 4th course for school leaders of elementary, compulsory secondary and special schools: the school departments of the respective country governments or the school leader association)
Target Group	Newly appointed school leaders in their first year in post
Aims	Preparation for an educational leadership task according to the characteristics of successful school leadership defined like this: relying on the resources and development capacities of each staff member; leading by developing goals and giving feedback; leadership and management as a service for staff, pupils and parents; motivating by providing development possibilities and positive reinforcement
Contents	Course I (basic course): self-reflection phase and role understanding (self-image, the role as school leader, gain and give trust, critical faculty); school organisation/school administration (class organisation, timetable, teacher employment, office organisation, budget planning); school law (legal basis, basic knowledge, problematic issues); Course II: moderating conferences and discussions (moderation and presentation techniques, leading conferences and setting up agendas); leading staff (basic issues of leadership functions, guidelines and styles, management techniques); communication training (basic issues, ways of communication, negotiation and discussion strategies); additional contents: team work, school concepts of an alternative school; Course III: leading and managing staff (conflict management and coping strategies); school development and school quality (model, school profile, model processes, Corporate-Identity-Models, TQM-techniques); eco-friendly school; Course IV: public relations work; parents' consultation; deployment of stand-in teachers; education of foreign pupils
Methods	Seminar papers; short lectures; group moderation; role plays; simulations of everyday school life and possibilities of learning by doing; reflection phase; excursion
Pattern	15 course days for school leaders of secondary schools and 20 course days for school leaders of elementary, compulsory secondary and special schools (timetabling: 3 one-week courses: the first course during summer break (before taking over); the second in November/December; the third in May/June; for school leaders of elementary, compulsory secondary and special schools a fourth course during the school term) within 1 year
Status	Mandatory state qualification program
Costs	Unknown; state-financed
Certification	Certificate of participation

Example: In-Service Program in North Rhine-Westphalia

In North Rhine-Westphalia (NRW) the Landesinstitut für Schule und Weiterbildung in Soest devised the conception for the qualification of school leaders. It also designs the materials used, selects the trainers and coaches, qualifies them, and assures the overall quality. The implementation of the program, however, lies with the regional institutions of the county governments.

The development opportunities include an orientation opportunity for teachers interested in leadership positions, a qualification programs for newly appointed school leaders, and events for experienced school leaders. For newly appointed school leaders the training is conducted immediately after taking over leadership. Participation in the qualification program is not obligatory, nevertheless the majority of newly appointed school leaders participate. The school leaders are only appointed for a probationary period of two years, and further appointment for three to five years is dependent on an assessment after this probationary period.

The concept of the qualification program is based on the principle of in-service support. The idea is that it provides process orientation, practice orientation, participant orientation, action orientation, system orientation, and academic orientation. Moreover, it is supposed to take the new concept of school leadership into consideration, and therefore learning is supposed to be understood as the willingness to deal with new tasks and new challenges, to deal with them on the basis of one's own experiences and to develop concepts, ways of behaviour and attitudes which contribute to the fulfilment of the new tasks (see Landesinstitut für Schule und Weiterbildung, 1998). The difference between a mere 'training' concept and a 'learning' concept, which aims at the development of professional personalities, is emphasised here. The development program therefore is to integrate subject-related, method-related, and social competence, in connection with a consciously reflected perception of leadership, and to support the transfer of what was learnt to one's own school and individual leadership practice.

The program covers the first 18 months of school leadership. During this whole time the participants attend several one- to three-day seminars and an overall number of ten to twelve collegial case counselling meetings. The seminar topics are based on a mandatory curriculum, yet do not follow a fixed chronological order. They are chosen from the pool of thematic foci and decided upon by the trainers and the participants, whose wishes and needs are taken into consideration.

The collegial case counselling meetings, which take place at intervals of six to eight weeks, enable the participants to work on questions closely related to their profession. The first six to eight of these meetings are led by moderators – the case counsellors – and serve the purpose of joint discussion of real cases and the development of action competence. These meetings are followed by four further ones, which are led by the participants themselves. If desired, case counsellors are available for these meetings. In many cases, the participants continue to meet even after the qualification program is finished. Overall, the duration of the program (which is strongly sequentially organised) adds up to 26 to 28 days plus the ten to twelve case counselling meetings, amounting to 31 to 34 days in total.

The contents of the program are focused on the newly developed school-based management model, which prescribes the conceptual aims for educational policy in the state of North Rhine-Westphalia (see Landesinstitut für Schule und Weiterbildung, 1998). The development of the school-based management model and the resulting demands on school leadership in North Rhine-Westphalia is determined by recent changes in public office as well as in business (see Ministerium für Schule und Weiterbildung, 1997). The ways in which these aims are achieved remain, however, within the responsibility of the individual school.

The following topics are emphasised for the qualification of school leaders for all school types:

- school development,
- staff management and staff development,
- quality development and quality assurance of school.

Each facilitator, trainer, etc. is asked to support the participants' understanding by connecting these topics with the following contents. The decision about the intensity and extent of engagement with these contents is made by the facilitator, trainer, etc. in cooperation with the participants, who may choose to carry out an additional focus on individual aspects by self-study. The contents include, among other options:

- cooperation with school external partners,
- the development of a vision of school leadership,
- management of schools,
- management of school development,
- leadership,
- assessment of staff,
- feedback after classroom observation,
- classroom development,
- the school as an organisation,
- leading groups, leading conversations, leading conferences,
- basics of school counselling,
- school law,
- communication and perception,
- organising administrative procedures,
- time management and work organisation,
- budgeting and school development.

The school is supposed to be discussed from an organisational-theoretical perspective during the seminars by dealing with recent research in related disciplines. Thereby, a focus on real school settings should always be ensured so that the connection between theory and professional practice becomes clear. The methodical considerations emphasise transfer learning for the practice and include varied teaching strategies. The course leader has to pay attention to a balance between systematic acquisition of decisive contents and process orientation. The following methods are used for this: input papers, specific literature, exercises, demonstrations, simulations, self-reflection and documentation in project reports, practical work in one's own school, work in small groups, partner work, role playing, group discussions, collegial case counselling.

Hence, learning takes place in three settings: in the continuous professional development seminar, at school, and during the collegial case counselling sessions. The material used is selected according to the specific learning method. In general, however, it consists of basic and concept papers of the Ministry for School and Further Education as well as readers dealing with different topics, which have been developed by the Landesinstitut für Schule und Weiterbildung itself.

The participation is of high professional relevance because of the probationary period. The costs for the qualification are paid for by the state.

Table 19: Outline of the In-Service Program in North Rhine-Westphalia, Germany

Training and Development of School Leaders in North Rhine-Westphalia, Germany The In-Service Program	
Provider	Landesinstitut für Schule und Weiterbildung through the district governments
Target Group	Newly appointed school leaders in their first 18 months in post
Aims	Development of professional personalities (according to the profile for school leadership oriented at the recently developed school-based management model of the state) by the integration of subject-related, method-related, and social competence, in connection with a consciously reflected perception of leadership
Contents	Main topics: school development; staff management and staff development; quality development and quality assurance of school; supplementary contents: cooperation with school external partners; the development of a vision of school leadership; management of schools; management of school development; leadership; assessment of staff; feedback after classroom observation; classroom development; the school as an organisation; leading groups, leading conversations, leading conferences; basics of school counselling; school law; communication and perception; organising administrative procedures; time management and work organisation; budgeting and school development
Methods	input papers, specific literature, exercises, demonstrations, simulations, self-reflection and documentation in project reports, practical work in one's own school, work in small groups, partner work, role playing, group discussions, collegial case counselling
Pattern	31-34 course days (timetabling: several one- to- three-day seminars plus 10 to 12 half-day collegial case counselling meetings) within 1½ years
Status	Optional; high relevance because participants are appointed for 2 years on probation
Costs	Unknown; state-financed
Certification	Certificate of participation

Example: Qualification for Newly Appointed Members of the School Leadership Teams in Rhineland-Palatinate

School leadership development in Rhineland-Palatinate is organised essentially by the Führungskolleg in Boppard, which is part of the state-run Institut für schulische Fortbildung und schulpsychologische Beratung, ifb. The programs offered by the Führungskolleg are organised in eight modules so that events for various target groups can be offered. Participation is voluntary. The overall aim of the qualification is the professionalisation of leadership for quality improvement of the school seen as a 'system'. The individual modules are structured differently and are of different intensity, some are organised in the form of a regional conversation circle (with representatives of school-external areas), a forum or a symposium, or as summer academy; the more intense modules contain a sequence of 'courses', or 'lectures', which are again planned in a modularised way.

In the course of the development of the qualification concept, particular emphasis was put on a continuous widening of the target groups for the sake of meeting essential pre-conditions for a distribution of leadership itself. With its events, the Leadership College aims to address teachers interested in taking over leadership positions, newly appointed members of leadership teams, senior management teams, or other teams at school, established school leaders, and members of local education authorities.

The module 'Qualification for newly appointed school leaders' is supposed to support the induction in leadership through training units, which take place during the first school term of the newly appointed school leader. All in all it includes 19 course days at the site of the Führungskolleg, two thirds of which take place during vacation time. During the summer break between appointment and taking over the post, a one-week information seminar takes place, which has the purpose of evaluating needs as well as discussing, planning, and fixing the main topics and their methodical realisation. After taking over the leadership position, three- to- five-day training sequences follow. Moreover, the participants meet four times between the course sequences for regional, one-day work groups and visit a school office for one or two days twice, as well as visiting business companies[24]on two occasions.

Taking this into consideration, the qualification program adds up to 26 to 34 full days.

Content-wise, the program is focused on five main topics, which cover decisive topics such as quality management (school development, staff development, teaching development), team development, school law, and educational leadership.

During 1999/2000 the courses had the following titles:

Course 1: Quality management (five days)
Course 2: Project management – tools and methods (three days)
Course 3: Evaluation of school development processes (three days)
Course 4: Classroom development (evaluation and quality) (three days)
Course 5: Thematically open (five days)

---

[24] There are cooperation relations between the ifb and the study circle "School and Business", which make projects like these possible.

The participants' training needs are taken into account by their active participation in the courses as well as by the modification of the courses on the basis of evaluation results.

Table 20: Outline of the Qualification for Newly Appointed Members of the School Leadership Teams in Rhineland-Palatinate, Germany

Training and Development of School Leaders in Rhineland-Palatinate, Germany The Qualification for Newly Appointed Members of the School Leadership Teams	
Provider	Führungskolleg des Instituts für schulische Fortbildung und schulpsychologische Beratung (ifb)
Target Group	Newly appointed members of the school leadership teams in their first year in post
Aims	Professionalisation of leadership for quality improvement in the school system
Contents	Main topics in general: school development, staff development, classroom development, team development; school law; educational leadership for 1999/2000: quality management; project management – tools and methods; evaluation of school development processes; classroom development (evaluation and quality)
Methods	Conference methods; management methods; systemic intervention; collegial counselling; school visits
Pattern	28 course days (timetabling: 24 course days at the ifb: during summer break before taking over office 1 one-week information seminar and afterwards 5 three- to five-day training sequences; between them 4 regional one-day work groups) plus ca. 4-8 days of internship (divided in 2 one- to two-day visits at schools and 2 one- to two-day visits at business companies) within 1 year
Status	Optional; nevertheless nearly all newly appointed school leaders participate as participation is expected; aim to make it mandatory
Costs	Unknown; state-financed
Certification	Certificate of participation

For established school leaders, the Führungskolleg offers various modules. Its first five training courses are held under the topic 'educational school leadership' and are divided up into:

Course I: Leadership action
Course II: Classroom development
Course III: Educational school management
Course IV: Developing school
Course V: Leading by agreeing on goals
Course VI will be conducted in a modularised way offering the topic 'quality management'. Each of its six parts takes half a school term and consists of two to five (each two- to three-day) central courses, additional regional

work groups (mostly taking two afternoons), and mutual school visits. This means that per part three to four teaching days, one Saturday and two afternoons have to be invested by the school leader (plus additional time for the mutual school visits, the required preparation, and the follow-up sessions). The titles of the modules state the main topics: understanding of leadership, quality of leading, evaluation, personal competence development, organisation development and classroom development.

In the courses of the Führungskolleg, a variety of methods are applied including conference methods, management methods, and systemic intervention. There is an effort to discover new learning settings, use new learning arrangements, and regard the individual school as a learning setting. The aim is to link school leader qualification and school development as well as support cooperation among the school leaders and collegial counselling. The cooperation with school-external experts (e.g. from business) and with other educational institutions (e.g. Kaiserslautern University) as well as providers of continuous professional development for teachers is an expression of the continuous modification of the program. A modularised structure will support a needs- and demands-oriented development, which allows individual choices and combinations.

3. Summary

In the Federal Republic of Germany, the responsibility for education is generally up to the departments of education of each of the 16 federal states known as the 'Länder'. However, the general structure of the school system and the role of school leaders are quite similar in all these states. In many of them there is a move towards decentralisation and increased self-management of schools, within a centrally fixed framework. School-based components are being introduced and school development initiatives at individual school level have become important issues. School leaders, who are increasingly seen as key figures in the process of improving the school, are employed by each state as civil servants and in general have non-terminable (lifelong) tenure. The role of school leaders and the necessity of developing adequate training and development models are currently being addressed.

The only common characteristics of school leader development models in the federal states are that all sixteen states offer training for newly appointed school leaders, and that this training is offered by the respective state's teacher training institute. Only to a very small degree are other providers involved in school leader training. However, the programs vary in several aspects: Firstly, in half of the states these development opportunities for school leaders, after they have taken up their posts, are obligatory. Only five states have introduced additional pre-service orientation courses lasting a few days, which, however, have not been made mandatory thus far. Secondly, they vary considerably as to the time-span, the total training days involved and the program structure. Some of the states offer a program that consists of only one basic course, lasting a week. Others have established development models of up to 40 days. In those states in which the training program comprises 20 days or more, these programs are either offered in intensive one-week periods (i.e. an induction weekend and then four one-week periods), or offered in frequent, short and sequential one- to- three-day training sessions, in which participants work in small groups to establish 'critical friendship teams'. Thirdly, the goals and guidelines of the development models are expressed in different ways, and in some states they have not been made explicit at all, however, a common trend is visible: some years ago, the emphasis was put on administrative competences, on management skills and on the knowledge of school law. During the 1990's there has been a shift towards the communication and interpersonal competences of the school leader, and towards school development. The emphasis is clearly given to the notion of 'leading by communicating', and there is a shift from regarding the school leader as a senior administrator towards seeing the role in terms of personal vision and influence. In the states of the former German Democratic Republic, however, topics like school law are still favoured by many school leaders, due to the political and organisational changes after the German reunification in 1989, which caused a good deal of uncertainty.

For reasons of illustration, three development models are outlined briefly, which represent an approach typical of the prevailing ones in Germany. In the federal state of Bavaria, the Akademie für Lehrerfortbildung und Personalführung in Dillingen is in charge state-wide of the central mandatory training program for all newly appointed school leaders of all different kinds of school types in their first year of leadership. The purpose of the program is to support the participants in their new roles as key figures for assuring that their schools are run effectively within the framework of the central guidelines and implementing educational reforms proposed by the State Ministry for Education. The program comprises a one-week 'Basic Course' (i.e. Course I) in the summer holidays between appointment and taking over leadership, and two more one-week courses (II and III) during the school term. For leaders in the compulsory sector a fourth one-week course (IV) is conducted decentrally by the regional school authorities or by the School Leaders' Association. Additionally, voluntary orientation courses are offered, in which teachers interested in applying for a leadership position can partake.

In North Rhine-Westphalia, the Landesinstitut für Schule und Weiterbildung in Soest has developed a strongly sequentialised in-service development measure for newly appointed school leaders, which is based on the new leadership concept for schools with a higher degree of site-based management and the respective school leader profile. It is defined as a learning concept aiming at developing professional personalities. Beside the seminar, the school of the individual participant itself has become a learning setting. Moreover, counselling meetings, collegial support, and counselling are provided. There are also development opportunities for experienced school leaders, and an orientation course for interested teachers is being developed.

In Rhineland-Palatinate, the Führungskolleg in Boppard, which is part of the state-run Institut für schulische Fortbildung und schulpsychologische Beratung, is in charge of the qualification. Its offerings are divided in modules for various target groups. Increasingly, a modularised form is aimed at as well as an attempt at dovetailing school leader qualifications and school development and the promotion of ways of cooperating and mutual collegial counselling among school leaders. Visits to and internships in schools and business are part of the offer for school leaders during their first year in the leadership position.

4. Recent Developments

In Germany, several interesting developments have started. Increasingly, school leader development has come on the political agenda. In more and more of the federal states, in addition to the state institutes, other providers (e.g. publishing houses or professional associations) are offering courses for established school leaders. There are cooperations with independent providers and with universities.

Moreover, a debate has started among educationalists and educational policy-makers on models of a more central provision and quality assurance of school leader development across the federal states. One consideration is to establish a College for School Leadership either nation-wide or – more realistically – for a few states. The idea is to bring different fields together, such as research, training and development, but also service offering. Moreover, it should bring different experiences and resources (human and financial) together to achieve synergy. It will be very interesting what role universities will play in the future landscape of school leader development in Germany.

5. Statement by Stephan Huber

Regarding the extent and the pace of numerous social, economic, and educational-political changes, the individual school today has to develop an understanding as being a 'learning school', that is to say, it has to adapt to social, economic, and cultural developments, even advance them in some regard, and counteract problems resulting from them. Hence, it would indeed be essential, before designing a school leader training and development program, to get a concept of the aims of the school today as well as of possible aims of the school in the future, to discuss them, and to derive certain working assumptions from these aims.

So far, Rosenbusch and Huber (2001) still conclude for the German programs that, in general, mainly a preparation for the status quo might be provided. In future, however, the programs will have to be based on an adequate reflection on the aims of school and on where school is supposed to develop. Questions should be asked like what is the aim of school and schooling today? How can this aim be reached, and how do the organisation and administration of school and the school system have to be managed in order to achieve these aims? What is the specific role and task of school leadership in this process? That means that leadership activities in terms of administration, decision making, problem solving, as well as the communicative everyday practice of school have to be oriented towards fundamental educational premises.

The demand by Rosenbusch and Huber (2001) is that school leaders are to be trained explicitly to be 'organisational-educational managers and leaders', The principle 'school has to be a model of what it teaches and preaches' (Rosenbusch, 1997b) must have an impact on the training and development of school leaders: school leader qualification has to be based on a clear understanding of the aims of education and instruction, as well as on a leadership conception like that of 'organisational-educational management and leadership'. This understanding has to influence the development programs regarding their overall conception, their contents, and the teaching strategies and learning methods applied.

As stated above, many changes are taking place in the German 'Länder', so that what has been described in this country report will doubtlessly have changed within the next few years. It may be expected that due to the foci in the educational policy of the states (above all of those that continuously decentralise their school systems granting the individual schools more self-management), school leadership is given more attention and more responsibility for school development. This will have to be supported by a suitable selection and training of school leadership personnel, without which improvement would be nothing but rhetoric. International comparative tests, such as PISA or PIRLS, will also have consequences on school management and leadership. School leader training and development programs will have to take all this into account.

Switzerland: Canton-Specific Qualification for Newly Established Principalships

Stephan Huber and Anton Strittmatter

1. Background

The Swiss education system has a federal structure, which means that the responsibility for the school system lies with the 26 cantons. This is why there are differing kinds of school organization and leadership, and also different systems of education authority or school inspection in different cantons. Federal law, which guides agreements among certain cantons, can obtain only a limited influence with regard to the secondary school system (grammar schools, vocational schools or technical colleges) and universities, despite the fact that these are mostly run by the cantons.

The cantons have laid down their public school system by school law. A number of cantons hereby pass on the duty of establishing and maintaining kindergartens and compulsory schools (1st to 9th grade) to the local communities. By maintaining the schools, the communities obtain some freedom of decision-making, which is limited, however, by financial and legal stipulations. In each canton there is, therefore, a different number of levels of political decision-making (see Rhyn, 1998). The supervision of schools and, in general, the management of school leadership training is, however, run by the state, through its local representative of the canton. The canton-run school system is administered and run by the canton government and its office of education; in some cantons, furthermore, a council of education, which is elected by the canton parliament, is responsible for more specific responsibilities such as teaching material and curricula.

The state-run supervision and inspection of schools is usually exercised by the cantons via school inspectors. The school inspectors are generally responsible for a certain region, in which they supervise teaching practice and the application of the teaching curriculum. Additionally, in many cantons there are 'Pädagogische Stabsstellen' (educational offices), which are responsible for academic-scientific tasks, such as guiding processes of school reform or school development. On the community level it is mostly 'Schulpflegen', school councils or school commissions, which are responsible for the supervision of a single local school or several schools belonging to the community. They are usually assigned to their positions via democratic election and work as honorary civil offices. Part of their responsibility lies in organisational and administrative tasks, which means that they basically take over operative school leading functions in most cases. These civil offices consist of laymen; they are supported adequately, however, by the canton's office of education and the regional school inspectorate. By law, teachers hold the right to voice their opinions and they frequently utilise this right. The input of parents is not quite as developed (with the notable exception of Bern). However, as citizens of Switzerland, they can make use of their rights as voters and – according to the model of direct democracy – have an influence on school matters at community and canton level.

The status of school leaders varies to a great extent in Switzerland. The higher school levels have always known distinct, authorised school leadership in the form of headship. The lay supervisory boards of the grammar schools, vocational schools, technical colleges, or universities agitate discretely in the background. The situation is extremely diverse among elementary schools. Cantons like Lucerne have used a strategy for school leadership, in the form of headship, for 30 years, whereas other cantons like Zurich or Thurgau have only recently begun to approach the establishment of a strategy, or theory of school leadership. Even this statement is not quite accurate as yet: In many cantons large schools have known school leadership for a longer time, whereas in small communities in rural areas decisive school leadership functions have been taken over either by the lay office, or not all, or somehow by staff. Additionally, often only the communities themselves have a central school leadership while the individual schools within those communities only have committees, which provide merely administrative supporting functions.

Furthermore, major cultural differences may be noticed when comparing the German-speaking region of Switzerland and the French-speaking region. Western Switzerland is characterised by the French tradition and its subsequent influence on school leadership, which is rather directorial and obtains a relatively high social standing for school leaders (monsieur le directeur). On the other hand, the teaching staff enjoy no such status. They often display quite an individualistic, occasionally anarchistic behaviour. In the German-Switzerland region, the role play is more often characterised by a Germanic rationality. People tend to hold school leadership down, at least concerning etiquettes, and tend to haggle over responsibilities in the descriptions of school leadership functions or in regulations.

The fact that the educational system has worked fine so far – in spite of lacking or confusingly organised school leadership – can be traced back to the old tradition of high self-responsibility of individual teachers. Many committed teachers therefore point out difficulties in accepting the establishment of school leadership; they fear incapacitation, bureaucratic problems and a resulting loss of quality at schools.

From the 1990s on, however, changes have been taking place in Switzerland concerning the structure of areas of responsibility in the school system. Those changes are having a strong impact on the situation of individual schools.

In view of a situation in which expectations towards the school and education system are rising, while at the same time financial public resources are declining, there is a great need for more efficiency and effectiveness in the sense of the new public management. The public administration is intended to become more service- and achievement-oriented. For this reason management procedures are to be re-organised. This means that decision competences are being delegated to the bottom, the bureaucratic apparatus is being reduced, and people working in institutions are given more responsibility. Superior offices are supposed to reduce themselves to strategic leadership, meaning to the establishment of general guidelines, while inferior offices are supposed to hold operative functions, leading to a rather self-responsible realisation of regulations.

The implications of this for the school system are that schools are supposed to develop further into partially autonomous schools, which see themselves as action units and work units with extended competences. These schools are also supposed to decide on issues that have previously been the responsibility of a higher authority in the educational system. The operative leadership will be assigned to single schools even more often. In many cantons in Switzerland, school leadership is therefore established as an independent institution. This institution, then, is taking over leadership functions in staff management and pedagogical issues in addition to organisational and administrative tasks (among which those of the traditional 'Schulpflegen' or school councils and school house committees are to be found). The cantons differ in development: some cantons are still conducting pilot schemes, whereas in others school leadership has been established in the regulations for public schools[25] or is currently being introduced. Bern provides an example of this (for an overview see Rhyn, 1998). These changes include the following aspects: schools define themselves as pedagogical organisms, which are granted further liberties in designing their internal organisation; their school leadership and teaching staff then work together in developing their own educational profiles, they acquire a higher standard of quality awareness and begin to develop a process of self-evaluation (see Szaday et al., 1996).

The existence of an individual school leader in place is supposed to have a great impact on the development processes of individual schools. School leadership functions as a decisive growth factor and as guarantee for the transition from development to culture (see Bildungsplanung Zentralschweiz, 2000). The upgrading of school leadership in Switzerland can be seen – according to Abächerli and Kopp (1997) – among other in the following fields:

- in literature: by increasingly working on the topic of school leadership,
- in local school development: by the development of models of school leadership, the development of compilations of school leadership tasks and functions and an increasing orientation towards team work,
- in efforts of professionalisation: by joining together school leaders in groups representing their interests as well as by offering school leadership training.

2. Development of School Leaders

Pre-service and in-service training opportunities for school leaders go back to the 1970s. Rolf Dubs created a course, which still exists and has by now been frequently attended by school leaders of grammar schools and business vocational schools. This training program was established at the University of St. Gallen in cooperation with the Swiss Centre for CPD (Continuous Professional Development) for grammar school teachers in Lucerne. The Swiss Institute has initiated courses for leaders of commercial vocational schools for vocational education. In the field of public

[25] In some cantons there are – as already described – educational regulations that have included the model of 'partially autonomous schools'. Additionally there are bottom-up developments and the State interferes via law and regulations. The control of the cantons is supposed to be ensured by external evaluations in the future; first pilot projects have been initiated.

schools it is assumed that the Conference of Lucerne Public School Leaders was the first to start continuous professional development offers in the mid 1970s.

Ten years ago a course was established at the Academy for Adult Education (AEB) in Lucerne. This course was meant for school leaders of all levels and all (German-Swiss) cantons. The offer is run as a joint venture between the AEB and the pedagogical section of the umbrella organization of Swiss teachers (LCH). The courses have been developed step by step from 12 days originally, up to 25 days, more recently. Today they include over 100 graduates from the kindergarten to the university sector. One variant of the course is called Leading in Social Organizations; the participants are made up of one third school leadership personnel and two thirds leadership personnel from other non-profit organisations. This model of purposely mixed course groups of participants from different school levels, functions and cantons has remained unique until today, except considering short-time programs offered by the traditional Swiss Teacher Training Courses during summer break.

In the mid-1990s several cantons started to offer their own continuous professional development opportunities for school leadership personnel of the public school sector. Today the cantons Aargau, Basel-Landschaft, Bern, Lucerne, St. Gallen and Zurich are conducting such training courses. In Western Switzerland in-service training has so far only been offered sporadically, actual courses are planned. An in-service training program for leaders in remedial education institutions has existed for several years at the Remedial Education Seminar in Zurich.

Certification and Accreditation of School Leadership Training and Development Programs

In the spring of 1999 the task group 'Certification de la formation des directeurs et directrices d'établissements scolaires du niveau primaire et secondaire I' was established; it consisted of two members of each educational region[26] and was supposed to work out a scheme for the accreditation of school leadership training. The report of the work group consisting of eight people was published on February 8th 2000 and includes, among other information, a survey of recent training courses offered over the whole of Switzerland. The task group's comment on this survey emphasised the main characteristics of qualification conditions for Swiss school leaders:

1. The task group recognises a comparable position in all of Switzerland regarding the notion of training and pre-service and in-service development opportunities for school leadership personnel. At the same time, however, different regions show different stipulations of contents. The aims of the training are not formulated explicitly in all cases. If they are, the type, and extent of, information given tends to vary considerably, and what the provider defines as leadership cannot always be explicitly seen.
2. Different organisations offer qualification programs. Among them are institutions under public law on the one hand, and task groups or private

[26] An 'educational region' consists of numerous cantons; there are four educational regions in Switzerland.

providers on the other hand. Most of the Swiss providers are, however, state-run institutions of teacher training, whose program includes – besides continuous development programs for teachers – qualification courses for school leadership personnel.

3. The target group mostly consists of established or newly appointed school leaders (Lucerne is an exception as it offers its own modularised school leadership training for teachers who wish to qualify for school leadership tasks). The rationale underpinning this is that the qualification is supposed to be an in-service support and this, therefore, allows it to be modified according to the participants' everyday experiences and school leadership challenges.
4. Concerning the contents, the major topics are comparable: leadership, communication, organisation, development processes, project management. Law and administration are topics that tend to receive a different emphasis (in some cases they are not even mentioned). The providers from different cantons focus thematically on specific conditions of the individual cantons.
5. The duration of the qualification measures differs greatly as they are all organised in a different way. Often, there is a basic level, an advanced/intermediate level and a certification level, and mostly there is accompanying project work.
6. The degrees that can be obtained vary from certificates to diplomas and confirmations that are granted either by the provider itself or the respective civil office. In some cases the degree can be achieved simply by passing the course, in other cases additional tasks are necessary.

From its analysis of the recent situation in the Swiss model of school leader qualification the task group summarised the following result: An equal accreditation for all of Switzerland would be reasonable for school leaders as well as feasible for providers. It would not be necessary to carry out intensive conceptual changes to ensure the criteria for such an acknowledgement as suggested by the task group. It would be necessary, however, to make the understanding of leadership more explicit for each of the offers. This, again, would lead to a greater transparency. Furthermore, the task group developed a conception for this accreditation of school leader training courses for all of Switzerland. It then suggested this conception to the Swiss Conference of Cantonal Education Directors (EDK)[27]. According to this conception, the EDK should establish a commission of representatives from (educational) policy, of the teacher association and of the training providers. This commission would then put a suitable institution in charge of carrying out the accreditation procedure. The individual providers then have the option of submitting the material necessary for accreditation and the possibility of becoming accredited. A meta-evaluation (controlling the self-evaluations of the courses) is supposed to control and develop further the program each five years.

The accreditation procedure suggested is characterised by the following goals: quality assurance in the field of school leader qualification and continuous quality improvements (providers that do not meet the standards are supposed to receive

---

[27]The EKD and its commissions hold an advisory function and can only give recommendations. On the level of public schools (including 9th grade), education is managed by the cantons.

suggestions for improvement on condition that the changes required are made), and creating a basis for the mutual acknowledgement of different kinds of training in the individual cantons.

The task group lists guidelines for standards for an accreditation as follows:

- a transfer-oriented and intensive learning package striving towards the success of each participant,
- training offers which satisfy the educational demands and needs of school leaders,
- a transparent display of the institution's concept of training and continuous development and of its pedagogical vision,
- progress achievements which are participant-oriented, efficient and effective,
- instructors who hold subject-related, methodical and didactical competences and who show a high standard of social and self competence,
- a systematic quality management and development.

These standards are being operationalised to develop indicators that allow control of the standards in the training courses.

Example: Training Program of the Educational Region Central Switzerland

The committee Education Planning Central Switzerland is responsible for the coordination of training offers for school leaders in central Switzerland. It is the pedagogical section of the Conference of Educational Directors in Central Switzerland (BKZ) and works for the cantons that make up the Educational Region of Central Switzerland[28]. It is responsible for taking over planning tasks in the field of acquiring specialised knowledge, evaluations, outlook, advising, networking and information. In the field of school leadership it works in different ways. It offers counselling and coaching to school leaders in office, it is a member of the Swiss task group for accreditation of school leader training, it coordinates the school leadership training and further education in its region and it ensures the coherence between these programs and the project Quality Evaluation of the Educational Region of Central Switzerland[29].

The organisation and implementation of this qualification measure for school leaders is then subject to the Academy for Adult Education (AEB), which carries it out in cooperation with the pedagogical section of the umbrella organization of Swiss teachers (PA LCH). The target group for this program consists of established and newly appointed school leadership personnel. The prerequisite for participation is the existence of a model of school leadership as well as a description of functions in the respective community[30].

---

[28] The 'Educational Region Central Switzerland' includes the cantons Lucerne, Nidwalden, Obwalden, Uri, Schwyz, Zug and the German-speaking part of Wallis.

[29] This includes three projects: 1. Self-evaluation of individual schools, 2. external evaluation, 3. standardised evaluation of success. The first two projects have been included in the training program for school leaders.

[30] It is expected that an organization chart of the leadership levels has been developed in the community as well as a job description of the newly established function of school leadership.

The goal of the qualification measure is to enable the participants to conduct their tasks as leaders. A compilation of the following competences is emphasised (see Education Planning Central Switzerland, 2000, p. 3):

- leadership competence,
- social competence,
- personal competence,
- pedagogical competence,
- school development competence,
- administrative-organisational competence.

The training consists of the following:

- participation in three modules on 26 days,
- an accompanying supervision of 21 hours,
- cooperation in a learning group,
- production of a final paper.

The three modules comprise the following content:

1. Leading: leading the school
   This module is seen as the basic course and is the prerequisite for the two further modules. It takes 12 days and includes the following topics (see ibid. p. 4):

- leadership biography,
- leadership philosophies, styles, models and techniques as construction process, power, gender-specific aspects of leadership,
- notions, theories and models of organisation, organisation analysis and diagnosis,
- development phases of organisations, organisation development,
- model-oriented and goal-oriented leadership, project management, dynamics of development processes, opposition,
- communication models, social strategies, problem-solving and conflict-solving models, discourse analysis, counselling models,
- groups and teams, fields of intervention, conference techniques, feedback,
- selection of staff, introducing staff, accompanying and counselling staff, confrontation and collegial discourse,
- concepts and methods of quality assurance.

2. Leading staff, team development
   This module takes seven days and offers extended contents and training in the following fields:

- selection, introduction, accompanying and counselling of personnel, confrontation and collegial discourse,
- communication models, social strategies, problem-solving and conflict-solving models, discourse analysis, counselling models,
- means of team development,
- fields of intervention, conference techniques, feedback,
- moderating /guiding development processes,
- recognizing and leading processes in groups,
- project management, dynamics of developmental processes, opposition,
- organization and administration,
- school law.

3. Quality development
   This module also takes seven days and includes the following:
- quality evaluation and development,
- concepts and methods of quality assurance,
- initiating, planning, moderating and evaluating school development processes.

The accompanying supervision enables the conduction of case discussions under professional leadership. In the learning groups – made up of teams of five to eight participants and conducted over the duration of one year – the participants are supposed to further deepen the learning contents of the course, discuss literature, enable counselling intervision and plan and carry out hospitations. The members of the learning group are self-responsible for organising their group work. They decide how often and at which intervals they meet and take over leadership tasks themselves (as, for example, the moderation of meetings, decisions about the working method of the learning group, the mutual feedback). Functions of the learning group consist among others of supporting the transfer from theory to practical application, mutual counselling in leadership issues, mutual support with planning, conducting and evaluating the school development processes, which are then documented in a final paper, mutual visits to each other's working place including hospitations, increasing familiarity with different kinds of collegial support – such as, for example, practise counselling and knowledge management – and the establishment of a collegial network beyond the duration of the training.

Each participant is intended to begin his or her final paper after the basic course and parallel to the following training elements. She or he is supposed to document the school development process that has been planned, led and evaluated by him or her, as well as to reflect the process by including the theories, models and concepts worked on during the qualification. The final paper fulfils several functions: It should on one hand support the theory-practice-transfer, on the other hand serve as a learning control. Moreover, it should have an informational and animating function for an interested audience and encourage them to conduct similar projects at other schools. Possible topics as suggested by the providers include the following: elaborating a model for the school, conducting self-evaluation projects or developing concepts on a certain topic (for example personnel development). The formal basic conditions for the final paper (including formulating and elaborating a documentation) include a time span of 80 work hours and a range of 20 to 40 pages. For the evaluation of the paper the following is decisive: The final paper describes steps of a development process and documents the personal reflection process in the sense of reflecting on the arrangement of development processes at one's own school, the reflection with one's own leadership role and the clarification of the leadership self-understanding that has been developed thereby (see AEB/PA-LCH, 1999).

The qualification program includes 26 course days (whole days), of which 50% take place during lesson-free time and the remaining 50% (without lesson cancellation) take place during school time, plus learning group lessons and 21 hours of supervision. It takes at least one and a half, and not more than three years, to complete. The program can be entered once a year. The first qualification course according to this concept started in November 1999, the second in May 2000.

The modularisation of the qualification enables the participants to graduate at individual times. The participants apply for the certificate at the Academy for Adult Education as soon as the following documents have been received: the course confirmation for the modules 1 to 3, the confirmation of group supervision, the confirmation of work conducted in learning groups, the final paper and – for a possible acknowledgement transferable across Switzerland in the future – the proof of at least two years employment in a school leadership position (for the acknowledgement in the educational region of Central Switzerland the latter specification is not necessary).

Table 21: Outline of the Training Program of the Educational Region Central Switzerland

Training and Development of School Leaders in Switzerland Example: The Training Program of the Educational Region Central Switzerland	
Provider	Akademie für Erwachsenenbildung (AEB) in cooperation with the Pädagogischen Arbeitsstelle des Dachverbands Schweizer Lehrerinnen und Lehrer (PA-LCH)
Target Group	Newly appointed and established school leaders in whose communities a school leadership position has been established
Aims	Enabling the school leaders for their tasks as leading persons, emphasising the competences to be acquired
Contents	Leading: leading the school (among others leadership philosophies, styles, models and techniques as construction processes, power, gender-specific aspects of leading, dynamics of development processes, development phases of organisation, organisation development, opposition); leading staff and developing teams (among others selection, introduction and counselling of staff, conference techniques, feedback, moderating decision processes, recognising and influencing processes in groups, project management, organisation and administration, school law); quality development (among others quality evaluation and development, concepts and methods of quality assurance, initiating, planning, moderating and evaluating school development processes)
Methods	Seminars; learning groups; supervision; project work and writing a documented final paper; literature studies
Pattern	26 course days (timetabling: several two- to four-day courses for conducting the modules) plus 21 hours of supervision, as well as additional individually arranged meetings of the learning groups, in addition time for literature studies and for conducting and documenting the final paper 1½ to 3 years
Status	Optional; appreciated; perhaps soon mandatory
Costs	Ca. 4 000 euro (6 030 Swiss francs) per participant; financed by the community or the canton or the participants themselves
Certification	Certificate of participation recognised by the educational directors of Central Switzerland (a standardised Switzerland-wide recognition is currently aimed at)

The costs for each participant are 6 030 Swiss francs in total, (about 4 000 euro). They are partially met by the communities or the cantons and partially by the school leaders themselves. The program is supposed to be accompanied by internal and external evaluation.

3. Summary

As part of the reorganisation of public administration, schools in Switzerland are in the middle of a development towards partly autonomous schools, whose operating leadership should be more strongly placed on the level of the individual school. In this context, for the first time in many cantons, school leadership is established as an independent function that goes beyond administrative and organisational tasks and includes educational leadership. The local authorities are called upon to develop appropriate school leadership models and descriptions of functions for their own schools.

An example of the qualification of established or newly appointed school leaders is the development program of the Bildungsregion Zentralschweiz in Central Switzerland. The provider is the Akademie für Erwachsenenbildung, the Academy for Adult Education, in cooperation with the profession itself. Target groups are established and newly appointed school leaders. Based on a competence-oriented approach, a modularised program consisting of seminars, self-organised learning groups, supervision and project work is offered. Individually, it can be completed within one and a half to three years. By conducting school development projects with appropriate documentation, the school leadership qualification is to be tightly connected with the re-structuring of the schools in terms of enhanced school-based management and responsibility. A certificate that is recognised within Central Switzerland is awarded. A Switzerland-wide accreditation of the program is aimed at and currently under discussion.

4. Recent Developments

Meanwhile the guidelines for the accreditation of development programs for school leadership personnel of the public school sector have been approved by the EDK (in spring 2002). The procedure of realisation is being set up at the moment.

At the same time, the Federal Office for Professional Development and Technology, the Bundesamt für Berufsbildung und Technologie (BBT) and the EDK have started a project to harmonise and modularise the development programs for school leadership personnel of the upper secondary school sector.

School leader development, which up to now has mostly been linked to teacher training institutions, is at the moment (in summer/autumn 2002) being shifted to institutions of the tertiary sector. Since the training for teachers of the public school sector is being transferred to universities or colleges of education, the development programs for school leaders will in future take place at university departments as well. The following years will show what chances and risks will emerge out of this change.

5. Statement by Anton Strittmatter

Being a provider of school leadership training I am concerned with two problematic issues to be reflected on – the future professional image of school leaders and the problems of training.

With reference to the future professional image of school leaders there currently appears to be a strained, bad design within the school system. The whole of the school autonomy movement seems to get stuck half way through, a real empowerment of schools does not happen, work is still conducted within a framework of diffuse and overlapping regulations of responsibilities. What is being allocated as autonomous spheres at first is de facto being regulated and blocked by sublime procedures of external evaluations by school authorities. Moreover, there can be observed a strong tendency in Switzerland to tighten the evaluations of teaching staff (partially connected with performance pay schemes), this results in the school leadership having to take over the task of teacher evaluation and assessment. Hence, the pressures on the role of school leadership, as can be observed, are unmanageable and make the profession appear unattractive (see Strittmatter, 2002).

Concerning the training and continuous development of school leaders, the same difficulties can be seen in all models: The transfer problem from the course into the everyday practice of chronically overstrained school leaders remains unsolved. Ideally, the training would have to be accompanied by supporting supervision and by organisational development, which does not as yet happen, mainly for financial reasons. A second problem can be found in the discrepancy between the leadership concepts and philosophies of the trainees, and the constantly changing school autonomy and school leadership concepts of those in charge of maintaining the school, especially when, as in Switzerland, unpredictable lay offices on-site blow hot and cold with them. We often do not know during courses, how to define the professional identity towards which we are supposed to teach and train in a way that enables people to acquire at the same time a reasonable, practicable basic foundation for their present practice and an innovative perspective for future developments. And finally we are tortured by the unclear responsibility for the product of the training. We generally cannot take over responsibility for whether those who have completed the course are going to be good school leaders or not. This depends upon a great number of factors, among which the continuous development is only one. The initiators of the continuous development, however, expect something like quality assurance from us and often have expectations towards the training that we cannot meet. We can take over responsibility for the following, however: For enabling the participants to experience appropriate learning settings in relevant areas of school leadership, to get well-informed, to encounter concepts supported academically and tested in practice, to make use of places of self-reflection and critical exchange with colleagues, and to receive – if necessary – open and constructive feedback about behaviour perceived during the course by the instructors.

Austria: Mandatory Training according to State Guidelines

Stephan Huber and Michael Schratz

1. Background

The school system in Austria is traditionally centrally organised and possesses a high density of regulations. However, jurisdiction in the Austrian school system is quite complex. According to the Austrian constitution, legislation and the administration of the school system are central responsibilities of the Federation. Legislation, the execution of various basic tasks of school administration and school supervision, of internal school order (e.g. school admittance, examination system, private school system, teachers' rights, personnel representatives' rights), and of external school organisation (school system, forms of organisation, the school buildings, the setting up, maintaining and closing down of schools, number of pupils etc.) are all matters of the Federation. Thus the Federal Education Department, Bundesministerium für Bildung, Wissenschaft und Kultur (BMBWK) is the body responsible for schools of academic secondary education. Federal schools (schools of higher education and the vocational schools of general and higher education) are directly financed by the Federation and their teachers are civil servants of the Federation. At the same time, the nine federal states possess a limited cultural sovereignty, they are partly incumbent on the executive legislation and the implementation of federal regulations. The teaching staff of general compulsory schools, i.e. federal state schools (primary schools, secondary modern schools, special schools and poly-technical schools), are civil servants of the respective federal state. On a local level, the respective councils are the bodies responsible for compulsory schools. As far as the choice of teaching staff is concerned, the regional school authorities and the school authorities of the federal state cooperate.

Vacancies for school leader positions are made public in schools, and teachers who are interested in the position apply to their respective educational authority. The choice is then made by representatives of the education department according to professional and personal suitability and length of service: In Federal Schools, a pre-selective choice is made by the school inspectors of the federal states. In the case of federal state schools, this choice is made by the authorities of the federal state government, advised by the school inspectors of the region and those of the federal state. School leader posts are securely anchored, which means that a head of school can be transferred to another school only for reasons defined clearly by law.

In Austria, it is the duty of school leaders to advise the teaching staff in their work as educators and teachers, and they have the right to observe and supervise lessons at any time. They are the immediate professional superiors to their teaching and non-teaching staff, although they have no legal influence on the choice of the allocated teaching staff. They execute the law and the decrees of the school authorities and are responsible for the distribution of the school budget, which has been allotted to the school, to the various school departments. They are also responsible for the annual administrative report. One of their tasks is to prepare and chair the meetings of the school community committee and to secure the implementation of its decisions. The school leaders themselves are advised and supervised by the school inspectors. As Schratz (1997) states, the legal basis for the function of school leadership in Austria (see School Education Act, § 56) shows that their range of action lies in the field of tension between supervising and reporting on the one hand, and advising on the other hand. This legal guideline presumes that only one single person should be in charge at school – a fact which eliminates the possibility of shared leadership.

Since the school year 1993/94, Austrian schools have had the chance of being more independent in their school life[31]. In this way, schools can enjoy a wider scope of development in determining professional and educational priorities, especially the consideration of the interests and capabilities not only of the pupils, but also of the teaching staff. They can take into account the necessities, needs and offers of the community; they can administer part of their financial means in a more independent manner etc. The aim is to achieve a more varied school landscape with a more distinct profile of each individual school. In concrete terms this means that within a predetermined framework (a maximum of 16 periods per week in secondary modern schools and 8 periods per week in the lower grades of the general schools of academic secondary education) and under the condition of cost neutrality and set regulations, each individual school could, for example, change the number of periods spent in compulsory subjects, introduce new contents, vary the size of groups, convert optional subjects into compulsory ones, or include school-external experts in their projects, etc.

School leaders are not only strongly involved in these changes, but are supposed to provide motivation. However, their scope of action is a limited one since there are legally defined procedures of decision making: The school curriculum has to be agreed upon by a two-thirds majority in each of the groups involved in the respective committees – comprising the head of school (a non-voting chairperson), teaching staff (in secondary modern schools: all the form teacher, in general schools of academic secondary education: three representatives) and parents (in secondary modern schools: all the parents' representatives; in general schools of academic secondary education: three representatives) and additionally in general schools of academic secondary education: pupils' representatives. Newly introduced measures are then to be evaluated regularly (for experiences so far see Bachmann et al., 1996; Krainz-Dürr et al., 1997; Krainz-Dürr, 1999).

Hence, the role of the Austrian school leader is undergoing a change – towards being a source of inspiration, a moderator of decision-making processes and the person responsible for the implementation of decisions. Further development towards school autonomy according to the principle of new public management will add to the new tasks of the school leaders. At the present time, the BMBWK are

---

[31] See the 14th School Organisation Amendment (1993) on a federal level and the amendment of the educational laws of 1993 as well as the laws of execution on the level of the federal states.

considering an expansion of financial autonomy, and a simplification and improvement of administration, school-internal organisation and school autonomy.

2. Development of School Leaders

In the school year 1988/89, a court appeal at a constitutional court against the appointment of school leaders started off a public discussion about the method of recruiting school leaders in Austria. Among the public, the nomination system was described as 'Parteibuchwirtschaft' (a system whereby fellow members of a political party are prejudiced in favour of party-political friends when called upon to make decisions), since among the criteria for the selection of school leaders were such activities as being a staff representative or a member of a teachers' union. The candidate's personal political tendencies were taken into consideration – sometimes quite openly – and even the mere membership of some political party or other could be of relevance. More objectiveness was demanded and as a result, in several federal states, new models of appointing school leaders and school supervisors were developed, which were based on objectiveness, transparency and comprehensibility (e.g. The Viennese Model of School Leader Nomination, December 1993). At the same time, the professionalisation of school leadership personnel via an adequate qualification went on the political and public agenda. For many years, up till then, headteacher training had been limited to a preparation for tackling legal and administrative tasks. On the one hand, being a teacher was regarded as sufficient qualification for a leadership position, and on the other hand, within the centralised form of organisation in the Austrian school system, the priority tasks of heads of schools were regarded to be in the field of executive administration. Alongside an improved system of selection of candidates, however, an appropriate qualification was regarded as a pressing issue and the BMBWK set up various work groups who paved the way – for example in 1989, the 'Work Team – School Management' and in 1991 the 'Bundeskoordinationskommission für Schulmanagement und schulartübergreifende Angelegenheiten der Schulaufsicht', the Federal Coordination Commission for School Management and Cross School-Type Matters of School Supervision (see Müllner, 2000).

Among other things, the results of these efforts led to a conception of a six-week training course for prospective school management trainers and a curriculum was drafted for a professional qualification training for school leaders (Fischer & Schratz, 1993). This conception visualised an extension of qualification offers for persons of all educational leadership functions, i.e. as a pre-qualification and as an in-service professional development course. The pre-qualification is intended to be an optional orientation offer, which does not warrant any corresponding position and is to be open to the widest possible circle of people who are interested in educational leadership. Educational leadership can become apparent as a personal profession – potential candidates could map their expectations against imparted estimates of the tasks and the areas of competence required. Another aim is that the responsibility for the development of schools should be shared among a larger group of people and that qualities of leadership will become of importance not only for the school leaders, but for everyone involved in the school system. This would be a significant condition for making use of the chances of an enhanced school autonomy.

A design of a modular program for a part-time course for appointed educational leadership personnel contains equal units on:

- educational rights, school administration and school organisation,
- communication and leadership,
- conflict management,
- lesson supervision, teacher consultation, and teacher assessment,
- school development.

The legal basis for the appointment of school leaders was regulated in 1996 in the 'Landeslehrer-Dienstrechtsgesetz' (LDG) for schools of general compulsory education and vocational schools and in 1997 in the 'Bundeslehrer-Dienstrechtsgesetz' (BDG) for schools of general academic secondary education, vocational schools of higher education, federal boarding schools and academies. School leaders are appointed for a period of four years. A condition for staying in this position is – apart from the fact that one has proved one's capabilities in this function (school partners and school authorities should have no objections to raise) – to have successfully completed a course of professional management training within two years of taking over a leadership position (see Müllner, 2000). The intended qualification measure is of high significance, since a successful completion is a decisive criterion for keeping the post. As only the induction part of the draft has been realised so far, the qualification can have no influence over selection and appointment of applicants, since these decisions were made previously.

The legal basis for the induction part time school management training course are the regulations laid down by the BMBWK for the realisation of the project for schools of general compulsory education as from December 1996 and for schools of general academic secondary education as from October 1999. Until now there have been no compulsory regulations on a federal basis for vocational schools of academic secondary education. The regulations for the above-mentioned schools are, for the most part, identically worded and respectively list the initial situation, aim, structure (2/3rd course studies, 1/3rd self-study), contents, scope of topics of the basic modules and the extended modules, the range of self-study and seminar organisation, criteria for a successful participation and a confirmation of participation. The mandatory qualification programs are organised by the Pädagogisches Institut (PI, the in-service training institute) of each federal state. On top of this, additional offers have been made by other organizations for relevant continuous development for school leaders.

Example: Part-Time Complementary School Management Course of Salzburg

The School Management Course of the PI in Salzburg, which is used here as an example, is based on the draft of the Federal Coordination Committee for School Management and Comprehensive Matters of School Supervision and the regulations already mentioned. (A short comparison with Tyrol and Vienna as well as with the other federal states will be made, too).

The target group for the qualification program are the newly-appointed school leaders (i.e. for a period of four years) according to the LDG and persons in leading positions according to the BDG. The realization of the course in Salzburg is purposely aimed at a comprehensive mixture of all types of schools in order to bridge the existing gap between them. The aim of the course is to promote and develop the pedagogical, function-related, social and personal scope of competences of school leaders and, therefore, to increase the quality of educational institutions (PI Salzburg, 2000). The preparation for dealing with educational and administrative tasks is stressed, as is also competence in creating a location-related school development with the aim of assuring the quality of the school.

The program comprises of a course-based social phase and a self-study-based individual phase. The social phase consists of mandatory compulsory basic modules and a mandatory choice among extended modules. These seminar modules are adjusted to the professional experience of the participants and are not only to arouse reflection, but also, using concrete examples, to work out significant educational tasks and topics of school leadership. The fact that the participants come from different types of school is regarded as a chance to extend perspectives.

The basic modules are (see PI Salzburg, 2000):

1. Communication and leadership (with 38 periods[32] = 4 1/2 days): Basic psychology of leadership; a new understanding of leadership and guidance; an analysis of different approaches to dialogue; counselling and problem-centered conversations.
2. Conflict management (with 38 periods = 4 1/2 days): work on case examples; personal attitude towards conflict and solutions; nature of conflicts; diagnosis of conflicts; concrete strategies of solving conflicts; dealing with conflict – a leadership task.
3. Lesson supervision: analysis – consultation – assessment (with 22 periods = 3 days): analysis of educational (didactic-methodical) conceptions; categories of lesson observation and methods of observation; cycle of consultation on lesson observation; feed-back procedures; educational controlling.
4. School development (with 22 periods = 3 days): an analysis of personal conceptions; systemic attitudes and strategies; possibilities of cooperation and team development; simulation of a development process; supportive measures for planning, implementation and evaluation in quality assurance.
5. School law, employment law, budgeting (with 22 periods = 3 days): – an additional module offered for participants listed under the BDG (from general schools of academic secondary education and vocational schools of academic secondary education) – practice-orientated introduction to the basic issues of school law, employment law, budgeting; research strategies, decision-making techniques, phrasing decisions, issuing notifications; working out case examples taken from practical examples of the respective participants concerning school law and employment law.

Additionally, heads of school of the category LDG are obliged to participate in 2 modules of the extended modules to choose from. Heads of school of the category

[32] One period comprises 45 minutes.

BDG are obliged to participate in one such course. Choice can be made from the following modules[33] (with duration of 3 days each):

School law, employment law and budgeting (for heads of school of the category LDG) (contents as above)

School Administration:

Contents for beginners: data input; updating reports of school events; using extended search for questionnaires, statistics and registers; setting up certificates; changing and saving texts (e.g. protocol) on floppy discs. Contents of an advanced course: Connecting up extended search; distributing subjects, timetables for classes, allocation of rooms, supplier plan, list of inventory, working out Socrates-Lists in Winword; copying Winword texts into mailboxes; possibilities of transferring data by First-Class/Offroad

Personal Development and Collegial Discussions:

Methods and materials for the assessment of existing personnel potential; collegial discussion as an instrument to motivate staff not only to apply their abilities and talents, but to develop them further; accompanying development measures

Conference Techniques – Effective Discussions:

Appropriate preparation of a discussion; methods of discovering ideas, treatment of topics and decision-making; possibilities of visualisation; functions of leading discussions e.g. introducing topics, setting impulses, presenting comprehensive overall views, summarizing; taking the minutes; supportive material

Time Management and Self-management:

Personal inventory of time; internal and external time management; optimisation of management orientation; methods, strategies and techniques for the rational use of time; personal work schedule and work organization

Project Management:

Conceptual characteristics of project, phases of project, organization of project (listing of tasks, schedule of order of events and appointments, operational means and financial planning), project supervision, arranging organization of projects (roles and distribution of responsibilities, competences, and tasks); project communication (information and reports, documentation and presentation)

New Curriculum – the Role of the Head of School:

Basic philosophy of curriculum (central themes) and the practical implementation (e.g. by determining the central and the extended spheres); elements and steps to be taken in establishing a school program; possibilities of developing and assuring quality; necessary material for this; planning, realization and consequences of self-evaluation

Topics of Current Interest:

Suggestions are orientated towards the priority of the respective educational policy, current demands and needs of the target group.

Methodically, topic-related learning by experience, the application and promotion of positive group dynamics, the working out of case examples and the use of short information-input, which add to group discussions and team work, are regarded to be of particular importance. The material used, for example, consists of

---

[33] The 'Educational- and Teaching Tasks' i.e. the learning/teaching aims as well the contents of the individual modules are classified respectively.

brief documents on theory and of case descriptions, which the participants themselves bring along, cooperatively designed posters, flip-charts, material for moderation purposes, etc.

The individual phase, or self-study, is mandatory for approximately 80 study periods, 15 of which are intended for a study of literature, 25 for project work and 40 for further offerings of training which are relevant for heads of schools (about five course days).

Study of Literature:

In the course of the study of literature, it is expected that the texts, which have been chosen by the trainers, will not only inspire a deeper and more extensive analysis of the contents of the seminars but help to prepare new topics, too.

Project Work:

The project work is designed to link theory and practice and is to result in a written report on a project – of about 10 pages – which has been previously agreed upon with the trainers. Examples of suitable topics are listed in the study plan: Among them are an account of a school development process or a project in the field of school autonomy, a description of the practical implementation of the contents of a seminar and/or the application of literature in the daily task of teachers, or keeping a diary of the seminar (innovation diary). The project is to be carried out during the basic module and to be suitably presented to the other participants (with the use of appropriate media – photos, wall newspapers, videos and internet-homepages).

Further Development Courses:

The attendance of further development courses which are relevant for heads of schools and the choice of topics depend on personal needs and the demands of the schools themselves. Naturally, one can take advantage of offerings by other providers as well, if one so wishes. The study plan names topics from which the participants can choose i.e. courses for leaders of primary schools on the topic of school entrance; for leaders of secondary modern schools on the topic of school autonomy and of certain aspects of management as well as training for special skills required for leadership.

31 working days are allotted for the School Management Course (240 study periods (UE), each of 45 minutes) as a whole, comprising 26 course days and about 5 additional days for literature studies and project work. The course begins each year in summer and should be carried through for three or four terms. Part of the course is held during school-free time.

The Study Commission of the respective PI decides whether the conditions of successful participation in the basic- and extended module courses are fulfilled and whether the self-study part has been appropriately realised. There are no final examinations. According to a nation-wide regulation, the graduates are awarded certificates that give them the right to the appropriate title. The program is state-financed.

Table 22: Outline of the Part-time Complementary School Management Course of Salzburg in Austria

Training and Development of School Leaders in Austria Example: The Part-time Complementary School Management Course of Salzburg	
Provider	Pädagogisches Institut Salzburg
Target Group	Newly appointed school leaders in primary and secondary compulsory schools and schools of academic secondary education
Aims	Promotion and development of pedagogical, functional, social and personal competence of heads of school and thus an improvement of the quality of educational institutions
Contents	Basic modules: communication and leadership; conflict management; lesson supervision: analysis – consultation – assessment; school development; school law, employment rights, budgeting; extended modules: school law, employment law, budgeting; school administration; personal development and collegial discussions; conference techniques – effective discussions; time- and self-management; project management; new curriculum – the role of the head of school; topics of current interest
Methods	Participant-orientated seminars on pre-determined topics (topic-related learning by experience); working out case examples; group discussions and team work; literature study and project work
Pattern	26 course days (timetabling: 204 study periods (SP) with 45 minutes each, including: 2 modules of 38 SP (4½ days per module), 4 modules of 22 SP (3 days per module), and relevant further education of 40 SP (5 days)) plus literature study of 15 SP and project work of 25 SP within 1½ to 2 years
Status	Mandatory; a precondition for further appointment after the end of the first four-year period as a school leader
Costs	Unknown; state-financed
Certification	Certificate according to a nation-wide regulation

Qualification of School Leaders in Other Federal States

The actual interpretation of the guidelines of the BMBWK is the responsibility of individual federal states. Various methods can be observed in the federal states:

1. Differentiation according to type of school: If there is a sufficient number of schools of a particular type of school in a federal state – which is often the case in the larger federal states – and if there are sufficient prospective participants, the PI offers a course which is differentiated according to the type of school in question, that is only for leaders of one particular type of school. In the smaller federal states, however, a comprehensive course is generally offered for leaders of all types of school (in varying degrees in Salzburg, Carinthia, in Burgenland, the Tyrol and Vorarlberg and partly in Upper Austria).
2. Differentiation according to position and function: In a few federal states, participants of varying hierarchical spheres of one type of school are trained as a group, regardless of their position or function. For example, in the

vocational school of academic secondary education, the law foresees the participation of two hierarchical spheres, that is the school leaders and the heads of departments, who then attend the same course (see Müllner, 2000). In Burgenland it is even possible for teachers who are interested to attend the courses. In Vienna, the courses for school management for schools of general academic secondary education are attended by school leaders, administrators, members of the school authorities, of the PI and teachers who are interested.

3. Summary

The Austrian school system traditionally boasts a density of rules and regulations. School leaders are appointed as life-long civil servants and are professionally superior to their teacher colleagues. With an increasing independence of schools they are to be faced with new tasks, however, only within a legally defined framework.

The recruitment and the qualification of school leaders has been reorganised in Austria during the last years. Part-time induction School Management Courses for newly-appointed school leaders have been developed on the basis of the guidelines set by the Bundesministerium für Bildung, Wissenschaft und Kultur (BMBWK). This centrally developed program is conducted by the Pedagogical Institutes of the federal states. The courses focus on 'educational leadership' and 'school development'.

The program, which is a combination of seminars made up of the so-called social phase, and of self-study and project work in the individual phase is to be completed within one-and-a-half to two years.

Participation is mandatory and a significant condition for staying in the leadership position after the original four-year appointment as head of school has been completed. For further development and training, however, there are no fixed regulations.

4. Recent Developments

The evaluation results of school management training programs and the monitoring of the present situation of school leadership in Austria have lead to further actions. The Ministry commissioned a study into the state of the art of innovation and development of school management policies within the German speaking countries. The key issues of the study were:

1. Where have change processes taken place within the framework of legislation/decrees/enactments during the past few years?
2. What was the basic intention of such modifications?
3. What theoretical/empirical assumptions concerning school development formed the basis for these intentions?
4. What relationship exists between the stated modifications and school development in Austria?

The results of this comparative analysis are to form the basis for further considerations in the profiling of heads of school in Austria and to facilitate the decision-making process of appropriate reforms.

The data of this study of normative guidelines and operative measures that were researched and analysed show on the whole a strengthening of the roles of heads of school and the educational bodies. It is now aimed to increase the dynamic force of development of the relevant school locations so that local and regional needs can be met more successfully. These trends suggest a need for change processes in Austria, the following spheres would appear desirable.

As far as heads of schools are concerned:

1. Tasks and Areas of Authority of Heads of School

- Since quality development in the individual schools depends to a large extent on the role of the school leader, it would appear necessary to transfer clearly defined areas of authority, responsibility and tasks concerning the development, the drawing up, the evaluation and continuation of the educational program to the heads of school.
- It would seem feasible for the responsibility for the budgeting of the monetary household funds, which are provided by the Board which maintains the school, to be transferred – at least in part, i.e. of the material and personnel costs – to the heads of school, to encourage further independence and autonomy.
- All-round talents, capabilities and knowledge are required of the heads of school in the fulfilment of their tasks. A compulsory personal in-service training in all matters concerning the areas of their authority would, therefore, appear to be absolutely imperative.
- In the interest of furthering quality development, the heads of school should be obliged to cooperate with institutions of teacher training and further education.
- With regard to the program and profile of the school in question, heads of school should be granted the right to co-determine the selection and assignment of the teaching staff.

2. Selection and Assignment

With the democratisation of schools in mind, it would seem necessary to allow educational bodies (educational forums and school community committees) more right to voice their opinions in the selection of school leaders.

As far as deputy heads and heads of school and extended/collegial school management are concerned:

The abundance of new tasks calls for a legal embodiment of the position of deputy heads of school as well as the possibility of arranging for an extended or collegial school management. Particularly with regard to the conscious involvement of female colleagues in leading functions, it should be possible to form a collegial school management for a limited period.

As far as educational forums and school community committees are concerned:

Considering the aspects of school autonomy, school partnership and a democratisation of schools, educational forums and school community committees, respectively, should be regarded as switchboards for school-internal decision-making processes. All the various groups involved in school life should be jointly responsible for cultivating and supporting the development of schools. It would appear appropriate, therefore, not only to involve pupils' representatives – as from the 5th grade – as well as representatives of the non-teaching staff as members who are eligible to vote in both of these bodies, but also to extend the decision-making areas of authority of these bodies.

5. Statement by Michael Schratz

Among German speaking countries, the training of school leaders in the Austrian school system started systematically at an early stage. Meanwhile, a mandatory foundation course has to be enrolled and graduated by all newly appointed leaders. Further training is usually chosen individually by the heads themselves or organised by the district or county school board. The strength of this system lies in its systematic training in the initiating phase of school leadership, its weakness in the fact that continuous development is more an individualistic venture. Nevertheless, school inspectors urge their heads to be updated on new developments within the system, which offers special occasions for re-training.

The main challenges for preparing school leaders in Austria for the 21st century lie in the following areas:

1. Personnel Development: the decentralisation of schools has shown that the heads need more capacity in leading their teachers and non-teaching staff towards the fulfilment of pressing tasks.
2. School Development: most heads were trained as teachers, so they lack the capacity to do work on the organisational level, e.g. preparing a development plan, doing evaluative work etc.
3. Opening up school towards its environment: heads still see schools as the main place of learning in today's society, they find it difficult to see the school in the social context of other actors around it.

The role of the school leaders will change as the role of the inspectorate is changing. This will give them more power on the subsidiary level, but they have had little training in that direction. Since heads represent the managerial line structure of the Austrian school system at the lowest level, they mirror the insecurities of the overall system. Further training activities will have to pay more attention to the vertical connectivity of the system.

Another problem arises from the fact that schools in Austria are characterised by very flat hierarchies, usually there is no management level between the school leader and his or her teachers. This might not cause problems in small schools on the primary and lower secondary level, but it is very disadvantageous for large schools. Leading such schools becomes more an activity of survival rather than a motivation to reach new visionary grounds. This problem has recently been due to new measures of restructuring the system on the employment level and severe cuts in certain areas, which has demotivated lots of teachers, especially the ones who previously were the more innovative ones. Heads of smaller schools suffer from the fact that they get very little administrative support at the school level, so that school leaders feel more involved in administrative tasks than in leadership activities.

South Tyrol, Italy: Qualifying for 'Dirigente' at a Government-Selected Private Provider

Stephan Huber and Michael Schratz

1. Background

In Italy, school administration is traditionally centrally controlled. The Ministry of Education in Rome, the Ministero della Pubblica Istruzione, is responsible for all matters of education throughout the country. The Ministry is, therefore, totally responsible for the supervision and coordination of educational measures throughout Italy. The Educational Departments – originally called the Sovrintendenze Scolastiche Regionale, now known as Uffici Scolastici Regionali – of the 20 regions and 94 provinces, respectively, are placed under the authority of the Ministry.

South Tyrol, or rather the province of Bolzano, is an autonomous Italian province. For South Tyrol, this far-reaching autonomy means that it has a right to primary power of legislation. However, in the field of education, only secondary rights of authority are allocated (though in some school–related matters it has primary power as well). In concrete terms, this means that Italian national law can be modified by the South Tyrolean Landtag – the provincial parliament – in accordance with the local situation. The South Tyrolean provincial law must then, however, be ratified in Rome. At the present time, the province of Bolzano and the neighbouring province of Trento still form a common region. This contact to Trento, however, is in a process of dissolution and the areas of responsibility for the region (for hospitals, land registery, real estate registery, laws of voting, municipal decrees etc.) are being handed over to the provinces themselves.

South Tyrol possesses a central administration, is financially secure and has become one of the richest provinces of Italy. Nevertheless, South Tyrol will continue to be an autonomous Italian province. Neither the plans of small German opponent political parties to achieve rights of self-determination for South Tyrol nor the state-level federalist ideas of the Lega Nord are particularly promising. Furthermore, the Südtiroler Volkspartei (SVP), the South Tyrolean People's Party, claims the absolute majority in Bolzano and thus rules the province supremely.

The specific political features of South Tyrol have a manifold effect:

- There is great commitment towards cultivating the German language[34]. Maintaining the image of Italy as a 'bogeyman' and continuing to have a separate educational policy is in sharp contrast to the attraction of Italian culture and the cohabitation of Germans and Italians in everyday life, which

[34] For many years the fear of Italianisation hindered the setting-up of a university in Bolzano. In the meantime, however, the project of the University of Bolzano/Bressanone has been realized.

is naturally taken for granted. One strictly regards proportional representation and a precise distribution of public offices among the three ethnic groups – Germans, Italians and the Ladin-speaking population – on a percentage basis, according to the representation of each ethnic group. A so-called 'immersion education' is impossible (which would mean that history could be taught in German at a school in which otherwise the Italian language is spoken).

- Until now, in the field of education – from kindergarten up to school-leaving classes – there has been a strict separation of the three ethnic groups of Italians, Germans and the Ladin-speaking population. As a result of this, there are three educational departments (each with its own Director) and three administrations. Only recently, the three educational departments have started to cooperate, particularly regarding questions of personnel: teachers and heads of schools are still salaried employees of the state but are taken over by the provincial region (at the present time, the trade union and the educational departments are negotiating with the province of Bolzano for a collective agreement to find a uniform solution).
- Provincial Law Number 12 (dated 29th of June 2000) has provided schools with many areas of authority. However, in reality, the provincial government of South Tyrol with its centrally organised administration tends towards stronger central supervision.

Since 1990 there has been a far-reaching process of change concerning schools – and consequently school leaders – within the context of a centrally determined reform in South Tyrol. A National School Conference determined the establishment of a school pact to improve the image of schools and to stress and strengthen not only their educational and social assignment but also their role as a transmission belt for occupation and development. School autonomy is stated to be one of the most significant factors involved. Schools are regarded as a public service or people's service and have been called upon to list their own service charts (Carte dei Servizi), the aim being to reform public administration according to the premises of effectiveness and efficiency and to re-determine transparency and subsidiarity (see EURAC, 1999).

The 1st Bassanini-Law, (Legge 59/97), secures the legal basis for the establishment of school autonomy. Article 21 determines the regulations of school autonomy and a step-by-step plan for its implementation. The most significant steps are as follows: firstly, it is determined, from which size on schools will be granted autonomy on the 1st of September 2000. Then the school leaders are up-graded, as far as rights of administration and salary are concerned, as 'Dirigente', that is as executive managers with corresponding tasks and areas of authority, but only after participating in a training course for executive managers. (The former was settled by the legislative decree on March 6th 1998, Nr. 59, and the latter by the ministerial decree on August 5th 1998). The next stage is the establishment of the didactical and organisational authority of the school. In the meantime efforts are being made to define educational aims and standards. These attempts are academically supported by the Servizio Nazionale per la Qualita dell'Ístruzione, the National Service for Educational Quality. A reform of collegial organs, of the Ministry, of educational departments and supportive organisations such as the Instituto Regionale di Recerca

Educativa (IRRE), the Centro Europeo dell'Educazione (CEDE) and the Biblioteca di Documentazione Pedagogica (BDP) is also expected.

2. Development of School Leaders

As can be gathered from these notes, a comprehensive process of change is taking place, of which only the main features have been realised until now. The qualification of school leaders, who are to be regarded as leading executives, is taking place within this field of tension. A whole list of queries remains unanswered as to the actual realisation of the scope of development of schools with school autonomy and the rules applying to political relationships, central and local school administration and the schools themselves. Consequentially, the qualification concept has been designed as relatively broad and flexible to compensate any change which could take place in the meantime. On the other hand, the participants have sufficient scope to make suggestions as long as such suggestions have been checked with the educational departments and political authorities. The framework conditions i.e. the aims, duration, scope and the mandatory contents of the training course for school leaders were, however, centrally determined for Italy in the ministerial decree of August 5th 1998. The autonomous region of the Valle Aosta and the autonomous provinces of Bolzano and Trento could have made exceptions, as they own the right to confer the order for training in their own interest. Nevertheless, in order to preserve an equality of all school leaders within the state territory, they agreed to bind to the conception set down by the ministerial decree (relating to aims and contents and their implementation as well as the quality of the qualification program).

Executive Leader Training for Heads of School

The Verwaltungsakademie der Europäischen Akademie Bozen (EURAC) was given the task of carrying out the executive leader training centrally in South Tyrol. As opposed to other Italian regions, in South Tyrol there was no public call for tenders to implement the ministerial decree, to which other interested persons or bodies could have replied (for example, personnel development departments of large firms or the University of Bolzano). The EURAC set up a planning committee made up of experts in the relevant fields, i.e. school law, school administration, school planning, school management and school development. Among these were longstanding EURAC partners as well as Professors of Education from Italy and North Tyrol, among whom were educationalists of the newly-founded Faculty of Education of the University of Bolzano in Bressanone.

After the planning committee had drawn up first suggestions for the structure of the training, a work-group was formed which consisted of representatives not only of the future participants, but also of the three educational departments as well as experts as advisors and discussion partners. This group worked together to modify conceptual ideas. A steering group was formed for the duration of the training out of members of the planning committee and the work-group. This proved necessary since – as mentioned above – significant aspects of school autonomy took concrete shape only after the executive manager training had begun. Continuous evaluation of the project is regarded indispensable for an adequate adaptation. The task of the steering group is to improve communication among the clients, planners, consultants and participants, to answer queries and to work out solutions to emerging problems.

The aims of the executive leader training were enriched by suggestions made by the regional school leaders' union. The following aims are listed in the EURAC conception (see EURAC, 1999):

- reflection on the field of action of the school in determining the new role of future autonomous schools,
- an extended imparting of skills and abilities which are necessary for the responsible leadership of an autonomous school and for the step from administering towards designing a new type of school,
- starting with an analysis of the actual current situation, to work out methods and instruments for realising the organisational and didactic autonomy of schools,
- a cooperative development of ideas of how certain aspects that have not yet been regulated by law could be settled.

The results for the participants one is striving to achieve are (see ibid.):

- They are aware of the radius of action of an autonomous school.
- They know the tasks and the responsibilities of a school leader of an autonomous school.
- Building on an analysis of the actual present situation, they have identified the capabilities necessary for the leadership of an autonomous school and have articulated the corresponding need for further training.
- They have developed the competences required and trained them in seminars.
- They have become acquainted with the instruments needed for the realisation of autonomy in pilot projects and have taken steps to introduce such instruments.
- They have newly defined the professional image of a school leader.

The South Tyrolean qualification of school leaders originates from a conception of training on three levels: content-, process- and transfer-related. The basic curriculum has three pre-defined fields of content:

School Planning and School Development:

To be concerned with school and its radius of action, with a vision, with an educational plan and with project work.

School Management:

To be concerned with topics such as the steering of the school, leading school as an organisational unity, staff and resources administration, leading staff, conflict management, organisation development, quality management, service quality, effectiveness and efficiency.

School Law:

To be concerned with topics such as the liability of the school leaders, contracts and agreements, procedures, supervision and accountability for the outcomes.

The basic curriculum comprises of 90 hours of seminar, of which 70 are the same for all participants and the remaining 20 are to serve as an extension of the topics. Participants are encouraged to bring forth suggestions, which are gathered at the very start of the training in an analysis of needs and requirements.

Additionally, the optional curriculum includes 48 hours of seminars in complementary topics. The topics to be dealt with are determined after the evaluation of the results of the analyses of needs and requirements, as described above. Originally, the ministerial decree for Italy planned the topic 'security in schools', however, since this topic had already been dealt with comprehensively country-wide in South Tyrol, it was omitted. On the other hand, the topic 'the role and significance of information technology' was of great relevance since the provincial schools information system LaSIS was introduced at the same time as the training, so that knowledge of information technology became important. The optional subject 'information technology' shows examples of how an electronic infrastructure can be put to use, not only for didactic means, but also for administration. The topics 'electronic data processing as a working tool in an autonomous school' and 'new media and education' were designed for this purpose. Ongoing school projects are built into the discussions. The participants are then encouraged to make use of the new media in the seminars, i.e. internet material is made available and email correspondence is set up between the participants and the instructors.

The action-oriented processing of the contents is to ensure the transfer from the training context into the actual workplace of the school leaders. This is reflected in the schedule of the course, the ministerial decree pre-determines that half of the intended periods, i.e. 150 study periods, should consist of 'attivita in situazione – learning by doing'. This contingent is divided into supervised self-study (40 periods), an exchange of experiences (30 periods) and project studies (80 periods).

The South Tyrolean training program explicitly formulates the process character of the qualification:

> The prerequisite for the introduction of school autonomy is a modification of procedures of thought and action of the manager leaders. In order to achieve this on the basis of a cooperatively structured culture of school autonomy, an organisational and personnel development process must be introduced and supported. (see. EURAC, 1999, p.9)

Obviously, on-the-spot coaching of such processes is far beyond the bounds of such a qualification program. An attempt to compensate for that is the supervised project work.

As far as the project work is concerned, a coaching approach is intended. Each tutor accompanies a group of heads of school and supports them in processing the imparted knowledge, in checking the practicability in job context and in adapting it to the specific demands of each individual school context. There are ten small groups, who work independently. Once the projects[35] have been determined, the

[35] An example of such an individual project on the topic 'School Model – School Reform' is the combination to a unity of the independent structures of two schools in Merano, a secondary trade school (GOB) and a secondary school of natural sciences (RG). Other projects are concerned with school autonomy, school development, school programmes, models, evaluation, new methods of teaching etc.

members of each group decide among themselves, when, how often and where they will meet. The (minimal) overall duration and a report of the current state of each individual project are predetermined. The tutor is available on six half-days for group meetings and on a total of a further four days for individual consultation (on-the-spot, by telephone or online).

Approximately 150 heads of school took part in the school leader training program (approx. 100 German, 40 Italian and 10 Ladin-speaking members) in the years 1999-2000. They were divided into four regional groups of 40 participants each (i.e. three German speaking groups and one Italian group. The Ladin-speaking members joined the German group). Plenary sessions alternated with group sessions. Additionally, there were small group sessions of a workshop character, where each group was divided into two sub-groups. Thus, ten small groups of approximately 15 persons were established for project work sharing one of the thematic priorities.

Table 23: Outline of the Executive Leader Training for Heads of School in South Tyrol, Italy

Training and Development of School Leaders in South Tyrol, Italy The Executive Leader Training for Heads of School	
Provider	Verwaltungsakademie der Europäischen Akademie Bozen (EURAC)
Target Group	Established school leaders prior to up-grading as Dirigente
Aims	Preparation to face the tasks of school leaders under the conditions of an extended school autonomy (1st Bassanini-Law), in detail: reflection on the field of action of the school in determining the new role of future autonomous schools; an extended imparting of skills and abilities which are necessary for the responsible leadership of an autonomous school and for the step from administering towards designing a new type of school; starting with an analysis of the actual current situation, to work out methods and instruments for realising the organisational and didactic autonomy of schools; a cooperative development of ideas of how certain aspects that have not yet been regulated by law could be settled.
Contents	School planning; school development; evaluation; school management and school law
Methods	Seminars; regional groups; small groups; workshops; project work with coaching; self-study; supervision and networking
Pattern	Ca. 23 course days* (timetabling: 138 hours of seminars: 90 hours for basic curriculum and 48 hours for an optional curriculum) plus an exchange of views and experiences (30 hours); project work (80 hours), including 6 half-days of tutor-coaching group sessions; self-study (40 hours) within 1½ years
Status	Mandatory; a prerequisite for the right to an administrative and salary up-grading of the school leaders to the position of Dirigente
Costs	Unknown; state-financed
Certification	Certificate of participation (thus a promotion to Dirigente)

** If there was no specification by the provider as far as the number of days is concerned, we converted the contact time in hours into the unit course day, taking 6 hours training as one day.*

The training centres involved were, on the one hand, the training centre of the Pedagogical Institute for the German speaking participants, the Rechtenthal Castle in Termeno, and, on the other hand, regional centres. The sessions of the project groups took place alternately in the schools of the respective participants. The plenary sessions of the entire group took place in Bolzano.

The consultant team consisted of members of the planning team and consultants of their choice. Among others were professors of the Universities of Bolzano/Bressanone and Innsbruck. Additionally, heads of school from other parts of Italy as well as other countries such as Switzerland, the Federal Republic of Germany and the Netherlands were invited to present an exchange of views.

Following participation in this state-financed program, the participants were up-graded to 'dirigente' according to Article 21 1st Bassanini-Law.

3. Summary

A far-reaching process of change is taking place in the field of education in the province of South Tyrol within the context of a centrally predetermined reform of public services. School autonomy has been legally introduced and is to be implemented in a step-by-step plan.

In this connection, heads of school experience the right to administrative and salaried up-grading to a position of 'dirigente' and are thus executive managers with corresponding tasks and areas of authority. A condition for this is their participation in a centrally planned Executive Leader Training Course within South Tyrol, for which the Verwaltungsakademie der Europäischen Akademie Bozen was commissioned to draft the concept and to realise the project. Mandatory contents are school planning, school development, evaluation, school management and school law. Methodically, there is a combination of seminars, regional groups, small groups, workshops, project work with coaching at the school of the respective participant, self-study, supervision and networking.

The existing field of tension and a certain paradox lie in the fact that in South Tyrol the increase in school autonomy is centrally decreed and that the heads of school are to be trained as leaders of autonomous schools in a development program which is mandatory and state-decreed.

On the other hand, school leadership in South Tyrol exists within a multi-cultural triple-lingual context, where there is little institutionalised interaction among the language groups. A shared central development program bridging the divide may indeed offer opportunities for a cross-regional exchange of experiences.

4. Recent Developments

Because of the change of governments the Italian school system is under new pressure. The merger of former primary and middle schools into one managerial unit has brought new challenges for school leaders with one main consequence – the same allocation of time for the responsibility of more people. Models of collegiality have been introduced in order to ease the time pressure.

Moreover, while a lot of attention has been paid to re-training school leaders, the local school administrative units, which have to carry a fair share of the new work loads involved in the decentralisation and restructuring of schools, have had little support within the system (see Tschenett, 2001). This seems to be a major problem school systems are confronted with, they have a lot of experience in implementing new policy making horizontally (e.g. via in-service training courses), but little know-how in working vertically on new grounds, as has become necessary in a more autonomous school system.

5. Statement by Michael Schratz

The Province of Alto Adige/Bolzano (South Tyrol) has a particular role within the Italian school system. Because of its history (it was part of Austria up to World War I) and its socio-cultural background, the province has been politically active to keep some independence from the rest of Italy. This has now materialised in the legal accreditation of the status of an autonomous province, which gives the provincial government a strong position influencing what matters in schools. On the other hand, the Italian school laws have to be obeyed, which sometimes causes fields of tension with the system within the system.

School leadership in South Tyrol has to be seen in a multi-cultural context of schooling. There are three language and culture groups represented by their own administrative bodies, which do not interact frequently. Only when the executive training for school leaders was organised to satisfy qualification requirements set by the Italian Government, a major in-service activity eventually succeeded in crossing the language borders within the province. This was only possible because it was not managed by one of the regional school boards but by the Verwaltungsakademie der Europäischen Akademie Bozen (EURAC).

Policy-making in Italy changes direction quickly compared to other countries, which puts much pressure on schools and their leaders, in particular. They have little time to develop strategies fulfilling policy matters because they might be changed quickly again. This does not coincide well with the more traditionally oriented culture of South Tyrol and therefore puts a lot of pressure on school leaders. They might have to convince teachers and parents today for one thing and tomorrow of the exact opposite.

At the moment, the Italian school system is being restructured by putting together the former primary and middle schools into one organisational structure, which is another challenge for heads. Making one out of two in a state system is probably more difficult than merging two firms.

Asia

Singapore: Full-Time Preparation for Challenging Times

Stephan Huber and S. Gopinathan

1. Background

Singapore's population numbers approximately 4.5 million. The city-state is made up of 78% Chinese, 14% Malay and 7% Indian and 1% from a variety of races. The four official languages of Singapore are Mandarin, Malay, Tamil and English. English is the dominant medium of instruction at all levels in the system and all pupils are required to learn their mother tongue (which in the case of the Chinese is Mandarin). As every pupil is expected to become proficient in English, it is the language which links different cultures and enables people to communicate with one another. The majority of Singaporean pupils leave high school with the ability to speak, read and write two languages.

Singapore can only look back upon three and a half decades of experience as a politically independent state. Being convinced that Western democratic values were not transferable to Asia without adaptation, Singapore's government under the former Prime Minister Lee Kuan Yew took a tough and strict line in leading Singapore. Strict regulations of social behaviour frequently backed up by publicity campaigns, oversight over the media, an 'Internal Security Act' are all limitations in personal freedom which seem to be accepted by most Singaporeans. Those who experienced the tough years after 1965 when Singapore struggled to become a viable state feel so grateful for the stupendous economic growth and political stability that they are obviously willing to pay the required price for it. This gratefulness, however, is less evident in the younger generation, which can be attributed in part to the growing use of new technologies (such as the internet) and hence the greater accessibility of information. Since Lee Kuan Yew resigned as Prime Minister in 1990 and Goh Chok Tong took over power, a more relaxed and liberalised approach has become noticeable. Despite the homogenizing effects of globalisation, the realisation of the value of identity as an Asian society where the community will always count more than the individual means that Singapore cannot be compared easily with western societies. Singapore continues to seek ways to strengthen traditional Confucian values (loyalty, diligence, and hard work) while integrating them in a new global view of life.

The school system is embedded in a meritocratic policy approach with strong emphasis on achievement, efficiency and economic success. On a national level the Ministry of Education (MOE) formulates and implements education policies, is responsible for the design of the curriculum and allocates resources. Furthermore, it controls the development and administration of the Government and Government-aided schools and also supervises private schools. The School Division of the Ministry of Education ensures that schools are effectively managed and that the education provided is in accordance with national objectives. The Superintendents of the School Division provide professional services and supervise, guide, support and assess school management teams.

Singapore's education system is extremely competitive, which might be the reason for the high pupil achievement results which Singapore's schools are known for. Those pupil-outcomes are tested in standardised tests that allow the ranking of all pupils within a class as well as classes within a school and also schools among each other on a national level. The results are published in ranking lists, like the ones known from Great Britain. The children's abilities are assessed at an early age. Exams place pupils into courses deemed suitable for their needs. Primary pupils are firstly streamed at the end of their fourth year and again at the end of Grade 6 when they sit for the nation-wide primary school graduation exam. The results of this exam place pupils into a stream to suit their intellectual and academic abilities in secondary school. The more able pupils are placed in four-year special or express courses, and the remainder are put into a four or five-year normal academic or technical course. Teacher performance is reviewed every year as well and teachers are ranked among the other teachers in the school, as judged by the principal, vice-principal and heads of department. The reports are confidential and principals must discuss negative reports with their staff.

A Career Advancement Chart was developed to plot the training needs and career prospects of all teachers and functioned as a formal guideline for promotions. This Chart indicates the possible further career steps for teachers in Singapore, positioning a teacher within a school according to his or her academic achievements and teaching experience as well as the reports.

In 1987 the report 'Towards Excellence in Schools' was published by a group of school principals as a result of their visit and study of selected schools in the US and GB. The principals recommended the decentralization of educational governance. They recommended more responsibility for principals to be able to introduce initiatives and respond more flexibly to changes. Since the late 1980s, the government created independent schools within the public system. More and more responsibility has been given to these independent schools so that decisions can be made by the schools themselves. Such schools develop many school programs and can decide on the allocation of financial and personnel resources. Decentralization makes schools theoretically more flexible and free to innovate but also keeps them strongly accountable to the Ministry and parents since their pupils continue to sit for national school leaving examinations. There are now a total of 8 independent schools and some 22 autonomous schools.

In 1997 a new initiative, the creation of 'school clusters', was launched to enable the transfer of autonomy to small school clusters. About five to seven schools are grouped into a cluster, in some cases involving both primary and secondary schools, and headed by an experienced school principal. Such a cluster operates like a small autonomous entity which has the flexibility to make certain decisions without having to refer to the Ministry of Education, and its members also have a greater say in the deployment of teachers within each cluster. The devolution of decision-making to the cluster level has allowed resources and expertise to be shared and used according to the needs of the schools in the clusters. Sharing experiences and self-evaluation on the cluster level through the other members is meant to lessen the control of the Ministry of Education. A continuous professional development for teachers is being pushed, teachers are entitled to a 100 hours per year of professional development time and a Teachers' Network promotes the sharing of good practice.

In 1997 the Ministry of Education introduced the 'Thinking Schools, Learning Nation' concept, which is, according to the Ministry, the new formula for Singapore's success in the new millennium. Schools need to become 'thinking schools' by creating a learning environment that helps pupils to become committed and creative citizens. Teamwork and the willingness for life-long learning are seen as desired attitudes and skills that will have lasting influence on the individual's adult competences. The life-long commitment to learning is to be strengthened through the school in order to put the second part of the formula 'learning nation' into action and to make sure that a commitment to lifelong learning and innovation is secured at all levels. A School Excellence Model has been developed, which allows schools to be more self-directed in the assessment of their programs, and gives them some autonomy in the setting of goals.

The school principal will obviously play a key role in this transition from a very result-oriented approach of viewing schools to a more process- and learning-oriented one. The principal has to make sure that the school reacts to varying needs and challenges, and she or he supervises the development of school programs. The main emphasis will be on character building, motivation and innovation, creative and committed learning. This could mean even more pressure to succeed for the single principal, since there will still be ranking lists and competition among schools while the range of criteria for all that has changed and increased.

It could thus be argued that school principals in Singapore have to cope with conflicting demands. On the one hand they need to holistically drive forward the vision of a thinking school: developing into a more organisationally independent and self-reflecting entity, even as they are ranked. In fact, schools are supposed to develop contrary to what has shaped them for decades. The school principal, therefore, plays an important role in this politically propagated societal change:

> The school is the basic operational unit for producing human resources for nation-building, and the principal is expected to be the person in charge of it. (Yip & Sim, 1990, p. 32)

2. Development of School Leaders

The origins of the program described here date back to July 1984. Since then, successful participants have been awarded a Diploma in Educational Administration (DEA). In terms of content, this early version focused primarily on theories and practices of management and action research to be applied to school situations during the participant's school attachment under the supervision of an experienced principal. In 1983, a pilot program for aspiring heads of department was initiated with the objective of developing a competent and professional leadership team able to efficiently manage schools. Time was ripe to try bottom-up approaches and the idea of supporting school principals by a competent team to help them fulfil their various tasks within a school (school improvement, school policies, resource management) became more and more established in the following years. At the end of the 1980's the dominant opinion was that for the complex task of leading a school, a top management team which can make use of the various strengths of its members – like in the business world – was the best solution.

In the following years the program was further developed and findings from research projects (especially those from the participants' action research projects) were added (see Wee & Chong, 1990).

Today, the vision of 'Thinking School, Learning Nation' is to be realised by a team of instructional leaders, who support the principal. That's why the heads of department get a 17-week management training, whose emphasis is on the management of departments at school, on team work and on the design, implementation and evaluation of the curriculum. Every school leader – in the course of his career – is expected to undergo this training for preparation to be a head of department or vice principal.

Diploma in Educational Administration

We describe below the Diploma in Educational Administration (DEA), which prepares school leaders in a full-time training mode for the tasks and responsibilities that lie ahead of them. The outline of the nine-month program has been jointly developed by the National Institute of Education (NIE) of Nanyang Technological University (NTU) and the Ministry of Education.

The target group of this program is primary and secondary school principals before they take over their post. The admission criteria for participants is a minimum of 5 years' trained teaching experience, favourable reports on teaching and administrative ability as well as professional conduct, and having served as heads of department, vice-principal, or specialist education officers at the Ministry of Education headquarters. During this full-time program participants receive the salary they had before entering the course.

The emphasis of the program Diploma in Educational Administration is on the preparation of school principals for designing learning organisations, which are able to react fast and efficiently to contextual changes. It aims to prepare future school principals in a full-time program (lasting one academic year) to act in accordance with the vision 'Thinking School, Learning Nation', and ensure that the education provided brings about committed citizens, who are able to think creatively and critically and are good problem solvers. The Ministry of Education has realised that good marks can no longer be the sole criteria for good pupils and schools, and that in a fast-changing world, mere knowledge can no longer be the only aim of education, but that a willingness for a life-long passion to learn needs to be fostered.

The aims of the Diploma in Educational Administration, as stated in the program outline, are to prepare principals to become leaders able to:

- create the conditions for their schools to become learning organisations,
- to manage time, human and material resources for value-added pupil learning,
- to encourage all members of the school to work collectively to achieve organisational goals,
- to respond effectively to the demands made on the school by both the internal and external environments,
- to strive to improve themselves and other members of the school in order to achieve a higher level of excellence, especially in line with national policies (e.g. National Education, IT Masterplan) in their work,
- to be conversant with financial and cost management,
- to understand and utilise information-technology for school management and learning.

The program consists of coursework in four subject areas and a school attachment. During the two semesters, the participants will work on various topics of those subject areas, each of which are made up of two or three compulsory modules. Each module will take up about 26 contact hours and can be completed and assessed in one semester by research reports, essays or school-based projects. The school attachment aims to link theory to practice.

The four subject areas are:

1. Subject Area A: The School as a Learning Organisation

- Module 1: Principles of Management
- Module 2: Systems Leadership
- Module 3: Workplace Learning

The modules of the subject area 'The School as a Learning Organisation' aim to facilitate the participants to transform their schools into Learning Organisations. The module 'Principles of Management' provides the participants with management know-how for planning, organising, staffing, leading and controlling, and facilitates discussion. The next two modules pick out the disciplines of learning organisations as central themes. In the module 'Systems Leadership' the participants are introduced to the concept of mental models and system thinking. Advantages of those ideas for the management of schools are discussed and analysed for a better understanding. The content of 'Workplace Learning' is directly linked to this and deals with personal mastery, shared vision and team learning. The contents of both modules are assessed together by a two-hour written examination.

2. Subject Area B: Action Research and Evaluation

- Module 1: Problem-based Practice
- Module 2: Marketing
- Module 3: Educational Evaluation

Another three modules focus on 'Action Research and Evaluation'. The module 'Problem-based Practice' will help participants to understand the importance of the application of problem-solving techniques as central to their task and to be able to identify and solve problems strategically and professionally. Necessary problem-solving strategies and skills are discussed and participants are encouraged to put them into action using their schools as case studies. The module 'Marketing' provides the participant with the basics of marketing and helps them to understand the importance of good communication with parents and the community. In order to successfully practise marketing and to improve the quality of schools, participants have to be knowledgeable about approaches to evaluation, which are being dealt with in the module 'Educational Evaluation'.

3. Subject Area C: Management of School Programs

- Module 1: Curriculum Development and Change Implementation
- Module 2: Professional Development of Staff

'Management of School Programs' is the title of the third subject area and comprises the contents of the modules 'Curriculum Development and Change Implementation' and 'Professional Development of Staff'. Module 1 provides the participants with an overview of the curriculum field, and with the role of a principal within the development and implementation of curricular change in their school. Module 2 stresses the relevance of professional staff development in schools and looks at different models of staff development. Future principals not only need to be familiar with the importance of continuous learning and self-renewal but also should be able to design programs and to respond to the changing needs of their staff.

4. Subject Area D: Governance of Singapore

- Module 1: Educational Policy Making
- Module 2: Financial Management in School
- Module 3: Ethics of Management Decisions

'Governance of Singapore' touches on specific issues concerning education in Singapore. The Module 'Educational Policy Making' aims to familiarise the participants with processes and policy dilemmas in education that face policymakers of the Ministry of Education and schools in Singapore. 'Financial Management in School' deals with the effective management of financial resources. As schools gain more control over resources, the principal's role in utilizing and managing financial resources becomes more important. He or she also has to have the know-how to organise and get funding elsewhere. Specific moral issues of school administration are discussed in 'Ethics of Management Decisions'. Principles including the authority of the school principal and fairness in staff appraisal, team work, promotion, assessment of pupils, morality of punishment, etc., as well as their meaning and justification in the school setting, are central to this module. Here the participants learn to anticipate and solve sensitive issues and situations.

In addition to the seminars, participants are being placed at a suitable school for four weeks, working together twice with an experienced school principal as their mentor. The first attachment will be directly after the first semester and the second attachment is planned to be during the second semester. The participants will have the opportunities to observe and role-model their mentors' skills in the participative management of their schools, carry out specific school management tasks and receive feedback, coaching and counselling by their mentors and their supervisor from the National Institute of Education. The strategy of mentoring has proved to be very effective and is highly appreciated both by DEA participants as well as by the principals themselves, who profit from the innovations being introduced to the schools through the students under the guidance of a NIE lecturer. Both attachments are meant to be completed at one school only, but school visits and observations at other schools are deemed necessary in order to provide participants with additional insights.

The range of teaching methods of each module includes lectures, tutorials, seminars and workshops. The central method being used in the school attachment is mentoring. Participants who have successfully completed and performed in all the modules as well as the school attachment are awarded the Diploma in Educational Administration.

Table 24: Outline of the Diploma in Educational Administration in Singapore

Training and Development of School Leaders in Singapore The Diploma in Educational Administration	
Provider	National Institute of Education (NIE) of Nanyang Technological University in cooperation with the Ministry of Education (MOE)
Target Group	Teachers aspiring to a school leadership position, i.e. before application
Aims	Preparation of school principals for the creation of school as learning organisation or Thinking School
Contents	School as learning organisation (principles of management; systems leadership; workplace learning); action research and evaluation (problem-based practice; marketing; educational evaluation); management of school programs (curriculum development and change implementation; professional development of staff); governance of Singapore (educational policy making; financial management in school; ethics of management decisions)
Methods	Lectures; seminars; workshops; tutorials; mentoring; project work
Pattern	Ca. 58 course days* (timetabling: 21 weeks of seminars (286 contact hours per semester)) plus 2 four-week full-time school internships and additional readings within 9 months full-time
Status	Mandatory
Costs	Unknown; state-financed; participants get release time from school for the duration of the first phase, full salary will be paid throughout this time
Certification	Diploma in Educational Administration

* *If there is no specification by the provider as far as the number of days is concerned, we converted the contact time in hours into the unit course day, taking 6 hours training as one day.*

3. Summary

The school system in Singapore is driven by government policy, whose most important aims are achievement, efficiency and economic success. The supervision and control of schooling in Singapore is done directly by the Ministry of Education. The schools and the teachers, as well as the pupils, are in competition with each other and are therefore under a great deal of pressure. Efforts to give the schools a higher degree of autonomy result in even higher amounts of such pressure. Today a slogan often used says Singapore is striving for thinking schools to develop creativity and lifelong learning.

The preparational training program for all school leaders, the Diploma in Administration Education (DEA), was developed by the Ministry of Education and the National Institute of Education (NIE) of the Nanyang Technological University. The explicit purpose of the training program is to enable the participants to become leaders of these thinking schools. The program consists of modularised courses and two four-week internships with a mentoring attachment. During this nine-month full-time program the participants receive their full salary. Another explicit aim is to regard schools as learning communities and to establish management teams supporting the respective school leaders. Hence the heads of department have gone through training on their own. This also means that every potential school leader – as in Singapore there is a very standardised promotion procedure – had training when they were heads of department, before in their professional career.

Though no formal evaluation appears to have been done of the DEA program and not all DEA graduates went on to become principals, it is certainly the case that the DEA is held in high regard. The Singapore education system achieved international recognition for curricular excellence and pupil achievement in the late 1990s through the publication of the results of the Third International Mathematics and Science study, and school leadership was a crucial element. Participants regarded it as an honour to be selected and the MOE continued to provide material and symbol support; many key officials made themselves available to brief participants.

4. Recent Developments

Notwithstanding the contributions made by DEA graduates it was clear by the late 1990s that the external environment had changed and would continue to change, posing new challenges for schools and school leaders. Decentralisation had resulted in the appearance of school clusters, independent and autonomous schools. The Edusave Fund was giving schools more financial discretion and outsourcing of services was becoming the norm. The 'Thinking Schools, Learning Nation' and the IT Master Plan both demanded and legitimised curricular innovation. Beyond schooling, Singapore itself was wrestling with the challenges of globalisation, having to restructure its economy, cope with an influx of migrants, both skilled and unskilled, and cope with massive social change as well. Singapore leaders argued that Singapore needed to innovate, to adopt and adapt best practices and to be entrepreneurial to survive.

The Leaders in Education Program (LEP), introduced in 2001, fashioned in conjunction with the Ministry, is, at its core, an executive program conceiving of the principal's role as that of a Chief Executive Officer (CEO). It is shorter in duration from the previous DEA, adopts an innovative process-as-content model to place the emphasis on learning, on problem solving and decision making, draws on the expertise available in industry and provides opportunities for field trips abroad. The following modules comprise the heart of the program, 1) Managing competitive learning school organisations 2) Marketing and strategic choice 3) Strategic information technology integration in schools 4) Achieving excellence in teaching and learning 5) Building human and intellectual capital 6) Leadership for the new millennium 7) Personal mastery and development for principals.

Admission to the new program continues to be rigorous, the program is resourced at a high level and senior officials take a deep personal interest in the program. Early indications are that the program has a sound conceptual base and participant response has been enthusiastic.

5. Statement by S. Gopinathan

Given that the DEA program is well regarded and produced capable schools leaders it might be wondered why it changed character so dramatically. A variety of reasons, some cost-related, others conceptual, played a part. It is expensive to have vice-principals spend a year away from school on full pay attending the DEA. More significantly however, policy makers saw change needs for school leadership in a context of extensive school reform. The idea of schools leaders as Chief Executive Officers had gained acceptance and it was felt that a model based on executive training was a more intensive appropriate model to prepare leaders for challenging times when they would have greater control over personnel and other resources. There was an acceptance also that there was much to learn from best business practices and indeed that closer contact between students and leaders in commercial/industrial firms would open up school leaders to new perspectives. Finally, there was a view that schools needed to be more nimble and innovative in meeting needs with their pupils, thus innovation capacity building is a core element in the program.

In itself there is nothing wrong with using tested commercial strategies in educational management. The danger lies in not seeing education institutions and educational processes as fundamentally different from processes involved in producing goods to sell to maximise profit. How will a school leader deal with pressures to do better in school rankings, or to attract more able students if the core discourse in leadership preparation programs is in terms of an education marketplace? How are future educational leaders to learn about and commit themselves to values of inclusion, respect for difference, equalising opportunity, valuing each child and seeking ways to maximise his or her potential if these are not centrally addressed? It is precisely because competitive pressures are great in education at the present time that discussion and understanding of the philosophical and ethical traditions underpinning education is so vital. It is to be hoped that excessive instrumentalism will be avoided in the program.

Hong Kong: A Task-Oriented Short Course

Stephan Huber and Huen Yu

1. Background

In July 1997, Hong Kong was handed over to China. After this political transition from a former British colony to a Special Administrative Region (SAR) of China, the Hong Kong SAR Government has enjoyed a high level of autonomy, including decision-making power in education and making policies regarding the educational system and its administration, as stated in the Joint Declaration of the Government of the United Kingdom of Great Britain and Northern Ireland and the Government of the People's Republic of China, 19 Dec 1984.

The problems emerging from the handover and the administration as an SAR, as well as the demands on school education and the school system, arising from the transition of the economic system due to Hong Kong being an international financial and business city, disclosed a need for reforms on several levels of the school system. Education had to not only be sufficiently available for all but also to guarantee high quality and effectiveness in order to make sure that qualified people were educated for the job market. The starting point of most of the reform initiatives was the idea that the educational aims of schools were perceived differently by all stakeholders and therefore the cooperation of teachers, parents and the public would become necessary.

In 1984 the Education Commission (EC) was established to support and advise the Education and Manpower Bureau (EMB) and Education Department(ED). The Commission was also given the responsibility of coordinating the necessary reforms in education. Since that time, the Commission has almost biennially published Education Commission Reports (ECR) on the current state of the art, on new developments and on problems in the sector of education. Initiatives that have been implemented so far are: a national target-oriented curriculum, a new language policy (regulating the use of English and Chinese) and in 1991 the School Management Initiative (SMI). The SMI was introduced to decentralise power and to give more responsibility and decision-making power to the school itself in order to make it more flexible and able to work in a target-oriented way. In return, schools had to keep to formal procedures by planning, implementing and evaluating their activities and to be accountable for the pupil achievements. The SMI was meant to be a framework for the initiation and implementation of school improvement at schools.

In spring 1999, the EC started to review the educational system of Hong Kong and consulted the public on their opinion regarding the aims of education in Hong Kong as well as the structure of the school system, curricula and the examination system. The findings from 14 000 contributions in this first phase of the Review of Education System formed the basis for reform proposals. In the second phase, ideas and proposals were published again and discussed. In the third phase, the EC made concrete reform proposals regarding the structure of the school system, curricula and the examination system, and suggested priorities. The public was consulted again and the final report was handed in to the chief executive, Tung Chee Hwa, by mid 2000[36].

At the same time, a new extensive reform was implemented, which had been recommended by the Education Commission in their Report No. 7 in 1997. It was here, that they demanded that school-based management had to be implemented in all schools by the year 2000. In Hong Kong, schools are usually provided by government-aided non-government organisations (Aided Schools). Only a small number of primary and secondary schools are directly managed by the government (government schools) (see Information Office, 1997). In the meantime, every school in Hong Kong actually administers its own school management committee, according to the Education Ordinance (Cap. 279). This means that a large part of the responsibility, authority and accountability was given to the individual school sponsoring bodies, though within a centrally fixed framework.

Chair of the school management committee of a government school is a directorate officer of the Education Department, and in the case of aided schools the committee is usually led by the person in charge or the representative(s) of the non-profit organisation which is the sponsoring body of the school. Besides those, the government aims to include other stakeholders of the school, such as the principal, representatives of teachers, parents, alumni and the community to form members of the committee. According to the Advisory Committee on School-Based Management (2000), the Committee has the following tasks and responsibilities:

- to ensure that the Education Ordinance is complied with and the vision of the school sponsoring body is fulfilled,
- to set the mission and the goals of the school,
- to determine policies on teaching and learning,
- to be responsible for program planning, budgeting and human resource management,
- to establish a community network and support system.

All this brought about a paradigm shift in the role of school leadership: the traditional view of principalship in the sense of maintaining the status quo and following a hierarchical allocation of tasks and responsibilities is to give way to team work and cooperation as well as an improvement- and effectiveness-oriented view. Besides the rather traditional tasks oriented to maintenance and the smooth running of the school, new tasks have been added oriented to change and innovation as well as communication and cooperation with all stakeholders.

2. Development of School Leaders

In January 1999, the ED assigned a task group on training and development of school heads to the task of conceptualising a training program for school leaders. The important role of school leaders for the effective implementation of reform

[36] Obviously there is great effort to base the reform initiatives on a broad level of acceptance among the public.

initiatives and the school-based management had been recognised, and the knowledge, skills and attitudes required had to be facilitated. Half a year later, in June 1999, the task group under the ED published a consultation paper on the Leadership Training Program for Principals. This proposal outline had been put together after the intensive study of programs in England, Scotland, Australia (State of Victoria) and Singapore. Besides this, the needs of school leaders themselves were ascertained. The training program planned by the task group on training and development of school heads covered topics ranging from leadership training, curriculum studies, IT, personnel management, and budgeting. Elements such as needs assessment and the use of action research would link theory and practice as well as take the actual needs of the participants into account (see Wong, 2000). In this 1999 consultation, there was great support for the government to put effort in the upgrade of principals' competence and to promote principals' continuous professional development. However, there were reservations about the proposals of running a uniform program for all principals and the requirement for all serving principals to obtain a certificate of principalship by 2007. Further description of Hong Kong principals' continuous professional development is provided later in the section of recent developments.

At present there is not as yet training for aspiring or potential principals in Hong Kong. With the training always taking place after appointment, it cannot play any part in the recruitment and selection process. Up to now, it must be admitted, there has been no culture of a preparatory qualification for school leaders in Hong Kong. The idea of a preparational training, however, is still pursued, so maybe the preparation course already planned will be realised after all.

Induction Course

There was one course in Hong Kong preparing school leaders that aimed to impart basic management skills and support for the school management tasks. This Induction Course was compulsory for all school leaders of all school types. It took nine days for secondary school principals and ten days for primary school heads and could be completed within a period of 2-3 weeks. The training was provided, coordinated and organised by the ED and took place in the training unit of the ED. The coordinator invited guest speakers from tertiary institutions, internal experts from the ED or experienced school leaders. The topics and course contents were formulated by the ED and therefore individual needs could not be taken into account to a high degree.

Those topics included:

- Introduction to management
- Communication
- Delegation
- Conflict management
- Finance
- Law and ordinances

Each of the nine days was organised in a similar way. Two to four topics were discussed during the day. The lectures and workshops could be roughly classified under the above-described topics. The method of instruction, mainly case studies, group discussions and lectures, was left to the individual teaching style of the lecturer.

The following list will give an overview over the Secondary School Administration Course from October 1998:

Day 1:
- Hot issues on education policies in Hong Kong
- Roles and functions of secondary school heads

Day 2:
- School vision and mission
- Communication
- Application of IT in education
- School visit

Day 3:
- Performance management
- Prevention of bribery
- Managing change
- Empowerment

Day 4:
- School finance and accounts I
- Working with staff having teaching/emotional problems
- Education ordinance and education regulations
- Working as a secondary school principal

Day 5:
- School head as a leader
- Selection of staff I
- Selection of staff II

Day 6:
- School finance & accounts II
- Curriculum leadership

Day 7:
- Relationship between school heads and mass media
- Quality assurance inspection
- Code of aid & annual estimates

Day 8:
- Employment ordinance
- Crisis management
- Team building

Day 9:
- In tray exercise, post-course action plan & evaluation
- Open forum and presentation of certificates

The courses were organised by the ED and therefore entirely free for the individual participant. After the completion of the course, participants received a confirmation of attendance. There was no formal accreditation or examination. These short courses were discontinued in the academic year of 1999/2000.

Table 25: Outline of the Induction Course in Hong Kong, China

Training and Development of School Leaders in Hong Kong, China The Induction Course	
Provider	Education Department (ED)
Target Group	Newly appointed school leaders
Aims	Introducing newly appointed school leaders into their tasks and responsibilities; promoting a re-conceptualisation of the role and the relationship with staff
Contents	Hot issues on education policies in Hong Kong; roles and functions of secondary school heads; school vision and mission; communication, application of IT in education and school visit; performance management; prevention of bribery; managing change; empowerment; working with staff having teaching/emotional problems; education ordinance and education regulations; working as a secondary school principal; school head as a leader; selection of staff; school finance and accounts; curriculum leadership; relationship between school heads and mass media; quality assurance; inspection; code of aid and annual estimates; employment ordinance; crisis management; team building; in tray exercise; post-course action plan and evaluation
Methods	Lectures by guest speakers; discussions; case studies
Pattern	9 course days (timetabling: 9 sessions with 3 - 6 hours each) within 2-3 weeks
Status	Mandatory
Costs	Unknown; state-financed
Certification	Certificate of participation

The Professional Development Program
School Management for Principals (SMP)

Another training program for serving principals was the program School Management for Principals (SMP), which was conducted at the Hong Kong Institute of Education (HKIEd), since school-based management was introduced. It was outlined as a direct response to the decentralization of the school administration.

The target group of the SMP was serving primary heads. The overall aim of the program was to equip the participants with perspectives and a new understanding of leadership in the context of school-based management. Participants would be stimulated to reflect and reason their current practice critically. The SMP brought in the practical experience of the individual participant in a constructive way throughout the course, and the philosophy could be reflected through the module contents, the teaching methods (mostly team teaching by two to four trainers of the management group of the lecturers of the HKIEd) and the final evaluation procedure.

The SMP required 45 contact hours spread over five weeks and divided into 15 seminar sessions taking 3 hours each.

Contents of the 15 seminar sessions:

- Overview (2)
- School-based management (3)
- Leadership (1)
- Team building (1)
- Communication (1)
- Motivation (1)
- Home-school relation (1)
- Conflict management (1)
- Management of change (1)
- Staff evaluation and appraisal (1)
- Staff development (1)
- Financial resources and management (1)

In the beginning of the course this structure was discussed with the participants and could then be modified. The assessment was a case study or a project aiming to encourage the participant to link theories of school management to school practices. After that, attendance was confirmed. However, this development program was discontinued in September 2000.

Before the initiation of a continuous professional development framework for principals in June 2002, the ED endeavoured to try out different development programs for newly appointed principals and serving principals. Programs for the two types of principals had a slightly different focus. The newly appointed principals had development opportunities through needs assessment, induction seminars, leadership development programs, and extended programs which included three distinct phases: workshops held in Hong Kong, overseas study trips, and post-trip reflection and action planning. For serving principals, their development focus was shifted to activities such as educational administration seminars, overseas study trips and post-trip reflection according to different developmental needs. Details of the principals' continuous professional development framework are elaborated in Section 4 on 'Recent developments'.

3. Summary

In the course of the reforms in the educational sector, and above all due to the establishment of school-based management, the self-responsibility of Hong Kong's schools has increased since the mid 1990s. However, this operates within a centrally determined framework. For the role of school leaders there is supposed to be a paradigm shift. It focuses now more on teamwork and cooperation, improvement and effectiveness.

There is no tradition of any training for school leaders before they take over principalship in schools. However, since 1999, a task group on professional development of principals has been set up to develop a conception of a comprehensive program including preparatory components. The Education Department is developing this new program. So far, there have been two programs. The first was a nine-day or ten-day induction course, offered by the Education Department and aimed to provide newly appointed school leaders of all types of schools with basic knowledge in school management. The second program was a continuous professional development program for experienced school leaders in the primary sector. This course, School Management for Principals (SMP), financially supported by the Education Department and run by the Hong Kong Institute of Education as a supportive measure after the establishment of more autonomy for schools, sought to help school leaders come to terms with their new context. Decentralisation has meant a major re-conceptualisation of the principal's role and relationships, with soft or people skills being given much more attention. A particular issue has been that of re-orientation of school principals without any parallel re-conceptualisation of the teacher's role.

4. Recent Developments

Subsequent to the 1999 consultation on Leadership Training Program for Principals, the ED set up another group, which is called Working Group on the Professional Development of Principals, in June 2001 and invited tertiary experts and experienced frontline practitioners to develop a continuous professional framework and different requirements for aspiring principals, newly appointed principals and serving principals. The Working Group worked out a proposal, which was published in the form of a consultation paper in February 2002. Since then, it was under public consultation until mid-April. After considering various views and suggestions, the ED announced the full implementation of the principals' continuous professional development framework at the end of June.

The framework comprises three interrelated components: beliefs, four leadership domains and six core areas of leadership. The statement of beliefs provides basic underpinnings for principals' continuous professional development. The leadership domains describe opportune forms of leadership that schools in the 21st century require principals to demonstrate. These domains include strategic leadership, instructional leadership, organizational leadership and community leadership. The core areas of leadership are values, knowledge, skills and attributes needed by Hong Kong principals. These areas encompass strategic direction and policy environment, learning, teaching and curriculum, teacher professional growth and development, staff and resources management, quality assurance and accountability, and external communication and connection to the outside world.

There are different requirements for aspiring principals, newly appointed principals and serving principals. Aspiring principals have to meet the requirements of Certification for Principalship (CFP). From the academic year 2004/05, aspirants will have to attain the CFP before they could be considered for appointment to principalship. Newly appointed principals are required to undergo designated courses throughout the first two years of principalship. The courses comprise four parts: the needs assessment for principals in Hong Kong, an induction program, a school leadership development program, and an extended program. Serving principals are required to undertake 50 hours professional development activities per year or adding up to a minimum of 150 hours every three years. The activities they undertake should cover all three activity modes, namely, structured learning, action learning, and service to education and the community.

5. Statement by Huen Yu

The training of the nine-day induction course for secondary principals or ten-day induction course for primary principals designed by the Education Department of Hong Kong was able to provide newly appointed principals with basic and survival skills. Also, these courses could establish networks for principals so that they were able to communicate and enhance mutual support with one another. However, this kind of short course was definitely not sufficient for preparing contemporary school leaders to cope with a host of educational innovations in schools in Hong Kong. The main problem was the content of the principal training courses. It emphasised more on management and less on leadership, focusing mainly on the practical skills of managing school. The themes related to management were: roles and functions of secondary school heads, selection of staff, application of IT, prevention of bribery, annual estimates, school finance and accounts, code of aid and ordinances, crisis management, quality assurance inspection, relationship between school heads and mass media, etc. The part of leadership content was much less in comparison with that of the management, such as school head as a leader, curriculum leadership, school vision and mission, communication, managing change, and empowerment. Conger (1999) pointed out that sometimes the inability to adapt to the changing world might be caused by organizations having too much management and too little leadership. In this changing world, organizations need more leaders than managers. It is no surprise that many principals may have had problems in leading teachers because they are trained as managers more than as leaders. Leaders and managers are said to be different. Bennis and Nanus (1985) distinguished the difference: "Managers are people who do things right and leaders are people who do the right things. The difference may be summarised as activities of vision and judgment – *effectiveness* – versus activities of mastering routines – *efficiency*" (p. 21). The heart of the problem was a severe shortage of leadership talent (Bennis & Nanus, 1985). The same problem may apply to Hong Kong principals. According to Yu's research (2000), Hong Kong primary school principals had difficulties in exercising sufficient transformational leadership to influence teachers' commitment to change. Dimmock and Walker (1998) found that the Hong Kong principals' greatest challenge was to get teachers to become genuinely involved in decision making. Thus, there may be a great shortage of leadership knowledge and talent in regard to school leadership in Hong Kong. In order to better equip principals for more effectively running schools, some leadership theories and concepts such as transformational leadership, moral leadership, value leadership, instructional leadership and participative leadership, could be introduced. With all these included in the content of principal training courses, the duration of nine to ten days of training proved to be too short.

As for the program of School Management for Principals for primary school heads, its duration of training was comparatively longer, being 45 contact hours spread over five weeks. Its content was comprised of themes relating to management and leadership, and current issues in education reforms such as school-based management, team building, home-school relation, and management of change. More theoretical concepts were provided. However, the part of leadership training was still not enough. Sufficient leadership concepts could be included to strengthen principals' knowledge base of leadership. In addition, for serving principals, problem-based learning could be used so as to raise their interest of learning and to suit the individual need for solving problems for his or her own schools. This type of learning is an appropriate method to prepare school leaders in this changing era (Murphy, 1992; Bridges, 1992; Bridges & Hallinger, 1995; Hallinger & Bridges, 1997).

Australia/New Zealand

New South Wales, Australia: Development of and for a 'Learning Community'

Stephan Huber and Peter Cuttance

1. Background

New South Wales (NSW) is one of the six federal states of Australia. Australia's federal structure of government assigns most of the responsibility for schooling to the six state and two territory governments. The Federal Government through the Commonwealth Department of Employment, Education, Training and Youth Affairs (DEETYA) provides national cohesion across the various school systems, a system of vocational training, funding for universities, which operate relatively autonomously and a policy framework linking education to the economy, society and culture of the nation. Approximately one-third of pupils attend schools that are governed and managed by entities other than the state and territory governments, mostly affiliated with various churches. Each state and territory has developed its own system of educational administration within this framework, New South Wales is the largest public school system, with 2 200 schools, 750 000 pupils and 46 000 teachers.

In New South Wales, as in most other states of Australia, in the course of the 1990s, reforms in the field of educational policy took place. The central administration was reduced and schools were given more self-management in terms of site-based management, by which local school committees and school leaders were delegated an increased level of responsibility. To some extent, individual schools and their leaders have become more accountable. Parents were given a greater say in decisions within schools[37].

Arising from these developments were a number of challenges for school leadership. Firstly, schools like other public and private institutions have become somewhat more accountable. Secondly, the expectancies of communities and the general public towards schools and the educational system have increased. Thirdly, pupil achievement as outcome of schooling has been made a more central interest. Fourthly, there has been some restructuring of positions within schools.

---

[37] While, this is representative of a general trend across the nation, NSW has one of the lowest levels of devolution to schools. For example, individual schools still have relatively little control over or direct input into decisions about the selection of new staff, who are still allocated centrally to schools.

Similar to other countries in the world, schools in New South Wales are confronted with swift technological changes, an explosion-like increase in information and easier access to sources of information. Additionally, a discussion on Australia's cultural and national identity and its increasing cultural diversity, which results in a re-evaluation of basic principles and values, as well as the increasing gap between rich and poor in Australia has started.

Through the Training and Development Directorate, the New South Wales Department of Education and Training[38] has developed a leadership program to address the enhanced demands on school leadership. The underlying aim was to re-structure schools as learning communities. The key features of such learning communities are grouped in the areas "organisation, practice, purpose, roles and relationships, core beliefs and values" (see table below).

Table 26: Key features of learning communities

Core Beliefs & Values All People can learn Diversity is honoured Learning is valued Learning how to learn is valued Quality of everyone's learning is important	
Organization - Diffuse internal and external boundaries - Open communication and information – flows - Networks and partnerships - Structures, time and space provided for - dialogue and discussion	Practice - Continuous learning – individual, team and organization - Trust and risk-taking - Participative decision-making - Balance of inquiry and advocacy - Experimentation - Critical reflection
Purpose - Sense of identity purpose and power - Sense of belonging and connectedness - School is a centre of inquiry - Focus on students and their learning - Learning focused work environment	Roles & Relationships - Principal as leading learner - Parents as learning partners - Teachers as learners and leaders - Students as self-directed and committed learners - Collaboration and collegiality

(from: Dawson, 1999, p. 12)

Seven important principles for leaders of learning communities result from this conception of school as a learning community (see Dawson, 1999):

- leaders are responsible for learning,
- leaders model effective learning,
- leaders lead teams,
- leadership is a function of ability, not position in the hierarchy,

[38] Henceforth, the Department.

- leaders exist throughout the school learning community,
- leaders are creative,
- leaders are ethical.

2. Development of School Leaders

The conception of schools as learning communities was supported by a School Leadership Strategy, implemented in March 1989, that aimed to develop the capacity for educational leaders to lead learning communities. The program was developed on following extensive consultation with school leaders and school leadership associations. Responses to this survey were numerous and the resulting report particularly emphasised the following concerns of the profession: Access to existing qualification programs and resources needs to be facilitated and incentives for continuous training need to be provided in order to improve school leader development in NSW. A clear-cut definition of the responsibilities on behalf of educational policy, the profession itself and the schools and the individual school leaders is called for as well. Equal access of all school leaders to the qualification programs is to be guaranteed. Inhibitions like distance, gender roles, family obligations, varying access to technology and the unequal level of collegial support are meant to have little or no influence, if possible. To disclose the individual needs, the selection or assessment of the candidates needs to become more effective (see NSW Department of Education and Training, 1998a). The focus of the Leadership Strategy was to be the cultivation of a professional commitment to ongoing development among school leaders and school executives.

School Leadership Strategy

The starting point of the School Leadership Strategy was the culture of schools. The program sought to develop school leaders as leading learners in school cultures that promoted effective learning for all members. The strategy sought to develop competences and leadership skills in the areas of educational leadership (to enhance learning outcomes for all pupils), cultural leadership (to articulate the values and beliefs that underpin the work of the school), strategic leadership (to negotiate purposes, goals and strategic directions for the school with the educational and societal context), organizational management (to create and articulate a shared purpose and vision for the school and to implement learning strategies to achieve the purposes and visions), and educational management to create an optimal learning environment and focus on resources to ensure optimal learning outcomes for all pupils (see NSW Department of Education and Training, 1998a).

The School Leadership Strategy was comprised of five elements, covering three stages of professional educational leadership in schools in the context of distributing the responsibility of leadership. Therefore, participation in one of the programs depends on the job position and the development needs of the respective participant (see table below).

- The School Leadership Preparation Program (SLPP): a multi-phase preparation program for teaching staff aspiring to leadership positions in the school.
- The School Executive Induction Program: for teachers who had been recently promoted to a new leadership position in the school.
- The Principal Induction Program: to support principals in their first year.
- The School Executive Development Program: for teachers already in a leadership position (but not principalship).
- The Principal Development Program: in-service development for existing principals.

Table 27: Survey of the qualification School Leadership Strategies of the Department of Education and Training

Preparation	School Leadership Preparation Program (SLPP)	
Induction	School Executive Induction Program	Principal Induction Program
Development	School Executive Development Program	Principal Development Program

All parts of the School Leadership Strategy Program will be shortly discussed in the following paragraphs, while only the School Leadership Preparation Program will be described in greater detail.

School Leadership Preparation Program

The School Leadership Preparation Program (SLPP) was designed to prepare aspiring school leaders and school executives for formal leadership responsibilities. The target group is consequently teachers aiming at such positions. The prospective participants apply for entering an Inter-District School Leadership Group (ISLG).

Since the entire qualification program is mainly organised and implemented by the ISLGs, these groups are of major importance. They have been formed out of two to three individual school districts respectively. The main task of each ISLG is to disseminate information about the programs, to coordinate the implementation, and to facilitate mentoring opportunities and the development of local collegial networks. The SLPP makes decisions on the recognition of prior learning or current learning and elements of individual programs to support the participants in developing their learning portfolios (documenting professional activities and qualifications), validates the learning outcomes for the portfolios and cooperates with the state-wide school leadership development reference group to assure consistency in the implementation of the program. The ISLGs make recommendations on the certification of participants and identify resources, offers and initiatives made available by external providers.

All participants in the SLPP take part in an initial two-day School Leadership Preparation Seminar provided through the ISLGs. This seminar provides an overview of the program, establishes an understanding of the concept of school leadership in the Department, provides a basis for participants to assess their own leadership qualities and plans the components of the program in response to this information. Participants are provided with guidelines for the development of learning portfolios, collegial support groups and networks.

The program consists of two further seminars of two-days each. Universities or other providers are engaged in the implementation of one of these seminars. The objective of the three seminars is to meet the leadership requirements for achieving the learning targets of the Department and the personal development needs of the individual participants.

The three seminars focus on:

- Leading Learning Communities
- Leadership for Enhanced Learning
- Leadership for Effective Management

The seminar Leading Learning Communities explores leadership in a broader context and the role of the leader as an educator in a learning community. The seminar encourages participants to reflect critically on the significance of an overall leadership vision. Learning options for teams or individuals in the context of this theme include understanding school cultures, team building, leading change processes, providing cultural and ethical leadership, systems thinking and strategic planning and leading educational change.

The seminar Leadership for Enhanced Learning focuses on the role of school leaders for the improvement of pupil learning outcomes. It addresses issues of improvement of the conditions for pupils' learning, the curriculum, pupil welfare and equity.

The final seminar, Leadership for Effective Management, focuses on the management function of school leaders. It includes the translation of educational ideas into practice, management of curriculum, relationships between staff, supervision, communication and interaction within the school community, personnel resource management, effective decision-making processes, and stress and time management.

In addition to the three seminars, the participants follow a program of self-study or in small learning teams. The material for this component is made available on CD-ROM, as print material and over the Internet. Participants may also undertake alternative programs by other accredited providers as part of the program.

Importance is attached to how the participants demonstrate their application of what they have learned. The preparation, presentation and analysis of a professional learning portfolio is an important means for documenting the skills and understanding of leadership and its application. There is, however, no requirement, to undertake a workplace action research project, as it is in some other leadership programs. The learning portfolio additionally serves to identify areas in which there are further learning needs and provides the evidence of professional learning that is the basis for the certification of the candidate. The learning portfolio remains the property of the individual participant.

After completing this program, the participant is awarded the Certificate for School Leadership. This certificate is, however, not a mandatory requirement for appointment to school leadership positions. If a participant is unable to complete the program and if their reasons are legitimate, they may be provided with a record of achievement for the elements that have been completed successfully.

The cost of participation is 1 300 euro (2 400 Australian pounds). One-quarter of this is paid by the participant or their school and three-quarters by the Department.

Table 28: Outline of the School Leadership Preparation Program of the Department of Education and Training in New South Wales, Australia

Training and Development of School Leaders in New South Wales, Australia Example: The School Leadership Preparation Program of the Department of Education and Training	
Provider	NSW Department of Education and Training through regional school leadership groups and external providers
Target Group	Teachers aspiring to any leadership position at school or directly to the position of a school leader
Aims	Preparation for school leadership and other leadership roles in learning communities
Contents	Leading learning communities (e.g. cultural and ethical leadership, system thinking etc.); leadership for enhanced learning (e.g. create optimal learning conditions for the school); leadership for effective management (management tasks of the school leader)
Methods	Seminars; small team sessions; networking; various use of electronic media; preparation and presentation of a learning portfolio and literature studies
Pattern	Ca. 14 course days (timetabling: 1 school leadership preparation seminar: 2 days; 3 school leadership excellence seminars: 2 days each; additional integrated individually selected program components for self-study or for small learning teams: 3 times 2 modules with 4 to 6 hours work time each) plus literature studies within 1-2 years
Status	Optional; recommended for the application to a leadership position
Costs	Ca. 1 300 euro (2 400 Australian dollars) per participant; one quarter (600 Australian dollars) is financed by the participants themselves or their schools and three quarters (1 800 Australian dollars) are financed by the NSW Department of Education and Training
Certification	Certificate of School Leadership after successful completion, otherwise records of achievement for elements completed successfully

School Executive Induction Program

The program of School Executive Induction was developed as a four-phase program for the induction of teachers into leadership positions, for example, as head of department, head of the year, member of the school leadership team, vice deputy school leader/vice head teacher, yet not for school leaders themselves. The responsibility for the conduction of the program was shared by the individual school itself and by the District Office. The program was comprised of four phases:

- school-based induction,
- District seminars,
- District Office orientation,

- ongoing support.

A member of the school's leadership team was responsible for introducing the teacher into their new task in the school with the assistance of support materials developed and provided by the Department. Part of this induction was to introduce the teacher to staff, to familiarise them with routines, to inform them about essential school policies, to give them essential information, to negotiate a role statement and for their responsibilities, to induct them into the school's planning processes, to negotiate an appropriate individual program of training and development, and to introduce them to school community organizations.

The Department provided materials on: setting priorities, leadership and culture, conflict resolution, team work and supervision, clarifying and negotiating your role, legal and industrial issues, self-management, and school self-evaluation.

The District Office orientation was designed to familiarise newly appointed leadership personnel with the responsibilities and support services provided by the Department through the District Office.

Ongoing support during their first year for newly appointed leadership personnel was provided by the principal and the District Superintendent, through participation in District or training and development opportunities, by participation in curriculum and support networks, by shadowing opportunities, or by taking part in mentoring programs.

Principal Induction Program

The Principal Induction Program was designed to support newly appointed school leaders. It was introduced in 1997 and conducted in consultation with the NSW Primary Principals' Association and the NSW Second Principals' Council. It provided theoretical information and concrete support for adjustment to the new role. The program was comprised of four phases:

- a start-up conference,
- assignment to a colleague,
- district orientation,
- a follow-up conference.

A two-day conference conducted during school holidays was provided for all newly appointed principals. The program consisted of core sessions and optional sessions focusing on practical strategies for newly appointed principals.

Orientation colleagues were assigned to provide newly appointed principals with an opportunity to establish collegial networks. These regional colleagues cooperated to offer their support, introduce the new principal to the existing network and assist them in using the resources available.

The new principal was also provided with a program to familiarise them with District operation and support mechanisms. This was organised and implemented by District office personnel in consultation with the respective District Superintendent and with assistance of the orientation colleagues.

A follow-up conference was held during the school vacation, focusing on issues associated with leadership and management, theory and practice, establishing networks, providing access to experienced school leaders, and to the personnel of the Department of Education and Training, with the Department.

School Executive Development Program

This program provided development of individual in-service training and development for school executive level (all leadership positions below those of Deputy Principal) personnel at school. The program addressed individual school and system leadership development priorities. The Department offered courses in the fields of leadership and management and curriculum leadership. The following development options were made available: executive interchange, teacher exchange, education fellowships at universities, tertiary research support, study leave, and short-term study tours.

The Principal Development Program

Ongoing support and development opportunities were offered to principals within the Principal Development Program. The following programs were offered:

- The Principal and School Development Program (PSDP), through which school leaders of different schools support each other in school development activities.
- The Team Leadership Course (TLC), a school based team-learning program, which developed skills in team leadership.
- Participation in the School Excellence Seminars of the preparation program and its in-depth learning modules.
- Participation in the tertiary course Certificate of School Leadership and Management (CSLM) at the Master's degree level.
- Training in mentoring and coaching and the opportunity to be assisted in identifying appropriate mentors and coaches to work with principals.
- The Peer-assisted Leadership Program (PAL), a collegial support and shadowing program, which developed the skills to provide professional advice to peers.
- The Certificate of Teaching and Learning (CTL) and Quality Teaching and Learning Materials (QTLM), in-depth programs that focussed on quality teaching and learning practices.
- The Rural Areas Principal Interchange Program, which provides the opportunity for school leaders in small primary and central schools the opportunity for short-term exchange with executives or principals in larger schools.

Additional seminars conducted in collaboration with Principals' associations were offered in the areas of: Personnel management, resource management, using technology, industrial relations, anti-discrimination and anti-racism.

3. Summary

In the course of the reforms in education, the concept of self-managing schools was introduced and as a result the individual responsibility of each school in New South Wales increased and the school management's spectrum of tasks expanded.

A comprehensive program of leadership, the School Leadership Strategy (SLS), was centrally developed and implemented by the NSW Department of Education and Training, with support being provided through local Inter-District School Leadership Groups (ISLGs). The program is systematic and addresses all levels of leadership in schools. It is based on an understanding of schools functioning as learning communities, with leadership distributed widely within each school. The program tries to take the learning needs of each phase of the career into consideration.

The School Leadership Preparation Program addressed the needs of future school leaders, and the broader leadership group within each school. The School Executive Induction Program and the Principal Induction Program are designed to induct new appointees into these leadership functions. The Principal Development Program and the School Executive Development Program provide continuing professional development for established school leaders, and for faculty with other leadership roles. Individual learning needs are taken into account and prior learning recognised. The programs comprise a combination of seminars, small-groups, individual as well cooperative research, shadowing and peer-assisted learning, and the opportunity for temporary exchange among leadership positions.

It is interesting to see how the Department for Education and Training is responsible for a – voluntary – qualification program which incorporates central conception (and with it a closed conceptual approach) as well as regional implementation (and with it a definite orientation towards the participants' needs within a rather decentralised framework in NSW.

4. Recent Developments

The above programs have continued to operate in NSW, however, a major revision and restructuring of programs is currently being considered. This has been occasioned by the changing context of leadership in the school system. The major changes are driven by the demographics of the teaching population and, within that, of school leaders. The average age of the teaching population has continued to increase as the baby boomer bulge has moved towards retirement age. The strategic focus of future programs will be driven by a work force planning orientation and the needs of aspirant, beginning and continuing leaders. A capabilities framework is being considered to assess the needs of individual leaders and to adjust provision to meet their needs. Information and communication technologies will form a central focus of the way in which new programs will be delivered.

Four key elements are being investigated as part of the revised leadership strategy. A web presence to deliver access to the knowledge base required by leaders and resources in the form of templates and support documents. A capabilities framework for assessing and diagnosing the learning and development needs of individual leaders. Learning partnerships focused on professional learning through mentoring and coaching strategies. Information from a capability (self)assessment for individual leaders will provide the basis for strategic access to these strategies at appropriate stages of a leadership career. A rebalancing of the local/central focus in developing initiatives to meet the needs of leaders in their local contexts. This will require an adjustment to the budget to be devolved to Districts which will develop initiatives to meet the needs of leaders based on the specific socio-economic profile, recruitment profile and teacher demographics and particular geographic and local variables of each District.

5. Statement by Peter Cuttance

The strong focus on leadership programs that were evident across Australian school systems has been dissipated somewhat since the mid-1990s. This is in part a consequence of a reduced focus on the devolution of responsibilities to the school level – referred to as school self-management. In some systems, notably in the State of Victoria, self-management has been extended about as far as possible within the constraints of the system of governance that prevails in Australian government school systems. For example, Victorian schools determine the mix and number of staff at each level of seniority at the local level. Further devolution in Victoria would require substantial changes to the legal and legislative status of schools. Nevertheless, the preparation of leaders is a significant issue, with declining numbers of applications for leadership positions and an expected turnover of 50% in the principalship over the next five years as the current incumbents reach retirement age.

NSW presents a contrast to the push towards devolution evident in Victoria over the last decade. Although NSW was in the vanguard of self-management, following the Scott Report of 1989 (Scott, 1989), there has been a centralising tendency since the reforms were impeded by a deliberately obstructionist bureaucracy from the early 1990s. The leadership programs discussed in this chapter were developed in the last days of the initiatives towards a more devolved system.

Further curtailment of the degree of self-management at school level was managed by withdrawing funding for teacher professional development from schools and returning it to central departments of the bureaucracy as the decade proceeded.

The Department became enmeshed in a Royal Commission of Inquiry into Police Corruption which investigated past practices in the handling of sex abuse allegations and charges by pupils and their parents against teachers. This ultimately led to a further withdrawal of the capacity of schools to make decisions locally. The net effect of these developments was to substantially reduce the focus on the need for schools to develop the capacity to provide local leadership, and hence a reduced emphasis on programs for the development of future school leaders.

By the late 1990s, total professional development funds had declined to the equivalent of 0.05% of the salary for each teacher. The funding for the leadership program had declined to about a quarter of the budget that was envisaged when it was developed in 1998. Current initiatives are aimed at significantly restoring the funding required to meet the professional learning and development needs of leaders.

New Zealand: Variety and Competition

Stephan Huber and Jan Robertson

1. Background

In New Zealand, as in other countries, there were important changes in educational policies in the 1980s. Up to the year 1989, school governance matters lay mainly with the regional Education Boards and the Department of Education. Due to decentralization from 1989 onwards, the former Education Boards were dissolved and replaced by a Board of Trustees at each school comprising five elected representatives of the parents, the school leader, one representative of staff, and – in the case of a secondary school – one representative of the pupils. This Board of Trustees then had responsibility for the administration of the budget, the maintenance of school buildings and equipment, legal matters, and the curriculum. They also had the responsibility for the employment of school leaders and teachers, as well as achievement assessments (see Robertson, 1998 and Harold, 1998). Developing a local school program is also part of the tasks of this committee. Each school in New Zealand is run as an autonomous institution and is equipped with a budget covering everything besides the income of the teachers, which is today still paid directly by the Ministry of Education. There has been continued controversy related to schools having the income for staffing. A change of government meant that at one stage in the last decade, schools were allowed to volunteer to be 'bulkfunded schools' which meant they had complete control of the staffing budget for the school. The Labour government from 1999 on reversed this decision of the previous national government and schools no longer had control of their staffing budget. The school's operational grant is allocated to the individual school on the basis of a formula related to the number of pupils and some equity factors, such as socio-economic area, geographical location and size of school. The organizational shift of responsibility follows a market-oriented approach which considers consumer orientation and competition to be preconditions of effectiveness and efficiency. It is implied that education is a good that the user buys (see Codd and Gordon, 1991). As, for example, in Great Britain, decentralization in New Zealand has been followed by the introduction of centralising control mechanisms, such as the requirements from the audit agency, the Education Review Office, and further centralised developments and presentation of the nation-wide curriculum.

The effects that these profound changes in the structure and ideology of the education system have had for the role, function and the tasks of school leaders, are great. School leaders are at the interface between centralised preconditions and local self-administration. The new emphasis is much more administration- and management-oriented, and the benchmark for their actions has to be, besides the educational effectiveness of their schools, a favourable cost-benefit relationship, i.e.

economic efficiency. School leaders in New Zealand have come into a situation much similar to that of their British or American colleagues. Self-management means for them a so far unknown increase in administration- and management-oriented workload. School leaders feel that this additional administration is at the expense of their educational leadership role (see, for example, Robertson, 1991 and 1998) and results in some alienation from their real educational target activity, i.e. teaching and learning.

Obviously school leaders are concerned that the great additional work load by administrative and management tasks makes it more difficult for them to react innovatively to the learning needs of the pupils in their individual schools, whereas this would be their personal desire (see Wylie, 1994). From this emerges a paradox obviously typical of school leaders in New Zealand. They have a far-reaching authority to introduce innovation, but lack time for it.

Moreover, they are commonly less accessible for staff and pupils, a task that seems to be deeply rooted in the professional ethos of principals in New Zealand. There is a lot of discontent about this (see Wylie, 1997). It seems to be increasingly difficult for school leaders in New Zealand to find a balance between collegiality, in relationship to staff, and accountability to the public, which sometimes may be to the disadvantage of staff relationships. School leaders in New Zealand have never ceased feeling they belong to the teaching profession. This is also mirrored in the fact that the majority of them are still members of their teachers' associations, though there are, of course, special school leader associations as well. According to Harold (1998), the government educational policy aimed at the exact opposite. School leaders were given different employment contracts, intending to prevent them from being organised in teachers' associations, which are mostly very critical of the government educational policy. This, however, was not successful. School leaders have always actively used their chances of making open comments to educational policy in New Zealand via their associations and unions.

Many school leaders try to cope with the new tasks and the new role by adding to their personal working time. This has resulted in overload and stress and has led to a high number of school leaders quitting their jobs (in 1995, 416 from 2 300 primary school leaders quitted their jobs). Increasingly, fewer and fewer teachers are interested in school leadership positions (see Harold, 1998).

The local school boards of trustees, the real decision making bodies, have restricted the school leaders in their new freedom of decision-making since 1989. School leaders see themselves in a conflict-rich situation: on the one hand, belonging to this committee and holding one of the votes, but on the other hand being at the same time the educational leader of this committee, and commonly the only one experienced in school administration. Ironically, they have been employed by this committee and can also be dismissed by it. Thus, they feel crushed between the committee, staff and different lobbies in the community. "You can only feel secure until the next meeting of the board of trustees", says an ironic slogan among school leaders in New Zealand, who feel like "being between two sandwich slices", and have the impression "whatever they do, it is wrong" (see Robertson, 1995).

Among the entrepreneurial activities which a school leader has to develop are: looking for sponsors, looking for grants for extra funding, seeking partnerships with local business companies and advertising for additional privately fee-paying pupils from neighbouring Asian countries. The ethical and moral dilemma that may emerge here lies in being torn between market conformity and competition on one hand, which the business-manager within the school leader has to show, and the ethos of cooperation and support on the other hand, which the pedagogue within the school leader would like to follow.

This competition and market orientation partly results in covering up difficulties within the school, for example, in school discipline, instead of actively tackling them, for fear of an image loss. Impression management is the motto; a facade of being successful is built up in the public and, according to studies, even in the interaction of school leaders among one another (see Wilson, 1996; Robertson, 1997). Recently there has been a lot of effort to help school leaders overcome their isolation. Support and challenge from trusted professional colleagues has been vital in this climate and the partnership programs with academic supervision have gone some way towards achieving this. This professional dialogue, using coaching techniques, is one way that has been established to support principals to develop new ways of working (see, for example, Robertson, 1998).

2. Development of School Leaders

In New Zealand there is no obligatory state qualification offer for school leaders or teachers aspiring to an educational leadership position. That means that all potential applicants have to look for an adequate qualification on their own, in order to be able to convince the particular Board of Trustees, which decides on the employment of school leaders, of their qualification. The applicants can choose from a broad variety of different providers. Among them are universities, so-called In-service Associations, consultancy firms and other private consultants. The structure of courses, the selection of themes, the conditions of access and so on, vary broadly and are dependent on the particular provider. That means that the participants can choose from a broad spectrum of qualification programs or those that are for professional development purposes only. Consequently, the various providers are in competition and are obliged to make efforts to guarantee an adequate quality of their programs (or to advertise them accordingly). Not surprisingly, the quality of training programs varies. Due to the fact that there are no standardised requirements for the development of the qualifications for school leadership, individual participants cannot always assess whether it is a good program or not before they make their commitment. They are faced with the problem of having to make a reasonable choice from a great number of offerings, oriented towards their own needs. For this reason, and because too few individuals aim at a further qualification, an obligatory, standard-oriented qualification for school leaders is demanded more and more.

As an example, the qualification programs of the School of Education of the University of Waikato in association with the Educational Leadership Centre shall be described in the following section.

The Educational Leadership Centre (ELC)

The Educational Leadership Centre is a non-profit institute, linked to the University of Waikato through the School of Education, which offers various nation-wide and international activities for the training and development of educational leadership personnel. For example, it cooperates with international institutions and organizations that train educational leadership personnel, is represented at national and international conferences, and engages in research in the area of educational leadership.

The starting point for the preparatory qualification and the in-service professional development of educational leaders is the day-to-day professional practice of the participants. The institute aims at long-term counselling and guidance as well as cooperation with practitioners. The Leadership Centre's objectives are the development of the ability to reflect, of interpersonal competences, and of basic educational values of the participants, all of which are fundamental conditions for their task as leaders of instruction in schools.

The ELC, through the Departments in the School of Education, offers various academic qualification programs. While the Doctor of Education Course certainly is the most time-consuming program, which focuses on writing a doctoral dissertation, the Master's Program (MEdLeadership) is a very attractive qualification for individuals, aimed at enhancing their professional qualities and improving their chances when applying for a school leadership position. It links themes relevant to school leaders with the attainment of an acknowledged post-graduate university degree. A shortened variation is the Post-Graduate Diploma in Educational Leadership (PGDipEdLeadership), which is a common entry point. Eventually however, many of them decide in the course of the program to continue on to a Master's Degree. The target group of the Graduate Diploma in Educational Leadership consists of teachers in leading positions or those who aspire to such a position. The program aims at inspiring critical reflection on the role of principals and their responsibilities and working conditions. Due to the development of a Bachelor of Teaching degree, the interest in the Graduate Diploma in Educational Leadership has been clearly reduced as most applicants decide on the higher-level post-grad programs. This program has no longer been offered from 2001 onwards.

Table 29: Survey of the programs offered by the Educational Leadership Centre

1. Whole academic programs	2. In-service offers / workshops
- Doctor of Education (EdD) - Master of Educational Leadership (MEdLeadership) - Post Graduate Diploma in Educational Leadership (PGDipEdLeadership) - Graduate Diploma in Educational Leadership (GradDipEL) (*no longer offered*)	- Professional leadership program - Workshops / seminars - Further opportunities: Leaders-Net, newsletters, publications, research, international and national contacts

Master of Educational Leadership

The Master's Degree (MEd Leadership) is designed for educational leaders and individuals already in educational leadership positions such as in Ministry of Education, the Education Review Office, through all types of school up to the Early Childhood Education. Although is not necessarily tailored for school leaders, it gets more and more accepted and even favoured by the members of the Board of Trustees, the committees deciding upon employment, as well as by the participants themselves, because they recognise its worth.

To qualify for entry the applicants must prove the necessary educational knowledge either by a Bachelor degree in Education or by equivalent professional qualification in teaching, as well as experience in the field.

There are eight papers in total, including a research dissertation or thesis. The following four papers are compulsory:

- Resource Management and Issues in Educational Administration
- Educational Leadership: Issues and Perspectives
- Educational Leadership: Organizational Development
- Educational Research Methods or Kaupapa Maori Research (aiming at the original inhabitants of New Zealand).

In order to link theory strongly to practice, the School of Education makes an effort not only to employ educational faculty in the teaching as course leaders, but also establish cooperation with lecturers from the School of Law and School of Management.

Besides the obligatory core papers, there are four more optional papers from which the candidates can select according to their personal needs and in consultation with the program coordinator. Up to two of these fours papers can be taken in other Schools in the University. Some examples for these optional papers are:

- Educational Assessment
- School Leadership and the Community
- Educational Leadership for Social Justice
- Developing Educational Leadership
- Professional Education Seminar
- Education Policy

Teaching methods used are lectures, seminars, workshops, email platforms, as well as international study tours. By also offering the courses in summer schools and by online offerings, the design is flexible and thus can be integrated into the professional life by those participants who go on working as educational leaders throughout the program, which most of them do.

The program can be finished fulltime within two years. Those participants who choose the part-time form mostly cope with one paper per semester (and perhaps Summer School or Winter School) and thus need about three to four years, including finishing their thesis.

For eight papers there are costs of about 500 euro (1 119 New Zealand dollars), which totals about 4 000 euro (8 952 New Zealand dollars). There are no extra costs for the Master's thesis.

Table 30: Outline of the Master of Educational Leadership of the University of Waikato in New Zealand

Training and Development of School Leaders in New Zealand Example: The Master of Educational Leadership of the University of Waikato	
Provider	Educational Leadership Center, School of Education of the University of Waikato
Target Group	Educational leaders and individuals holding leading positions in different areas of the educational sector
Aims	Development of the ability to reflect, interpersonal competence, and basic values as prerequisites for instructional leaders
Contents	Mandatory modules: resource management and issues in educational administration; educational leadership: issues and perspectives; educational leadership: organisational development; educational research methods or Kaupapa Maori Research Additional modules: educational assessment; school leadership and the community; educational leadership for social justice; developing educational leadership; professional education leadership
Methods	Lectures; seminars; workshops; email platforms; international study tours
Pattern	Ca. 48 course days* (24 credit hours across 12 weeks = 288 hours) (timetabling: 8 three-hour seminars, either in the late afternoon or on Saturdays (2 per semester in full-time or 1 per semester in part-time; individual scheduling for part-time students is possible due to the online offer)) plus ca. 1 600 hours of individual study and participation in online platforms and conduction of school projects within 2-4 years
Status	Optional; seen as adequate qualification by the employing committee, the board of trustees of the school
Costs	Ca. 4 000 euro (8 952 New Zealand dollars) for eight units per participant (1 119 New Zealand dollars each paper); financed by the participants themselves
Certification	Master of Educational Leadership

* *If there is no specification by the provider as far as the number of days is concerned, we converted the contact time in hours into the unit course day, taking 6 hours training as one day.*

Doctor of Education

The target group of the Doctor of Education are individuals who may have been away from academic study for some time and now want to take a doctoral degree in modules or stages, perhaps because they are distance learners and are dependent on working on their studies in an open university way. This program is suitable for people holding jobs as education professionals who aspire to careers as academics or researchers as well as for people who want to focus their studies directly on their work situation. Special criteria for admission are the same as the academic prerequisites for the PhD and there are also professional criteria such as holding a professional qualification in teaching and a minimum of three years satisfactory experience as an education professional.

The degree structure is twofold:

1. There is a course phase in which the participants enrol in four courses. Two courses in research methodology are obligatory, i.e. those are advanced educational inquiries and a directed study. Finally the participant selects two more courses from the following areas:

- Advanced studies in educational leadership
- Advanced studies in education policy
- Advanced studies in curriculum: theory, policy and development
- Advanced studies in equality, equity and social justice in education
- Advanced Studies in counselling

2. The second part consists of a 60 000-word thesis.

Participation in the program Doctor of Education is possible fulltime or part-time and takes from three to six years to complete. Many participants decide upon taking the course phase part-time and then adding a fulltime phase for finishing their thesis. In this case, four years are a realistic time span.

The costs for the complete program are about 1 800 euro (4 000 New Zealand dollars) for the thesis and about 500 euro (1 119 New Zealand dollars) for each of the four papers, about 4 000 euro (8 796 New Zealand dollars). They have to be paid by the candidates themselves.

Post-Graduate Diploma in Educational Leadership

The Post-Graduate Diploma in Educational Leadership (PGDipEdLeadership) has the same entry prerequisites as the Master of Educational Leadership. Many participants who feel they could not cope with the workload of the Master's Program as well as their present professional situation, and who do not want to conduct a research program on their own but are interested in the content of the courses offered, first choose this program. Here, a thesis or dissertation is not included. Participants attend the same courses as candidates for the Master's Degree. However, only half of the papers are required (i.e. four papers instead of eight). The Post-Graduate Diploma in Educational Leadership can be finished part-time in less than two years.

The candidates choose four papers, three of which are core papers, namely:

- Resource Management and Issues in Educational Administration
- Educational Leadership: Issues and Perspectives
- Educational Leadership: Organizational Development

The optional paper is chosen in consultation with the program coordinator.

The costs are altogether about 2 000 euro (4 476 New Zealand dollars), i.e. about 500 euro (1 119 New Zealand dollars) per paper.

Graduate Diploma in Educational Leadership (GradDipEL)

The program Graduate Diploma in Educational Leadership (GradDipEL) has four obligatory core papers from which the candidates have to choose three. Additionally they have to choose three optional papers. A choice of the obligatory papers is among the following:

- Foundations of Educational Leadership
- Leadership within Education
- Educational Review and Development
- Professional Practice: Teachers as Professionals

Three more papers can be chosen from various themes, for example: Women and Educational Leadership, Intercultural Communication, a Socio-Historical Analysis of Maori Education and so on. This qualification has been discontinued.

Further In-Service Offers

A free electronic information and discussion platform for educational leaders, teachers, psychologists and educationalists in general, is the Leaders' Net[39]. It facilitates an exchange among interested individuals and provides online material for various themes, as well as a newsletter. Moreover, one- to two-day seminars and workshops with individual topics related to educational leadership are offered through the Educational Leadership Centre throughout the year, both at the University site, and in the local regions.

Very much practice-oriented is the Professional Partnerships Leadership Program. The purpose of this program is in terms of collegial support and challenge. A principal works with another principal as a critical friend and coach, who can observe, give evaluative feedback and coach the leadership practices and achievement of goals of their partner. Principals often experience similar issues and problems and can work together in this way, to assist each other with dialogue and support. The personal needs of development thus can be recognised and focused on in a very specific way. A handbook serves as a guideline for the critical partnership and as a database on personal development needs.

School Support Services

The so-called school support services of the University of Waikato represent yet another possibility for in-service training through workshops or short courses for a broad target group of persons working at schools (teachers, principals, members of the boards of trustees and parents). A great variety of courses on the topic of Management and Leadership enable the applicant to enrol in a suitable course that best suits his or her position and needs. Experienced teachers and lecturers of the School of Education, who can be contacted as external advisors on school development processes, work at various support centres in the greater Waikato area (Gisborne, Hamilton, Tauranga etc.).

---

[39] http://www.waikato.ac.nz/education/elc/leadersnet/leadersnet.html

3. Summary

In the course of decentralisation each school in New Zealand has become an autonomous institution, administrating its budget independently. The basis for this is clearly a market approach. School leaders are torn between their educational tasks and economic pressure, between the local school Board, the staff and different groups of interest within the community. However, there is general consensus that school leaders do not wish to return to the prior administrative regime, and the amount of educational autonomy has led to much innovation in New Zealand schools.

Market orientation also characterises the training and development opportunities for leadership personnel. There are no obligatory qualifications for school leaders or aspiring teachers, nor are there any governmental guidelines, prerequisites, or centrally agreed standards for these qualification programs, apart from the graduate programs offered by Universities which are nationally and internationally recognised. Every potential applicant has to choose from the high amount of diverse development programs offered by the different providers. More and more school leaders are turning to the Universities for their professional development and further qualifications.

One example of a comprehensive, academically-oriented university based program is the one offered through the Educational Leadership Centre, a unit in the School of Education at the University of Waikato. Besides the academic programs, such as the Doctor of Education, the Master of Educational Leadership or the Post Graduate Diploma in Educational Leadership, there is a broad variety of in-service workshops and seminars, professional internet discussion groups, and counselling services.

4. Recent Developments

Meanwhile, a National Induction Program for newly appointed principals, as described in the Statement by Jan Robertson (see below), is now in its first year of operation. Each principal attending has been given a laptop computer and has a combination of seminars, school-based support and web-based support throughout their first year of principalship. This is a voluntary program but the majority of newly appointed principals have chosen to be a part of it. There is no qualification or credentialing attached to this program.

The contracts for the Development Centres have been put on hold at present.

The Ministry of Education is now offering scholarships and sabbaticals for teachers to complete qualifications, particularly in the area of mathematics, science and Maori education.

The National Elections on 27 July, 2002 should return a Labour Government, but there could be many changes to education policy if this is not the case.

We have just completed another major International Institute, with our Canadian partner, the University of Calgary. This Institute launched our Boundary Breaking Model of Leadership Development (Robertson & Webber, in press; Webber & Robertson, 1998), where graduate students in Canada and those enrolled in New Zealand, worked together for the month of July. The learning spaces included a

Graduate Conference, an online web-based Conference; and international listserver discussion group, a two week Institute, a public seminar, video conferencing, and school-based enquiry.

The demand for the Master of Educational Leadership is increasing daily, particularly with the ability to be able to apply for a scholarship for release from school duties and fees.

Also increasingly sought after are the Educational Leadership Centre's yearlong programs of professional development in middle leadership and principal development which are linked to a paper for the Bachelor of Teaching or Master of Educational Leadership qualifications.

5. Statement by Jan Robertson

As I write my part in this comparative study, exciting developments are occurring in principal development in New Zealand. Heeding the many calls for a consistent, quality program of principals' development in the last decade, the current Labour Government has just invested 11.2 million dollars in school principals' development.

One initiative will be a Principals' Net – an online facility for all principals to be able to access and communicate with principals elsewhere nationally and internationally, on key topics of importance. Another initiative will be Principals' Development Centres, where principals will be able to get in-depth feedback on their practice and formulate a professional development plan.

The third major undertaking is the design of a nation-wide induction program for newly appointed principals. The Educational Leadership Centre and School of Education at the University of Waikato, in conjunction with the Principal and Leadership Centre at Massey University, have won the contract to design this program. We are currently developing the program and hope to be the team monitoring its delivery throughout New Zealand in 2002. It is a voluntary program, and principals may elect to do this program during their first year of principalship, not prior to appointment. We believe this will build on their concrete experiences of principalship and provides support and challenge during this most vital year of learning. New Zealand principals, like their counterparts in England have always been regarded as headteachers – that is, they move up through the teaching ranks, through positions of responsibility, until they earn the right and responsibility of being the principal of the school. This is why New Zealand leaders have generally been held in high regard by the teaching profession, here and internationally. They are educational leaders. They are committed to pupils' learning. We have designed this new program on the premise that the principal must continue to be the educational leader, with responsibility for successful pupil achievement in the school. This, we believe, is the primary role of a principal and principals must be assisted to hold onto this role, even in a climate more conducive to competition and managerialism. Through case study work, problem-based learning, professional partnerships' coaching, and knowledge of key tasks and responsibilities, and a link to Master's and doctoral degrees, we hope that we will continue to build the ethos of New Zealand principals who are able to demonstrate stewardship to the vision of developing a learning community, for staff and students alike.

North America

Ontario, Canada: Qualifying School Leaders according to Standards of the Profession

Stephan Huber and Kenneth Leithwood

1. Background

The British North America Act of 1867 placed the responsibility for education in the hands of each of the 10 provinces and two territories; only the education of native Canadians remains a federal responsibility. Consequently, there is no federal governing body for education in Canada. However, the federal government does allocate money to the provinces for tertiary education and also funds educational research. Apart from that, each province manages its own education system through a Department or Ministry of Education. These agencies serve the elected government, essentially advising government and overseeing the implementation and monitoring of government educational policies. Included among these responsibilities is the development of curriculum guidelines, funding of schools and school districts, and ensuring that schools and districts comply with educational legislation. Colleges and universities are also governed by the provinces. Besides these very similar structures, the provinces are quite similar in terms of regulations concerning compulsory schooling, the duration of primary schooling, the entrance age of children to school, the duration of the school years, and the structure of tertiary education. While such similarities are remarkable among the English-speaking provinces, the French-speaking Quebec education system is more dissimilar, especially with respect to the structure and funding of secondary and tertiary education.

Within each province, major responsibilities for the operation of primary and secondary schools are assumed by local administrative units called school boards or districts. These local and regional units, governed by an elected board of trustees, operate more or less independently within the framework of regulations and laws determined by departments of education. Within such frameworks, school districts are responsible for the construction and maintenance of school buildings, the refinement of curriculum, and the allocation of grant awards from the government. The current Ontario government, however, places substantial limits on the flexibility of both districts and schools to deviate from the provincial curriculum. One source of these limits is a system of province-wide pupil achievement testing in selected subjects and grades in both elementary and secondary schools. Developed and administered by an agency arms length from the government, these tests are administered annually, with results reported back to districts and individual schools and usually released to the media, as well. The tests are aligned to curriculum standards specified by the Ontario Ministry of Education, standards which also serve as the structure for a provincial report card which every school must use to report pupil progress to parents.

The largest Canadian province, Ontario comprises 78 school districts. While on average, these districts include about 32 000 pupils, the range in size is enormous, the largest district (Toronto) serving about 300 000 pupils. Another Ontario school district, the Durham Board of Education, has become well known in Germany as a result of being awarded the Carl-Bertelsmann-Prize 1996 for being 'The most innovative school system in an international comparison'. Located on the northern shores of Lake Ontario not far from Toronto, it includes 96 elementary schools (from five to twelve years of age, first to eighth grades) and 22 secondary schools (integrated comprehensive high schools – from thirteen to sixteen years of age). It employs more than 5 600 people (3 700 of whom are teachers) serving approximately 62 000 pupils.

In Canada as a whole, as well as in Ontario, schools are increasingly required to be publicly accountable and are the focus of considerable public and political attention. Provincial governments pursuing accountability agendas argue from a neo liberal ideology that stresses the contribution of schools to the economic welfare of the country. This has a substantial impact on the goals of education, stressing vocationalism, problem-solving skills, team-work, communication skills and self-management.

The recentralization of the Ontario school system in the late 1990s, was prompted by a number of specific reforms introduced by law through the government of Ontario that have been widely criticised. One of them was the rather sudden amalgamation of school boards reducing them from more than 150 to about half that number. In many cases, as in Toronto, amalgamation led to significant organizational and cultural challenges. Furthermore, the workload of teachers was increased by about 10% without increasing their salary at the same time. Teachers had been without an increase in salary since 1991. As a result, teachers went on strike for several weeks and then worked to rule for more than a year. Extra-curricular activities were dramatically reduced at most schools during the work to rule period. Other government initiatives during this period included, for example, initial teacher testing, requirements for teacher recertification, new curriculum, new provincial report cards, and an altered funding formula. It does not seem likely that the government of Ontario will turn away from the direction of these policies any time soon.

While the Ontario education system has become highly centralised in the past six years, districts are still responsible for hiring staff, negotiating teacher contracts, supervising all employees, and monitoring their own effectiveness. Principals typically have considerable involvement in such matters, perhaps especially now that they are no longer part of the teachers' bargaining units but are considered strictly management. Indeed, they have experienced a significant increase in their workloads and responsibilities and many express feelings of isolation. This affects the attractiveness of principal positions and the interest in such functions has decreased considerably. Principal posts are now difficult to fill as increasing numbers of them become vacant because of retirements, for example.

School principals are expected to respond to many masters: their district administrators, the provincial ministry, pupils, parents and business interests. The recent requirement that every school establish and work with a parent-dominated advisory council illustrates the need for principals increasingly to take account of parent perspectives in their work and to adopt participative management practices.

The altered conditions outlined above affect the tasks and roles of school leaders in the Ontario schools in two ways. Firstly, changed expectations of society for schools have made school management a much more complex task and requires new competences. Secondly, alterations in the organization like decentralization and accountability have brought about a shift the understanding of the principal's role, which those already in the role have to learn to deal with.

2. Development of School Leaders

In 1997 the Ontario College of Teachers (OCT) was created as a regulatory body of the profession equipped with full decision-making powers by the Ministry of Education. All 178 000 teachers of Ontario are required to be members of the OCT and are entitled to elect 17 of the 31 members of its Council. The remaining 14 council members are appointed directly by the government. The 'silo-principle' underlying the election makes sure that everybody is represented appropriately. Every elector votes for a candidate from his position (a secondary school teacher would vote for a secondary school candidate, a principal for the principal's candidate, a teacher from a private school for private schools). The OCT represents the profession also to the public and is responsible for matters that were in the hands of the Ministry previously. It deals with disciplinary issues (such as complaints, revocation of the teaching license), the licensing of teachers from other provinces and abroad and all other professional issues. It has full controlling, regulatory and licensing authority for school personnel as a whole. Furthermore, the Ontario College of Teachers Act states that the OCT is also responsible for the continuous professional development of teachers and the formulation and enforcement of appropriate standards. These standards are valid for the qualification programs for teachers and for the qualification of school principals as well (see Regulation 184).

The guidelines for principals' training, the Ontario Principal's Qualification Program (PQP), have been developed by the College and it is now up to the respective providers to develop a qualification program for interested candidates. In order be accredited by the Accreditation Committee of the Ontario College of Teachers, a training program has to prove that it meets those guidelines. Until now only about ten educational departments of universities have been accredited to offer a PQP, among the largest of them the Centre for Leadership Development of the Ontario Institute for Studies in Education, University of Toronto (OISE/UT) and the Faculty of Education of York University. However, other providers are now allowed to offer the PQP. The aim of establishing common guidelines for providers is to ensure a basic level of consistency across offerings insofar as essential knowledge and skill for school administrators are concerned.

The guidelines for the PQP aim to prepare school leaders to:

- uphold the Standards of Practice in the Teaching Profession and the Ethical Standards of Practice in the Teaching Profession,
- build and sustain learning communities that support diversity and promote excellence, accountability, anti-racism, equity, partnerships and innovation,
- assume accountability for the achievement of all pupils and promote student success and life-long learning in partnership with staff, parents and the community,
- align and monitor programs, structures, processes, resources and staff to support student achievement,
- manage and direct the human, material, capital and technological resources for efficient and effective schools,
- initiate and facilitate change, and operate successfully in a dynamic environment that is characterised by increasing complexity,
- understand and apply education and student related legislation in Ontario and district school board policies that have an impact on the school, pupils, staff and community,
- liaise with educational stakeholders concerning all aspects of provincial and district school board issues and initiatives.

The PQP consists of two main parts each calling for about 120 hours of contact time. Between these two parts is a Practicum often carried out in the candidates' schools with supervision and evaluation shared by PQP staff and a local administrator.

Part One addresses the following themes:

- What it means to be a principal
- Human Resources
- Interpersonal Skills
- Decision making
- Students with Exceptionalities
- Legal Issues
- School Operations
- Communication Strategies
- The Practicum Experience

Part Two, extending several of these themes and adding new ones, focuses on:

- Leadership
- Human Resources
- Community
- Decision Making
- School Program
- Assessment, Evaluation and Reporting
- School Planning
- Resources Management
- Communication Strategies

The Practicum: The guidelines describe in detail the design of the 60-hour practical experience, which has to be completed between Part One and Part Two. Candidates, for example, take a leadership role for some project in their school, supervised by their principal to meet the practicum requirement. The practicum is intended to enable candidates to put the knowledge from Part One to the test in a real school setting and to conduct a specific leadership project independently. They often work as a member of an administrative team and apply appropriate legislation, district policies and related research and theory. The guidelines also demand the extensive use of modern communication technology and stress the importance of getting to know all aspects of school leadership through readings, reflection, discussion and practical experience (see 'attachment').

The following descriptions of the PQPs offered by OISE/UT and York University do not differ in terms of content – this is clearly defined by the College's guidelines. However, they vary in the applied teaching methods and organization. Those elements are left to the individual provider who can decide on the mode of delivery and modify the program in order to meet the demand (seminars on weekends to encourage and promote women, online-tutoring for distance learners, etc.). Nevertheless, it is still possible for candidates to change the providers after completion of Part One.

Example: Principal's Qualification Program of the Ontario Institute for Studies in Education/University of Toronto

The Ontario Institute for Studies in Education (OISE) is the education faculty of the University of Toronto. Several research centres are a part of it such as the Center for Leadership Development and the Center for the Studies of Values and Leadership. The Center for Leadership Development (CLD) was among the first to develop and implement a qualification program for school principals in 1985, with approval of the provincial Ministry of Education. Like each PQP program, the CLC program consists of two parts. Part One is called the 'The Principal as Leader' and Part Two 'The Principal and Change'.

Application requirements, set by the Ministry for the candidates, are a minimum of five years' teaching experience in primary or secondary school level, certification in three divisional areas (one of which must be intermediate), an undergraduate degree and either a Master's degree or the completion of two appropriate development programs and the competed course work equivalent to half of a Master's degree.

Part One, 'The principal as Leader', consists of four modules:

- Module 1: Social Context
- Module 2: Staff development and Teacher Supervision
- Module 3: Management
- Module 4: Leadership

The four modules of Part Two 'The Principal and Change' deal with:

- Module 1: The School and its community
- Module 2: Initiation of change
- Module 3 Implementation of change
- Module 4: Institutionalization of change

Particularly interesting about the OISE/UT program is how they group their participants. About 100 candidates are enrolled in each part at one time and are divided into Home Groups each of which consists of 15 to 20 persons and one group leader. These 15 to 20 people in the Home Group form Modular Leadership Teams (MLT), each consisting of five people in order to work and communicate more effectively with one another. Groups are organised to ensure a balance between male and female, educationalists from the primary and secondary sector, from private, catholic and public schools and rural as well as urban areas.

The idea of forming this kind of work group reflects the situation in school, where various teams will work on school problems and discuss their proposals with their colleagues. At the same time this method supports content issues like problem-solving and the promotion of team-building. This process-oriented element is accompanied by a relatively substantial amount of reading in OISE/UT program. All assigned readings are compiled for each candidate in an individual folder. The deep insight in the relevant research literature in school leadership is meant to help the aspiring principal to explore their new area of responsibility and equip them with the necessary knowledge in order to fall back on relevant research findings in case of problems, conflicts or uncertainties of decision.

The first part of the program requires the candidate to take part in six weekend seminars spread over half a year – this means one seminar per month. Within three days after this weekend, an online-tutorial will be held: the candidates communicate via email with their group leaders and send them a reflective essay. This piece of reflective writing touches on new issues having been discussed during the session and work on specific questions like what kind of impact the newly learned content will have on their role and practice or how they would solve a certain conflict. The group leaders then comment on the text and deal with individual questions, disclose individual needs and develop new aims for their professional development. This kind of electronic tutorial may be supplemented by individual meetings of the participants with their tutors.

Within their MLT, participants also work together on an Interactive Electronic Communication Project in which they exchange information relevant to their school concerning budgeting, timetabling, etc. Through this, participants get a chance to know the situation in other schools and to prepare a presentation for their Home Group on this basis. Furthermore, the MLTs prepare a two-day module for one weekend session. Basically this means that every weekend one MLT in every Home Group is responsible for the preparation of a module. OISE/UT invites guest speakers who frame the program with speeches and workshops. The content of those talks will then be discussed in Home Groups. Thus, contents like 'team building' or 'problem-solving' are put into practice methodically.

After the successful completion of this first part of the PQP, the participants complete a school practicum, usually in their own school. They spend altogether more than ten weeks at their school observing a practising and experienced principal for about 40 hours and carry out a leadership project independently. Recently, the number of hours expected for the practicum has been increased to 60 and it has been suggested that candidates complete their practicums at a school other than their own. These changes have created several problems not yet resolved including the need for leaves of absence and missing their teaching assignments.

Table 31: Outline of the Principal's Qualification Program of the Ontario Institute for Studies in Education/University of Toronto in Ontario, Canada

Training and Development of School Leaders in Ontario, Canada Example: The Principal's Qualification Program of OISE/UT	
Provider	Center for Leadership Development of the Ontario Institute of Studies in Education, University of Toronto (OISE/UT)
Target Group	Teachers aspiring to a school leadership position, i.e. before application
Aims	Imparting to the participants the knowledge, skills and practices to enable them to uphold the standards of practice in the teaching profession; to build and sustain learning communities that support diversity and promote excellence, accountability, anti-racism, equity, partnerships and innovation; to assume accountability for the achievement of all students and promote student success and life-long learning in partnership with staff, parents and the community; to align and monitor programs, structures, processes, resources and staff to support student achievement; to initiate and facilitate change, and operate successfully in a dynamic environment that is characterised by increasing complexity…
Contents	Social context; staff development and teacher supervision; management; leadership; the school and its community; initiation of change; implementation of change; institutionalisation of change
Methods	Weekend seminars; reflective writing; interactive electronic communication projects; literature studies
Pattern	24 course days (timetabling: Part 1: 6 weekends (Saturday 8.30 am to 4.30 pm and Sunday 8.30 am to 5.00 pm); Part 2: 6 weekends (see Part 1)) plus 1 ten-week school attachment (altogether 60 hours) and literature studies within 1 year
Status	Mandatory; prerequisite for being employed as a school leader (the provider can be chosen by the candidates)
Costs	Ca. 1 100 euro (1 440 Canadian dollars) per participant (695 Canadian dollars for each part plus application fee of 50 Canadian dollars); financed by the participants themselves, sometimes by local school authorities
Certification	Certificate without grading; after successful completion there is a recommendation (recommended/not recommended)

Part Two of the PQP is delivered in a substantially different manner than Part One. While there are still Home Groups, lectures, workshops and readings, this part of the program is heavily problem based. One detailed case problem is presented to the candidates working in teams and their job is to develop a school improvement plan by the end of Part Two.

The workload for each of the program parts is about 125 hours, which splits into 100 contact hours (this includes participation in module units throughout six weekends every half-year, followed by reflection upon the new material with the group leader via email, as well as participation in an interactive electronic course), and 25 hours for individual reading and writing of essays, presentations, meetings with group leaders and with the Modular Leadership Teams. Holding the contact times on weekends and completing assignments online is meant to promote opportunities for those who would find it difficult to do such work in the evenings such as women, who usually find it more difficult to take up the course, as this might conflict with other duties such as child-rearing.

The fees for the program are 530 euro (695 Canadian dollars and additionally there is a registration fee of 40 euro (50 Canadian dollars). Until now, participants have had to pay for their course themselves, but recently there have occasionally been single school boards who paid half of the course fees, especially in the Northern provinces, where there is a considerable shortage of qualified principals.

Example: Principal's Qualification Program of York University

The guidelines of the Ontario College of Teachers for Part One and Part Two of the Ontario Principals' Qualification Program also build the framework for the aims of the program conducted by the Faculty of Education of York University. It has been developed in a joint effort by practising school principals and professors of York University. The leading ideas of the program – equity, reflective practice and critical action research – were formulated. The teaching staff and online mentors or program instructors are experienced educational leaders from school boards, who see themselves as a team. The central approach or the philosophy of the program focuses the work life practice of school leadership as a cooperation- and relation-oriented activity, which occurs amidst social communities and has to follow the principles of justice and equality. Emphasis is put on the role of the interactive, action-based dynamic approach of sessions, online-mentoring and the school attachment.

The admission requirements are the same as for the PQP at OISE/UT.

The PQP comprises weekly four-hour sessions from March to June, one one-week full-time session in July, and a minimum of 14 hours of online-mentoring. Both parts are quite similar in terms of teaching methods. There are evening session and online-mentoring, reflecting pairs, portfolios and a school attachment. In each part, about 100 candidates are admitted at the same time, which means that there are 200 people enrolled altogether. Twenty participants form a group and share one instructor. These instructors are responsible for the weekly four-hour sessions, which take place in one of the four venues in the Ontario area. According to the standards of the OCT, similar topics as at OISE/UT are worked on. The meetings are action-oriented and interactive, current issues are discussed, and the participants get an opportunity for reflective writing and group work on topics, which will then be presented to the others. For special topics, guest speakers are invited.

Another very important element of this PQP is the Online-Mentoring. Participants meet for an Online-conference once a week for about one hour emphasizing current issues relevant to the practice of the participants. For that 16 Mentoring Groups are formed with 15 persons each, who are connected online with one another. Members of those groups are supposed to come from all four regions, and it is taken into consideration that the group is properly mixed including male and female participants, those from more rural and urban areas, the private school sector and public schools, primary and secondary school, etc. Two experienced school principals, one from a primary and one from a secondary school, work online with this group of fifteen people and present to them a problem every 7-10 days which is relevant to their daily life in school. The participants then think about this problem, do some research and discuss it with their own principal. Then all of them give their answers online – what they would have done as a principal in that case. The other candidates can also put in their proposals and suggestions, read the ones of the others and comment on them. At the end of the 10 days the principal who presented the problem reports how he had handled the situation and whether he had been successful with it.

Table 32: Outline of the Principal's Qualification Program of York University in Ontario, Canada

Training and Development of School Leaders in Ontario, Canada Example: The Principal's Qualification Program of York University	
Provider	Faculty of Education of York University
Target Group	Teachers aspiring to a school leadership position, i.e. before application
Aims	The development of competent, impartial and just leaders who are developing their schools in a goal-oriented way and have a cooperative style of leading; the cooperation with the political community and the social context are to be taken into account
Contents	Part 1: what it means to be a principal; human resources; interpersonal skills; decision making; students with exceptionalities; legal issues; school operations; communication strategies; the practicum experience Part 2: leadership; human resources; community; decision making; school program; assessment, evaluation and reporting; school planning; resources management; communication strategies (see guidelines of the OCT)
Methods	Weekly evening sessions; online-mentoring; reflective pairs; portfolios; school attachment; individual counselling; literature studies
Pattern	23 course days* (timetabling: 28 four-hour group meetings and 1 one-week session) plus 1 ten-week school attachment (altogether 60 hours); ca. 28 hours online-mentoring and literature studies within 1 year
Status	Mandatory; prerequisite for being employed as a school leader (the provider can be chosen by the candidates)
Costs	Ca. 1 280 euro (1 700 Canadian dollars, 1 600 for early birds) per participant (800 Canadian dollars each part plus application fee 50 Canadian dollars); financed by the participants themselves, sometimes by local school authorities
Certification	Certificate without grading; after successful completion there is a recommendation (recommended/not recommended)

** If there is no specification by the provider as far as the number of days is concerned, we converted the contact time in hours into the unit course day, taking 6 hours training as one day.*

As stated by the standards of the OCT uniformly, this PQP also has a school attachment element of 60 hours. The participants design a proposal of what they want to be doing in their attachment, while doing Part One. Until they complete Part Two, they should have put this proposal into action. Usually the participants do their attachment at the school where they are teaching, but sometimes they choose another school. While their principal acts as their mentor, they get the opportunity to do action research and present their findings respectively. Additional school visits to meet with innovative leadership and teaching behaviour and action are supposed to provide the participants with a wide insight into different schools.

Two participants form a reflective pair and write a short reflection each week on a certain aspect. This again is done online so that the instructors can join in and sometimes get very useful insights.

In the course of the program, participants always have the chance to arrange individually scheduled meetings with the teaching staff.

An even more online-based variation of this PQP will be provided by the York University in August 2001 and is already advertised as Partial On-line Principal's Qualification Program. It is of special advantage to distance learners. All participants meet for an initial six-day session at York University and will be accommodated there. Most of the following elements can be done online.

All participants prepare an individual portfolio during the course in which they collect their reflective writings, their analyses of their professional aims and all their other assignments. It forms the basis for the final assessment, which only discriminates between 'recommended' and 'not recommended'.

The fees for each part of the PQP are 600 euro (800 Canadian dollars), there is an early bird reduction of 40 euro (50 Canadian dollars).Here, too, it is reported that partly the school boards share the costs with the participants.

3. Summary

In the school system of Ontario, school leaders have to develop a cooperative partnership with local and regional groups, with the community, municipalities and with industry, above all with the school community council. School leaders are facing high level expectancies, which also bring about a different role concept.

In Ontario there is a self-regulatory organisation of the teaching profession, the Ontario College of Teachers, within whose responsibilities are the accreditation of degree programs, the development of professional standards and the accreditation of in-service training following these standards. Moreover Ontario has established guidelines for the qualification of school leadership staff, which the respective provider is obliged to follow. The Principal's Qualification Program, which is accredited on this basis, is now offered by several universities, the new Ontario Principals' Council, as well as other agencies.

Hence the Principal's Qualification Program, for example, offered by the Ontario Institute of Studies in Education at the University of Toronto (OISE/UT) and the one offered by York University do not differ based on general objectives and content, pedagogically. The program of OISE/UT focuses on groups from different types of schools, which are subdivided into small teams. Great emphasis is put on process-oriented components. The small teams have a vivid exchange of the respective situations in their schools and current topics taken from their workplaces. This rather academically-oriented program nevertheless tries to guarantee a strong link with practice. Therefore a school leadership project should be taken over by the participants during the internship.

The Principal's Qualification Program at York University also includes interactive group meetings and applies action-oriented and problem-based methods. A particularly important method here is Online-Mentoring: Participants are presented with authentic problems by experienced school leaders, which then are commented on and discussed. Additionally, participants form Reflective Pairs, counsel each other online, with the support of their trainers. An even more online-based variation of this PQP will be provided by the York University in August 2001, which is of special advantage to distance learners.

4. Recent Developments

The primary and recent changes for Principals, especially in Ontario, are threefold. Firstly, greater demands for accountability have pressed principals to acquire new skills in the management and use of data as well as a new willingness to collaborate with the community and business groups. Secondly, these increased demands have come at a time of unprecedented rates of retirement – an aging administrator group and a hostile political context are the cause making it very difficult for districts to find adequately qualified replacements. Finally, the market for principal training has been opened up so that many groups other than universities are now eligible to get into the business. It is not clear yet what effect this will have on the quality of training but it is a promising response to the need for training larger numbers of potential school leaders.

5. Statement by Kenneth Leithwood

The direct preparation of school leaders in Ontario can be viewed as building on what is learned through a Masters degree program in Educational Administration or a closely related field such as Curriculum and Instruction. In this respect Ontario's model is substantially different from the dominant model in the United States where the Masters degree itself is the preparation for becoming certified as a school administrator. The Ontario model, as a consequence, does not run into some of the persistent dilemmas that have yet to be resolved in the US context. For example, university faculty in the area of Educational Leadership and Administration continue to pursue the development of theory and research in the field as a primary responsibility. Their graduate instruction is not intended as direct preparation for leadership practice. Rather it provides a broad foundation for such practice and alerts prospective school leaders to the types of contributions that theory and research might make to practice. It also has, as a goal, the development of critical consumers of educational research.

The direct preparation of leaders is largely the responsibility of skilful existing school leaders. Acting as instructors, these people offer their own models of how to use research in practice and help prospective leaders identify other forms of knowledge that will be valuable to them in the day-to-day work they do, should they become school leaders. There is little unproductive tension in this model between theory and practice yet the relationship between the university and the direct preparation programs remains close.

Of course this model is by no means a perfect one. One might easily argue that the amount of time available for direct preparation is quite limited and does not begin to address the full range of skills and competences that school leaders will need as they assume the role. But it also assumes that this is not the end of their training but rather the beginning. With this in mind Ontario's approach to leadership preparation seems promising in comparison with at least some other approaches.

Washington, New Jersey, California, USA: Extensive Qualification Programs and a Long History of School Leader Preparation

Stephan Huber

1. Background

In the United States the responsibility for education lies with the individual states, which are independent as far as educational policies are concerned. The American constitution does not recognise any federal authority in law making in the sphere of education[40]. There has indeed been a national cross-state Department of Education since 1982, but this federal office does not have the direct power to make mandatory directives. It can, however, exert influence on state and local educational decisions by its allocation of grants (Grants-In-Aid). Apart from this, the National Government can only achieve any level of uniformity in its education system by cross-state programs approved by Congress. Within any individual Federal State it is the respective State Department of Education that is the highest administrative authority.

The main tasks of this State Department of Education are seen above all as (see INES-data base/DIPF, 2000c):

- 'Leadership', understood as direction and guidance towards the setting up and improvement of both state and local educational programs through planning, research, advice, coordination and public relations,
- 'Regulation', in the sense of the laying down of standards, that is to say a minimum acceptable level applicable to the whole state,
- 'Operations', i.e. the administration and control function for schools, higher education, and educational projects.

The school board, or Board of Directors for each school district has a very powerful position in the education system. They are responsible to uphold the laws regarding education requirements and also to administer special bonds, levies, and other local fund raising activities to support schools. School board members are elected officials and the Superintendent of Schools serves at the pleasure of the Board. Important decisions about school reform, curriculum, instruction, finance, administration, and personnel are ultimately approved by the Board of Directors. Each state has authority to set its own rules governing the election, expectation, and function of the school board.

In the past 20 years in the United States, as in most industrial nations, education and particularly the school education of younger generations has been seen as the

---

[40] The Federal Government only has authority for setting up guidelines for educational institutions within the federal administration and the Armed Forces.

key to economic and social development and to the competitiveness of the nation in the market place. It was therefore a considerable shock when comparisons were made between the school standards of American school children and those of other countries and clear deficiencies became obvious. Besides the low academic level and failures directly experienced at the schools (such as a decrease in achievement levels, disciplinary problems and drug abuse), fundamental problems became evident, like the poor financial equipment of the educational sector and the lack of vocational qualification chances for High School graduates. There was loud criticism of schools across the country, and initiatives to improve and reform education began. The re-structuring of the school system in many American states followed in three 'pushes' or 'waves' (see Murphy, 1991; Elmore, 1991). As part of the first reform wave there were the top down measures introduced following the publication of 'A Nation at Risk' by the National Commission on Excellence in Education in 1983. These were administrative orders designed to increase the level of responsibility of teachers for the standards achieved by their pupils. These measures were aimed at the curriculum, the means of measuring achievement, and also teacher training. The initiative for these measures came from central government, but the responsibility for their implementation lay with the regions and the degree to which they were implemented was very variable. Critics in general noted that these reforms had very little influence on teaching and learning processes. What actually went on in the classroom, that is to say teaching and learning, became the centre-point of consideration in the second reform wave in the second half of the 1980s and particularly the concept of 'teaching for understanding'. In this the responsibility shifted more prominently to the local level and that of the individual school, whose teachers were more directly involved. Subjects such as 'professionalisation', 'collegiality and cooperation' and 'group decision-making' were of topical interest. Then in the third 'reform push' – the end of the 1980s into the 1990s – the experiments that had already been started were further strengthened and linked together.

The term 'restructuring' is a blanket title for the various main elements of reform. It is not easy to define it clearly, however. Across the patchwork of various attempts at reform in the United States Elmore (1991) recognises three general threads which he lists as characteristics of 'restructuring':

- efforts towards improvement which are aimed directly at the teaching and learning process,
- rather more structural reforms, for example relating to the training and recruitment of teachers and other members of a school's administration, to the school structure, to conditions of work in schools, and to the process of decision making,
- reforms in the allocation of responsibility and the authority to make decisions as far as this relates to educational authorities, to individual schools and to their clients.

Thus, 'restructuring' in the United States means for schools more than just a more independent status in the areas of finance and administration. It means as well that they must take responsibility for their own development and success. For the teaching staff, for school leaders, for senior management teams and for school administrators, this of course requires not only a change in how they see themselves, but also in their relationships with one another.

For school leaders, the administrators or principals, new roles and new areas of responsibility evolve out of this. While school leaders in the United States up until the 1960s were predominantly seen as administrative managers, in the 1960s and 1970s they were expected to become program managers; that is to say they were expected to organise and carry through the large number of support programs beginning at that time, which were financed by the state and aimed at the various minority and problem groups in school populations. Later in the 1980s, and linked to the results of school effectiveness research, instructional leadership[41] was in the forefront, that is to say leadership in the school, in the areas of teaching methodology and pedagogy. However, the large number of administrative tasks prevented many school leaders from doing justice to this aspect of leadership. Even greater were the demands made on school leaders in the 1990s in the course of increasing decentralization and school autonomy, for they are supposed to be the driving force in the dynamic process of school development, pursuing this goal by devising joint objectives with their teachers, by improving internal school communication, and by employing efficient, cooperative decision making and problem solving strategies. That is to say, they had to show a high level of ability in the areas of motivation and integration in order to liberate the potential of their workforce; that will integrate them and help them to bring about results (transformational leadership).

Above all, three important reforms practised in various states of the USA have had serious consequences for the role, function and range of tasks expected of school leaders: local self-government of schools (School Site Management), open choice of schools by parents (Open Enrolment) and school accountability (see Daresh 1998).

School site management

Since the 1960s and 1970s the decentralization of decision processes (that is, the shift of responsibility onto the level of the individual school) has been seen in some American states as a promising method of involving those local people connected in some way with the school in school work and decisions (for example, in Dade Country in Florida; in Chicago, Illinois; in South Carolina, Colorado and in Texas). Local self-administration has since then become the rule right across the USA, although the degree of autonomy varies greatly from state to state, district to district and sometimes from one school to its neighbouring school. It is clear that the steady transfer of a range of tasks which used to be tackled centrally now has become the responsibility of the individual school, this means a (substantial) change for these schools and particularly for the school leader. In the old centralised structures the school leader was indeed the leader of his or her school, but also had a role as middle manager in the hierarchical organisation of the district. His main job was to make sure that the instructions of the central authorities were carried out in his school, but now in local self-management he has acquired a new range of tasks

[41] In the US context, the notion of "instructional leadership" emphasizes the leading role in the pedagogical-didactical sense, while the term "school administration" captures the activities of deployment of resources.

which, for example, were once the responsibility of the Superintendent of the local education authority.

In many (but not all) schools one very specific new task is the school's own management of its complete budget, which means that the school leader must be competent in financial policy and business management. In addition, she or he increasingly has to work with local bodies, including business representatives and local politicians, who are usually all laymen in the area of education and who certainly represent many non-educational interests. This is particularly evident in what happens with the newly established School Council.

While the composition of a School Council varies across schools, districts, and states, consider the following example. Imagine that the Council consists of two local politicians, two representatives of the teaching staff, the school leader him or herself, a representative from the pupil body and six parents' representatives. This School Council decides on the appointment and the job description of the school leader and also indirectly influences further staff selection. The school leader is in fact appointed by the district, but this only following the advice of the School Council. So as to be able to get rid of school leaders more easily, their contract cannot be extended but must be renewed ever three years. The substantial power of this committee with its weighting in favour of the parents has often led to criticism, as it encourages a situation of dependency, and because teachers, who are the real experts in schools and teaching, are in their opinion under-represented, with only two out of twelve votes. The School Council meetings are no longer the forum for the traditional discussions between parents' representatives and those responsible for school expenditure on typical, largely administrative matters such as pupil safety on the way to school, equipment in classrooms or the layout of the school playground. It is far more often now a question of matters which only used to be discussed among teaching colleagues or with the education authority, for example, about curriculum content, the use of teaching resources, the evaluation of successful teaching, personnel questions, the nature and financing of in-service training for teachers, school discipline etc.

As a consequence new skills are required: financial knowledge and ability, political skills, skills in working with people, negotiation, the management of conflict, skills in adult education, in being able to present the school to the outside world, the ability to fit in with the aspirations of the locality but also to resist individual factions wanting undue influence, and much more. At the same time the school leader of course retains his or her traditional role and must remain true to his or her personal educational vision. He or she remains the pedagogic leader of the school and is as such responsible for the learning success and personal development of the pupils and for the quality of teaching, for the well being of the staff, for the school ethos, for the achievement of higher educational goals, for the integration and support of minority groups etc.

Open enrolment

The free choice of schools by parents and pupils exists in various forms (see Kowalski & Reitzug, 1993). It ranges from a freedom of choice within one school district, to the freedom to choose across a whole state, and in quite a few schools also includes the option for the pupil to choose to change his class and teacher. The basis for this open enrolment is that schools are inherently different from one another, that what they offer is different, for example, in terms of the curriculum, teaching methods, the general educational direction of the school and the school profile. Schools compete with one another and are therefore obliged to respond to the needs of their 'customers', that is the parents, and their 'clients', the pupils, in order to survive in the 'market place'. Critics of this market philosophy identify a danger in that not all parents are capable of making an informed choice of an appropriate school based on the talents of their child, but instead quote superficial arguments such as the performance of the school basketball team or the absence for them of 'undesirable' ethnic groups (see Glenn, 1989).

Apart from the school leader's internal conflict between her or his educational conscience and the need to please her or his 'customer parents', open enrolment means for her or him showing marketing skills that have traditionally been reserved for managers of commercial firms. That is certainly a challenge for the training of school leaders, both before and after they take up their position, for these are skills that must be learnt. Beyond this, American school leaders have to become familiar with the culture and norms of non-school organizations and have to break out of the ivory tower of their educational careers.

Accountability

A third element of reform in recent years in many American states has been the accountability of schools to the public, usually by the assessment of pupil achievement in standardised tests carried out in the whole district or across a Federal State. The champions of these tests see them as clear proof of which schools are preparing their pupils adequately and bringing them to a level at which they are ready for the competition with other countries (see O'Neill, 1992; Palmer, LeMahieu and Eresh, 1992; Steen 1991). Critics, on the other hand, doubt the practical and ethical value of standardised paper and pencil tests if it is a question of doing justice to the complexity of learning for individual pupils and the standards of complete schools as individual organizations (see Eisner, 1991).

School evaluations which are predominantly, or even exclusively, arrived at on the basis of such test results, seem just as thin as judging individual teachers on the same sort of basis (as is done, for example, in Texas, where a part of the state evaluation process of teachers consists of information as to how their pupils have achieved in the annual Texas Achievement of Academic Skills test. However sensible or questionable that might be, the regular use of centrally set state tests in various year groups means a further new challenge for the school leader. He or she needs to become knowledgeable about test procedures, the interpretation of test results and the evaluation of their validity and reliability, in order to use the test results profitably for his or her particular school and to present them appropriately and professionally to the general public, so that she or he can counter the inevitable tendency to simplification and unreliable exaggeration and mis-interpretation.

2. Development of School Leaders in the USA

Besides possibly Great Britain in the European context, the USA certainly holds the longest tradition and very extensive experiences in the area of school management and school leader development. It is highly instructive to dwell upon these experiences and their sequence of different approaches and increasingly greater differentiation in more detail than is done in the other country reports in this volume.

There is very little evidence from the decades between about 1820 and 1900 on what was said about school administration, school supervision and school leadership, and there were no formal induction or training programs for these positions. Murphy (1992) calls this time the 'era of ideology' (1992, p. 21). The first administrative office in the field of education, the school superintendency, began to make its mark in the last decades of the 19th Century, when it appointed personnel for school administration and supervision (see Gregg, 1969). These Superintendents learnt their profession on the job by trial and error. The minimum formal education expected at that time for teachers was also seen as adequate preparation for leadership positions in school supervision and administration. The first course in school administration was provided by Payne, himself a Superintendent in Michigan, in 1879 at the University of Michigan, but there were no other such courses until the early 20th Century. In 1875 Payne also wrote the first book about school leadership, 'Chapters on School Supervision'. As the title indicates, school leadership was seen primarily as supervision (of teachers and their lessons). The head of the school was seen as a teacher of teachers. This was imparted by anecdotes of experiences of exemplary school leaders who were depicted as great men. An attempt to define a theoretical framework was made by interpreting greatness as a personality trait. Another point of reference was the notion of school leadership as applied philosophy As an academic and moral authority, the principal possessed wisdom and truth in all areas of education, and he applied this wisdom in his actions, which granted him a quasi-clerical function (see Button, 1966).

Murphy (1992) calls the time from 1900 – 1945 the 'prescriptive era' (1992, p. 23). In this period the importance of school leader training gradually increased. By the end of the Second World War the number of institutions, mostly universities, which offered courses in school leader training had grown to about 125 and the providers of these programs had to compete with one another. Many states now formally required the successful completion of a training program for the higher educational jobs in school leadership, school supervision or school administration, and the graduates of these training programs gained an appropriate certificate. The trainers were often practitioners from school administration or school inspection. The participants, mostly male and white, took part in the programs in addition to their full time job at school. Regarding the content, the courses focused on the administrational and organisational aspects of school management, included knowledge bases from the area of finance and economy and were also based on the ideas of scientific management. The school leader viewed his or her role primarily as administrator of the regulations of the school authorities.

Between 1930 and 1950, this rather management-oriented perspective was heavily criticised. As a consequence, sociological components were integrated into this training and there were the beginnings of a concept of 'human relations'. Ideas such as job satisfaction and the motivation of the workforce were included in the agenda of training programs. It was expected that someone in a leadership position should show sensitivity towards human needs. It was recognised that the human factor plays an important role and that therefore communication and networking skills would need to be learnt, or at least developed.

The next phase, which began end of the 1940s or 1950s, was characterised by an attempt to improve the training by making it more academic. Murphy's designation for the time between 1946 and about 1985 is 'Behavioural Science Era' (1992, p. 36). First of all the behavioural sciences and above all the psychology of behaviour and learning offered a new basis for the training programs. On the other hand there was also a strong influence from the latest principles formulated by organisation theory. Those responsible for the development programs realised the complexities of educational and school administration and tried to cope with it through a solid, objective, academically founded basis for the qualification programs. The capacity of expressing matters in an abstract, theoretical way was seen as a core feature of academic foundation. This was attempted as well, when dealing with organizations, and school research was trying hard to construct a convincing theory of school which would dictate how school managers should act. Its aim was to develop a 'science of administration' (Murphy, 1992, p. 51). School management was hence seen as applied science, within which theory and research were directly linked to professional practice, and in which this theory and research were to be decisive for practical actions. The closer a program's link with contents of psychology, sociology, anthropology, political or economic science, the better and more suitable it was considered to be. Taking over research methods and instruments from these academic disciplines was regarded essential for the qualification. From this, conceptual frameworks were to be derived for the school leader's work in schools or other educational institutions. The focus of this interdisciplinary approach was the ideal of an administrator as decision-maker, who made use of academically founded insights.

The development programs usually were offered as Master or PhD courses at universities. In the late 1960's there were 212 such programs in the USA, and in the mid-1980s their number had grown to 500. They were provided by smaller institutions for the local geographical area, as well as by large universities with a country-wide catchment area.

A series of events at this time had a lot of influence on the training programs for school leaders. The first was the foundation of the National Conference of Professional Educational Administration (NCPEA). This association was composed of a number of educational scientists from various American universities. The second event of equivalent importance was the foundation of the Cooperative Project in Educational Administration (CPEA), a consortium of eight universities who had started training initiatives for leadership roles. In addition, in 1955 the Committee for the Advancement of School Administration (CASA) was founded. The eight directors of CPEA belonged to this together with a number of professors and practitioners. This committee endeavoured to develop standards for training in school supervision and leadership, and worked towards the introduction of a state certificate and accreditation for the graduates of such training programs. The cooperative CPEA project resulted in 1956 in the formation of the University Council for Educational Administration (UCEA). At that time this encompassed 34 training programs in educational leadership at leading American universities, and in the 1960s and 1970s the UCEA became the dominant authority for qualification programs in the training of educational leaders.

In the 1970s, two more developments could be observed: on the one hand an increased specialization and differentiation, i.e., in opportunities for the participants to attend different courses for different positions at a school demanding different competences. The second novelty was a tendency towards competence-oriented approaches and hence towards development programs based on a competence-approach. By the mid-1970s, the competence-approach had developed one of the most relevant foundations for school leader development programs (see Griffith et al., 1988; Silver, 1978).

Regarding the methods applied, as early as in the 1960s, there was a development away from purely lecture- or discussion-oriented methods towards a broader variety. In addition to traditional ways of working with course material, text books or lectures, case studies, simulations problem-based learning, action research and increasingly internships were used. Apparent was the effort to make use of principles of adult learning in the learning processes intended. This development went on in the 1970s and 1980s, so that the variety of methods grew continuously. Various media were used, such as audiovisual media and computer programs. As school and educational administration was regarded not only as a domain of practical actions but also of theoretical studies, there were efforts to motivate research-oriented thinking and questioning and to outline problems for which solutions had to be found.

Murphy attaches the label 'Dialectic Era' to the period after 1985 (1992, p. 67). This time was characterised by a general desire for the reform of American society and by an increasing dissatisfaction in many areas with prevalent values. As far as the training of school leaders was concerned there was a spreading disillusionment with the cognitive basis underlying the training programs. The attempt to present theoretical knowledge in such a way that it would be useful to practitioners in their daily work and give concrete help to school management and administration appeared not to have worked. The gulf between theory and practice had not been overcome. In addition the concept of the school leader had narrowed to the extent that he was really no more than a sort of chief administrator.

There was a lot of criticism of this sort of training. Thus a survey by Heller et al. (1988) established that 51% of the more than a thousand heads questioned thought that their training as a head had been moderate to poor. 46% thought that their training had not related closely enough to the realities of their everyday work. When they were questioned as to when and from what they had learnt their expertise as school leaders, only 10% referred to their training program and more than 60% said that they had learnt most by actually doing the job. The increasing criticism of the training principles was encapsulated in a UCEA general survey of the whole situation regarding initial training and inset, in which more than 50 leading university faculties were represented. This investigation was published in Leaders for America's Schools (Griffith et al., 1988; NCEEA/UCEA, 1987/1988), and this report concluded that the training programs were in urgent need of reform.

It recommended that new training programs should fulfil certain criteria such as

- define effective educational leadership,
- recruit high-quality candidates who have the potential to become future leaders,
- develop collaborative relationships with school-district leaders,
- encourage ethnic minorities and women to enter the field,
- promote continuing professional development for practising administrators,
- redesign preparation programs so that they are sequential, updated in content, and include meaningful clinical experiences.

(Milstein, 1993, p. ix)

During the academic year 1988/1989 and with the financial support of the Danforth Foundation, the National Policy Board for Educational Administration (NPDEA) was founded at the University of Virginia. This was composed of the Boards of Directors of the ten most influential groups concerned with school leadership and school leader development. The NPDEA published their widely respected report Improving the Preparation of School Administrators: the Reform Agenda.

This document recommended that certain fundamentals both of knowledge and skills should be agreed which would be basic training skills for school leaders before they took up position, i.e. in pre-service programs. These included:

- societal and cultural influences on schooling,
- teaching and learning processes and school improvement,
- organization theory, methodologies of organizational studies and policy analysis,
- leadership and management processes and functions,
- policy studies and politics of education,
- moral and ethical dimensions of schooling.

(Miklos, in press, quoted from Murphy, 1992, p. 74)

The content of these programs was to be based on and supported by practical experience.

Other proposals included the demand for a national system of accreditation, that is to say an official state recognition of training programs and a national certificate for those successfully completing the programs.

Out of this very varied landscape of development measures for school leaders in the United States, three programs will be focused on here:

- The Danforth Educational Leadership Program of the College of Education of the University of Washington in Seattle, in the State of Washington: it is directly rooted in the reform initiatives of the Danforth Foundation described above. Since then, the program has continuously been modified and enjoys an excellent reputation.
- The newly developed program of the William Paterson University in Wayne, in the State of New Jersey: this development program was conducted for the first time in 2000. It aims at combining topical knowledge from the school improvement research and the school leadership research as well as responding to the needs of the State of New Jersey.

- The qualification offered by the California School Leadership Academy (CSLA) in Santa Barbara, in the State of California: here, a comprehensive program is presented, which is highly systematic and aims at addressing different phases in the careers of a school leader, comprising a preparation program, an in-service-program and a qualification offer for teams.

These three programs will be presented in the following three chapters. A summary of three chapters will be offered after the last one.

Washington: Working Together to Prepare Leaders

Stephan Huber and Kathy Kimball

Example for the State of Washington: The Danforth Educational Leadership Program at the University of Washington

Background: The Involvement of the Danforth Foundation and its Criteria for Promoting New Qualification Programs

The Danforth Foundation (DF), a cross-state American foundation, has a long history of promoting education in the United States. It has given scholarships, initiated competitions, organised seminars, discussion groups, partnership programs with schools, educational districts, educational authorities, colleges and universities and supported all these financially. In the middle of the 1980s the DF decided to focus their support on the training of the staff in school leadership positions, thus reacting to a situation of widespread dissatisfaction with the training methods provided at the time. (For a description of how this was implemented, see Milstein, 1993.) Following a decision in the spring of 1985, activities were to be initiated which would have an influence on the training of those in a position of leadership in education. An essential starting point in this was and still is an attempt to bring together the various interest groups and the relevant people in authority, and to try together to draw up plans as to how one might prepare teachers to take up positions as school leaders before they even begin their appointment. On a budget of 150 000 US dollars a year, those working with the DF began to develop a catalogue of proposals and to pass this on to the Colleges and Schools of Education at various universities.

This document, titled "The Danforth Foundation Program for Preparation of School Principals" (DPPSP), was based on an analysis of what most of the training programs in operation at that time did not contain. It was thus a sort of list of omissions and gave seven deficiencies in principal training (Milstein, 1993, pp. 5-6):

1. There was a lack of communication and collaboration between the universities that were preparing principals for the consumer school districts and the needs of the schools where they would be employed.
2. There was a need to recruit high-quality candidates rather than taking those who had been self-selected.
3. There was an under supply of ethnic minorities and female candidates for the principalship.
4. Newly licensed principals had no or minimal field experience as principals prior to employment.

5. Pedagogy in the university classroom needed greater variation in approaches in order to respond to adult learning needs.
6. University departments of educational administration needed to enlarge the number of people involved in preparation programs by reaching outside to other schools in the university and to school-district personnel.
7. A curriculum audit was needed in the educational administration department to determine whether the content was relevant for the newly proposed reform initiatives and for the needs of the consumer school district(s).

Universities that were to be supported and which would develop the new training programs were chosen according to a list of criteria. It was decided that a group of three to six faculties would be selected every 18 months. Every selection for the 18 month period was called a cycle. Between 1987 and 1991, there were five such cycles. Every faculty received an initial payment of 5 000 US dollars with which to begin a new program. Then later every selected institution received up to 40 000 US dollars in order to implement their program.

The selected faculties and institutions were obliged to base their program developments on the following DPSSP components:

1. An Assessment of Potential Participants

On the one hand, assessment procedures were developed to screen appropriate candidates for participation in the program in order to guarantee appropriate entry standards, and on the other hand, they were also used for diagnostic purposes in the planning of appropriate training programs for these candidates. The faculties at each university carried out their own assessments. A variety of approaches were used for selection: a formal interview conducted by members of the Steering Committee; written statements by the candidates; a paper on a particular subject or a presentation on a topic selected by the committee. A number of universities put this procedure into the hands of special assessment centres, for example, the NASSP (National Association of Secondary School Principals) assessment centre. The goal, as also recommended by the DF, was to carefully identify well qualified candidates and also to think beyond the traditional training programs of the past, whose selection procedures, in as much as they had any, were not considered adequate for the task of choosing future educational leaders.

2. The Candidates

A lot of importance was attached to increasing the number of candidates from ethnic minorities and also the number of women. It was observed that the number of these applicants grew whenever relevant publications and application forms were widely distributed to all the teachers in a school system. This also encouraged teachers to recommend colleagues whom they considered particularly suitable.

3. The Curriculum

Those working with the DF recommended that the Curriculum Committee of the relevant faculty should consider the following content and/or themes:

- ethics,
- interpersonal relations,
- planning,
- speaking,
- writing,
- facilitating.

Another related aspect that the DF wanted to improve was the methods used in the seminars. Alternative teaching methods, such as case studies, simulations, role playing and Socratic seminar would be demonstrated to the faculty of each new program. Finally, each one of the selected institutions would work together with representatives of the cooperating school districts to analyse and evaluate their existing curriculum and methods. This work required a comparison of the existing curriculum with recommendations from current academic literature and advice from the field. Following this, the Educational Administration faculties were to begin a program of improvement. On the basis of this comparison and the related discussions with practising school leaders, plans would then be made as to how the training of principals should be carried out in the future at these universities.

4. The Implementation Cycle

Three to six universities took part in each implementation cycle, during which time regular meetings were established between the participating institutions, in order to create a forum for a productive exchange of ideas, plans and also results.

5. Practical Experience

The universities were requested to design a period of work experience for each principal candidate. The ideal work experience would last for at least a semester and take place in a school different from the one in which the candidate taught. Those working with the DF proposed that the candidate should experience both the elementary and secondary levels. This experience was seen as a firm foundation on which one could test out theory and practice and prove oneself ready to begin a career as a school leader. During his or her work experience, the future principal should have a mentor at his or her side, and as a third team member, the university facilitator or another member of the faculty. This trio was to be responsible for the activities and also the assessment of each trainee.

6. Mentors

One part of the work undertaken in cooperation with the school districts included the selection and training of mentors. Mentors in this program were to be principals or others working in school administration. In some cases, mentors were also chosen from those in leadership roles in public administration, business and politics. The aim was that the participants should work with high quality mentors representing multiple perspectives during their work experience.

7. Partnerships between Universities and Schools

It was intended that university faculties should develop long term working relationships with the representatives of the relevant education authority. In each case a steering committee would be formed which would include representatives from the university faculty and the regional school district leadership. In addition, representatives from the state offices of education, from business and industry, and the relevant individual schools would be included. Later it was decided that student representatives as well as their mentors should be involved. The first thing the individual steering committee would then do was to define program components best suited to the regional situation.

8. Responsibility for the Implementation of the Program

After the university faculties had reached a consensus about their participation in the DF program, a faculty member was selected to be responsible for this program. He was to be a teaching member of the faculty and to act as a link between the faculty and the DF. Over a period of 18 months, this person was to prepare her or himself for her or his role by attending a series of in-service training sessions that were sponsored and organised by the DF. Much was learned from the exchange of ideas during these meetings.

Looking back now, it appears that without the investment of the DF, the multitude of changes that took place in the training programs for school leaders throughout the United States at so many universities at the same time would hardly have been possible.

The Danforth Educational Leadership Program

The University of Washington (UW) belongs to the group of universities in the second implementation cycle supported by DF. It developed its Danforth Training Program in 1987 and began the first stage of the program in 1988, which is interesting because, in that same year, it received a prise for exceptional quality from the American Association of School Administrators for its existing training program. The fact that, in spite of this award, there was still interest in and energy for the development of a new model, parallel at first to the existing tried and tested version, is certainly evidence of quite exceptional innovative zeal. The following objectives were fundamental in this (see Milstein, 1993, pp. 146-147):

- developing strong partnerships with school leaders in the region and including these partners in the dialogue and decision making about overall values and program intent,
- delivering meaningful content and doing so when content related issues (e.g., budget) would be a high-priority topic in school settings,
- moving away from the 3-hour course approach and replacing it with flexible and variable-length educational units,
- establishing thematic approaches to guide the election and development of content,
- significantly increasing the proportion of the program devoted to field experiences and making those experiences more rigorous,
- focused recruiting, with the intention of increasing ethnic and racial minority participation in the program, given the increasing and shifting minority population in the Puget Sound area.

Following an evaluation of the program and an analysis of the success of its graduates in their applications for positions as principals, the Danforth Educational Leadership Program became, in Autumn 1992, the only program in the Area of Educational Leadership and Policy Studies in the College of Education at the University of Washington. In addition to the Principal Preparation Program, there is also a School District Leadership Program leading to a State Superintendent Certificate. By August 1999, 166 candidates had successfully completed the program and 87% of those graduates were appointed as principals, assistant principals, and program administrators or to other leadership positions in school administration.

The target group for this training program is highly successful teachers who seek to improve schools through formal leadership roles. Applicants must have at least a

Bachelor degree and current teaching certification. They must meet the academic qualifications of the University of Washington, including evidence of high achievement and capacity for scholarship. At least three years of successful teaching experience is required, and it must be possible for them to reduce their teaching schedule during the program to half time in order to fully participate in the work experience component. In addition, potential candidates must show evidence of their leadership competence, which is achieved by means of the three references required, one from the principal, one from a member of staff nominated by the principal, and one from a colleague selected by the candidate himself. The candidate also has to submit a four to seven page document in which she he sets out her or his personal goals for leadership. This will show her or his views on education and leadership qualities, how she or he personally sees these qualities being translated into practical decisions, her or his reasons for applying for the training program, a personal justification for her or his desire to take up a leadership role in education, and the way in which the reduction in her or his teaching schedule will be organised and financed. Candidates without a Master of Education degree must concurrently enrol in additional coursework to complete the degree (required for principal certification in Washington State). To be admitted, they must prove their academic competence by submitting appropriate papers for consideration or by completing an additional essay and providing further testimonials.

As the program was being developed, the following key ideas were taken as fundamental. They are still quoted today in only a slightly changed format as 'working assumptions' and determine the conceptual framework of the program (see Milstein, 1993, pp. 147-148):

- Equity and Excellence: High-quality education for all students is a viable and morally correct goal.
- Leadership: Management skills constitute an important component, but leaders also need to articulate, justify, and protect core values that constitute the purposes and functions of schools.
- Organisational Change: Schools need to be self-renewing. The leader's role is to help initiate, implement, and evaluate that process.
- Collaboration: Leaders must facilitate widespread involvement in the renewal of schools.
- Inquiry and Reflective Practice: Leaders must value and have the ability to reflect critically about their schools' dynamics and to promote this value and skill in others.

The training program uses a cohort model and lasts one full year. The curriculum is organised as integrated, instructional modules for which students earn 39 credit hours[42], twelve of the credits are granted for the very intensive internship, or work experience.

The training program begins with a one-week residential institute at the end of July. The participants are given the opportunity to develop into a learning community away from their everyday work in school. The central theme is the transition from the classroom perspective to a view of the school as an organisation. It is also necessary to be familiar with school district policies and the latest political

[42] "Credit hour" translates to one hour of university involvement per week over a period of 15 weeks

changes in education in the State of Washington. For example, Washington has content standards, defined as Essential Academic Learning Requirements and performance standards to measure achievement related to those standards (the Washington Assessment of Student Learning, WASL). These standards are required in all public schools in the state. Each school district has implemented procedures to improve teaching and learning related to these standards.

All through the year, seminars take place. These instructional modules deal with the following topics:

1. Ethnography – understanding the culture of the school

Understanding the school's culture is an essential skill for a school leader. The module is based on the view that the effective principal must be a strong 'instructional leader', that is the person responsible for giving momentum both to the teaching and learning processes and to the school in general). Participants learn how to use such ethnographic methods as participant observation, interviewing, taking and analysing notes, and assembling data from documents. These skills are practised in the internship to understand the principal's role and should then be put into practice at the placement school to discover and then shape school culture.

2. Leadership

Students learn to think about leadership from a variety of theoretical perspectives. They examine leadership styles and tasks and write reflective analyses of leadership incidents in the school based on personal experience. Thus, they examine how leadership is practised, its moral and ethical aspects, and the interaction between leading and being led.

3. Moral and political dimensions of school and educational leadership in a democracy

This module takes as its subject the high level of responsibility implicit in leadership in a democratic society. If leading implies that linguistic and rhetorical skills can give you influence over others, then awareness must be developed of how important it is to use language responsibly. The aim must be that those who are led and influenced should develop into independent thinkers and responsible, mature citizens who participate fully in the democratic society.

4. Inquiry, evaluation, assessment and organisational learning

This module deals with topics such as critical inquiry, learning and the acquisition of knowledge, organisational theories, innovation, and the characteristics of change processes. Students conduct an inquiry project by gathering and using information to make decisions. They learn how to understand statistics, standardised tests, and alternative methods of assessment relative to the required state assessment and accountability program.

5. Staff development

This module addresses practical models of staff development, including questions of motivation, of the standard learning processes applicable to adults, and the relationship between staff development and staff evaluation.

6. Supervision of instruction

Central to this module are questions relating to the topic of improving teaching through observation and supervision based on an analysis of what pupils learn. This module includes supervision methods, coaching procedures, the assessment of teachers, communication and giving feedback, but also staff selection. The methods used are role playing, the analysis of video lessons, and discussions about the role of the principal in checking to what extent the state's Essential Academic Learning requirements are being implemented.

7. Curriculum and teaching

This module includes increasing awareness of the school-wide reform models used across the United States (e.g., Success for All, Coalition of Essential Schools). In addition, participants learn about the principal's role in curriculum leadership, such as adapting the curriculum to conform to the guidelines of Washington State's Essential Academic Learning Requirements, and also investigating the implementation of these guidelines in individual schools.

8. Multicultural education

In this module, students develop awareness of the wide range of diversity found in American public schools (e.g., ethnic, religious, cultural, language, social class, and family structures). The broad range of students creates conditions that require understanding and respect. Issues of equity and excellence, frameworks for understanding social class and poverty, and unique learning needs and styles are woven into this module. A major goal is to help program participants towards creating equal opportunity and counterbalancing the disadvantages that some students bring with them.

9. Educational leadership and special needs

Using case studies of special needs children in Washington schools; the program builds skills in trouble shooting and case management for special populations of students. Participants learn how to work within the complex legal requirements for special education and also how to set up a school-wide system that supports pupils with various disabilities. The module develops skills in behaviour management and adapting instruction and assessment for special needs pupils.

10. Budgeting

This module raises issues about resource allocation within schools and districts. A basic assumption is that schools decide how to use resources based on their values. Students also learn skills to tackle the budget process at the school level, and get a primer course in school funding from federal and state sources.

11. School law

The legal system has strong implications for leading schools in America. Students learn a system for legal decision making and enough about the federal and state statutes, regulations and procedures to know how to find out the information they need to know. Legal questions arise on a daily basis in public schools. Students explore case studies to understand how to deal with legal issues (e.g., religion, free speech, discipline, special education, and evaluation).

12. The internship: Experiential learning

The experience-based internship is a key component of the program. Practical experience and theoretical knowledge should be tightly linked according to program philosophy and the internship serves as the place where practice and theory come together.

During the academic year, the cohort meets in blocks of six hours every Thursday from 1 o'clock to 7 o'clock at the University. Four hours of each Thursday block is scheduled for instructional content modules (e.g., inquiry, law, moral and political dimensions of leadership). Two hours are devoted to a reflective seminar which is designed to help candidates think critically about experiences from their internship placements, relating such experiences to the theoretical content of the program and benefiting from the exchange of views within the learning group. In addition, there are ten whole-day meetings on Saturdays scheduled across the year.

Particular importance is attached to the work experience during the internship placement. Participants teach half time to keep their school schedule free for this component of their training. This is a hardship for many students since they would only receive half pay. Students spend at least 16 hours a week at their placement school. While schedules vary, most interns spend part time every day throughout the school year at the internship site. Washington's requirement is 720 hours of work experience; most of the UW students serve well beyond the minimum, often as many as 1 000 during the year. Sometimes, District Superintendents support participants by arranging an administrative assistant position, which allows for the internship at full pay. At the internship school, students are supervised by their mentor principal and/or by another education leader serving in a program administrator position (e.g. Director of Curriculum). The trainee and the mentor also work closely together with a representative of the university. As much as possible, the work experience activities are coordinated to link in with the typical leadership tasks that occur in the course of a school year, from the preparation for a new school year right through to its conclusion in June of the following year. Students are also encouraged to get work experience in a variety of settings (e.g., elementary and secondary levels; rural, urban, and suburban schools). To this end one part of the placement should be completed at a school which is very different from the trainee's own.

A five-day concluding seminar gives the students an opportunity to present the results of their experiences through a portfolio, to clear up uncertainties regarding personal career goals, and to prepare for their desired leadership post.

In order to make the curriculum as relevant as possible, the program schedule follows the public school schedule and differs from the usual organisational structure of the university academic year (three quarters, or two semesters).

The faculty in the Danforth Educational Leadership Program are either academics from the university or school practitioners. For example, principals, superintendents, or program administrators from the schools often are hired as part time faculty. The success of the program is largely dependent on a close and equal partnership between school and university faculty. The philosophy of the Danforth program also expects that the participants will share their experience and expertise and remain involved as partners in the further development of the curriculum.

A wide repertoire of instructional methodology is used including case studies, reflection on experience, simulation, role playing, and interactive discussions. Problem-based learning, a method based on individual cases that is highly suitable to integrate the participants into cooperative problem solving processes, is of key importance. Theoretical perspectives about how adults best learn underlie the teaching of the seminars. The school internship placements are very important in providing concrete leadership tasks that can link back to the seminars.

The number of participants in each cohort was limited to twenty until the year 1999. Beginning at that time 32 students were admitted to help meet the increased needs for leaders in Washington schools.

Table 33: Outline of the Danforth Educational Leadership Program of the University of Washington in Washington, USA

Training and Development of School Leaders in Washington, USA Example: The Danforth Educational Leadership Program of the University of Washington	
Provider	College of Education of the University of Washington
Target Group	Teachers aspiring to a school leadership position, i.e. before application
Aims	To enable candidates to work effectively towards the key goals of quality improvement, educational leadership, the further development of organisations, cooperation, the expansion of knowledge, and personal reflection as a part of educational responsibility
Contents	Understanding of the culture of a school; the moral and political dimensions of educational leadership in a democracy; organisational learning and evaluation; staff development and in-service training for teachers; lesson observation and assessment; the curriculum and teaching; multi-cultural education; school leadership and support for special needs children; financial competence; school law; work based on the placement experience
Methods	Case studies; simulations; role plays; interactive discussion; the completion of concrete leadership tasks during work experience; personal study
Pattern	Ca. 98 course days* (39 credit hours of study over 15 weeks = 585 hours) (timetabling: blocks of several days of seminars at the beginning and the end; six-hour seminars every Thursday; all day seminars on 10 Saturdays; 16 hours per week practical experience (4 days with 4 hours per day)) plus 120 days of practical experience (720 hours = 360 hours per semester); additional study time for reading the relevant literature and time to complete the necessary documentation, such as reports and the journal within 1 year
Status	Mandatory; however, the candidates may choose between different approved programs by different providers
Costs	Ca. 9 200 euro (8 600 US dollars) per participant for a course ending with the Initial Principal Certification and 11 800 euro (11 000 US dollars) with the additional degree of Master of Education, financed by the participants themselves, sometimes scholarships and district support are available
Certification	Washington State Certificate for Principal, sometimes plus Master of Education

** If there is no specification by the provider as far as the number of days is concerned, we converted the contact time in hours into the unit course day, taking 6 hours training as one day.*

Great emphasis is placed on the fact that this group of learners remains unchanged and stays together throughout the whole program (no new members are admitted once the program begins). The cohort creates a secure and cooperative environment in which people feel confident enough to explore their own communication, problem solving, organisation and leadership skills and in which constructive feedback can be given and further development made possible[43].

The final assessment for the participants is based on written work, seminar presentations and facilitation, active participation in discussion groups, and on written reports within the context of their work experience. For example, written work from the internship might include evaluation studies, implementation of pupil support in the classroom, professional development plans. During the course of the program, each participant builds up a portfolio folder with her or his academic essays, the documentation and reports from her or his placement, other essays and a journal, which documents her or his personal thoughts on her or his experiences. In addition, in their personal leadership platform, the participants express an ideal model of their philosophy of leadership. All of this is then judged against the state criteria of Washington's Performance Domains. If the candidate satisfies the assessment criteria she or he is awarded the Washington State Initial Principal Certificate. A Master's degree is a requirement for application for the position of a principal in Washington. If the degree has not already been earnt, the academic degree Master of Education can then be obtained by upgrading the program.

Participation in the program involves costs of 8 600 US dollars (9 200 euro) for the principal certificate and 11 000 US dollars for the certificate and additional Master's degree (11,800 euro). There are a variety of tuition scholarships and other grants available to students.

Recent Developments in Washington

In 1999, based on concerns about a shortage of school principals, the university administration, and educators in the field, encouraged the program to add students. Thirty-two students participated in the 1999-00, and 2001-02 program. Also, in 2000, due to state budget shortfalls, the University of Washington administration began encouraging programs to change to a self-sustaining budget. Therefore, the 2001-02 program is operating on a new basis.

At the time of publication, the faculty have made adaptations to instruction to better meet the needs of 32 students. These changes include discussion groups of 10-11 each, additional field supervisors, the use of peer critical friends to give additional feedback on written and speaking assignments.

As yet, the faculty is unaware of what long-range effects these changes in budget and size will have on program quality and reputation.

---

[43] Of the 20 participants in the year 1999/00, 6 are working towards both the Washington State Certificate for Principal and the Master of Education qualification. The other participants already have a Master's degree.

Statement by Kathy Kimball

The Danforth Educational Leadership Program at the University of Washington has earned and maintained a reputation for excellence in preparation of educational leaders over the fourteen years of its existence. In my opinion, there are several reasons that sustain this excellent reputation, including: (1) willingness to improve based on on-going, formative evaluation that leads to responsiveness to the needs of the schools and new leaders, (2) a balance of program faculty (one half from the schools; one half from the university), (3) careful selection of candidates by requiring evidence of proven leadership capacity.

On-going, formative evaluation keeps the faculty constantly aware of whether content and pedagogy are relevant and result in pupil learning. There is frequent monitoring of instruction and many opportunities for students to make suggestions for improvement in all aspects of the program. A mix of faculty from the university and the schools also helps maintain relevance and timeliness in the curriculum. Faculty from the schools keep the program current, while faculty from the university bring research-based practice and critical thinking about enduring issues in educational leadership. There is a strong program commitment to developing high quality leaders and an insistence on careful selection of candidates. Future leaders are sought, identified, nurtured, and supported by the participating districts and the university. Providing high quality mentor principals, internship positions, opportunities to do important work, and a commitment to help students succeed in the program, does this. Further, students are expected to perform well, to contribute leadership to the schools, and to express their values for excellence and equity for all students.

A survey of all program graduates was completed in spring 2001. A preliminary analysis of these data suggest that graduates are considered well prepared by peers, highly satisfied with their preparation, and eager to participate in follow-up activities that continue discussion of the important themes of the Danforth program. There is an on-going faculty discussion about what is (and is not, working with emphasis on finding those elements that need improvement. On the whole, the faculty and program graduates are vigilant about increasing program quality. Some of the areas that are under consideration for this academic year 2001-2002, include a commitment to:

1. increase the diversity of each cohort,
2. increase sources for student support,
3. organise and implement follow-up activities for program graduates,
4. maintain high quality standards while responding to the need to prepare more principals/program administrators and move to self-sustaining budget,
5. improve curriculum of specific modules (e.g., special education, add information on leadership for 'English as a second language' learners; modify case study approach).

New Jersey: A New Paradigm for Preparing School Leaders

Stephan Huber and Michael Chirichello

Example for the State of New Jersey: The Educational Leadership Program at William Paterson University

The State of New Jersey will be facing a significant number of school principal retirements in the next five to ten years. To be able to fill the vacancies with qualified principals, new university programs have begun to emerge.

In New Jersey, too, the qualifications for the principalship are controlled by the State Department of Education. The NJ State Department of Education requires a Master's degree from an accredited institution in one of the recognised fields of leadership or management, such as educational administration, public administration, business administration, leadership or management science. Study must be completed within the master's program in each of the following topics: (1) leadership and human resource management, (2) communications, (3) quantitative decision-making, (4) finance, and (5) law. Additionally, candidates must pass a praxis exam in educational leadership that is administered by the Educational Testing Service in Princeton, NJ[44]. Since June 1, 1990, there is no teaching requirement for principal licensure[45]. In NJ the course contents for each of the required areas may vary from university to university but must be approved by the NJ State Department of Education prior to its implementation. William Paterson University of New Jersey through its Department of Educational Leadership offers a program leading to the Master's degree in educational leadership that has been approved by the NJ State Department of Education for principal licensure. This new program began in May 2000.

Applicants have to meet certain academic requirements to be considered for admissions, including a Bachelor's degree[46] and participate in an interview. Applicants may already have a Master's degree but those who do not must take the Graduate Record Examination or the Miller Analogies Test[47]. Other requirements for admissions include five years or more of successful school experience under a

[44] Testing requirements may be viewed at www.state.nj.us/njded/educators/license/ 1112.htm.

[45] Qualifications are listed on the NJ Department of Education's web site, www.state.nj.us/ njded/educators/license/1127.htm.

[46] A Bachelor's degree from an accredited college or university with a GPA of not less than 3.0 on a 4.0 scale. An applicant who has a graduate degree from an accredited college or university must have a GPA of not less than 3.5 on a 4.0 scale.

[47] A score of at least 45 on the Miller Analogies Test. Applicants who take the Graduate Record Exam must obtain a minimum of 475 on the verbal score and a minimum score of either 525 on the quantitative score or 550 on the analytical score or a total score of at least 1 000 on any two parts of the GRE.

teaching, educational service, and/or supervisor's certificate[48]. This sharply contrasts with the State Department of Education's own requirements, which does not require teaching experience as a prerequisite for the certification of principals. Candidates seeking admission to this program must prepare a portfolio, which supports the applicant's capacity to teach effectively, write clearly, and demonstrate competence in analytical and evaluative skills, and participate in an interview. The portfolio must contain evidence that the applicant has undertaken successful leadership roles in schools and/or communities. Applicants must include a writing sample that includes their reasons for applying to this program. This portfolio must support the candidate's competence in teaching and education, their communication skills, their analytical and evaluating abilities and their experiences with successfully accomplished leadership tasks within schools or in other communities; the portfolio must also contain two letters of recommendation, one being from the candidate's school principal.

The program at William Paterson University's Department of Educational Leadership mainly encourages aspiring principals to apply, but teachers who want to strengthen their leadership skills without intending to apply for a leadership position are also encouraged to seek admission to this program. This graduate program supports the belief that leadership in school organizations goes beyond the principal and should include teachers as leaders. Now more than ever, teachers must take on new roles and responsibilities in ever changing school environments. The transformation of school organizations cannot be successful without teacher leadership (see Barth, 2001). Teachers must lead in their roles and responsibilities with practicum and student teachers and as mentors for new teachers. They must lead in curriculum design and professional development programs. Leadership should become more collective that centrist (see Chirichello, 2001). Therefore, one of the goals of this program is to encourage career teachers who aspire to lead to apply for admission into this program. Teachers who are not attracted to the principalship may find this program supports and strengthens their role as teacher leader.

Applicants who have backgrounds in leadership outside education or who may be non-practising educators may apply for admission to this graduate program if they meet all other qualifications. In these cases, it will be the responsibility of the prospective candidates to establish active partnerships with practising school leaders so they may complete the field experiences that are required throughout the program.

Twenty-three participants were admitted to the first cohort in 2000; this number, however, is insufficient to meet the needs of the projected number of principals who may be leaving their positions due to retirements. Throughout New Jersey, there are other universities that offer approved programs.

To work within the context of the changing needs of American schools and, as the need for more principals continues to grow and the complexity of the principalship increases, we must create graduate programs that are designed to develop school leaders who have the capacity to meet these challenges in our

[48] Applicants with fewer than five but at least three years of successful school experiences or related experiences may be considered for admission contingent upon materials submitted in their portfolio that would support their leadership potential.

schools. There is no simple solution- but any solution must advance effective leaders who have the capacity to build influencing relationships, encourage followership, initiate substantive and lasting change, and create visions that will transform schooling. Within these programs there must be room for developing values, beliefs, and competences for aspiring principals. This Master's program will allow participants to re-imagine leadership and create personal metaphors for the principalship. We will strive to develop visionary leaders who focus on 'we' and build successful teams; who understand how to create a culture of learning and leading, and who believe in collaborating with the staff to improve pupil results.

At the same time, we are witnessing major shifts in what leadership is all about. Public schools need highly competent principals with vision, courage, and leadership to undertake significant change and improve pupil outcomes. School restructuring is creating a new role for principals in post-bureaucratic organizations (see Leithwood et al., 1992). Significant and continuing changes in our society have created shifts in roles, responsibilities, and relationships in our schools. Principals must understand their new roles if school restructuring is to be successful (see Bredeson, 1992; Peterson, 2001). To meet the expectations for these new roles, responsibilities, and relationships, schools need leaders who move beyond reforming and look towards transforming school organizations (see Avolio, 1999; Goens & Clover, 1991; Koehler & Pankowski, 1997).

The Educational Leadership Program at William Paterson University of New Jersey (WPUNJ) was designed to meet these challenges in the context of an M. Ed. in Educational Leadership.

The following are the goals of this program:

- to attract a diverse pool of highly able educators who aspire to leadership positions in education,
- to nurture highly competent educators to develop a personal vision[49] and capacity to lead,
- to build understandings about initiating and sustaining substantive and lasting change within complex organizational structures,
- to prepare and assist candidates to acquire leadership positions,
- to create a graduate program in which aspiring principals and teacher leaders work together effectively to understand their roles, responsibilities, and relationships in educational organizations,
- to provide support for graduates who acquire principalships through continued collegiality and professional development.

Additionally, by linking each of the course topics to the Interstate Leadership Standards from the Council of Chief State School Officers (1996), the program hopes to develop leaders who demonstrate competence in the knowledge, skills, and dispositions supported by each standard:

1. A school administrator is an educational leader who promotes the success of all students by facilitating the development, articulation, implementation,

[49] Each candidate will develop a personal vision of leadership and encourage the behaviours that will support and develop collaboration and highly effective teams both in the cohort as well as in one's school. Candidates should become architects of continuous change and improvement both during and subsequent to completing this leadership program.

and stewardship of a vision of learning that is shared and supported by the school community.

2. A school administrator is an educational leader who promotes the success of all students by advocating, nurturing, and sustaining a school culture and instructional program conducive to pupil learning and staff professional growth.
3. A school administrator is an educational leader who promotes the success of all students by ensuring management of the organization, operations, and resources for a safe, efficient, and effective learning environment.
4. A school administrator is an educational leader who promotes the success of all students by collaborating with families and community members, responding to diverse community interests and needs, and mobilizing community resources.
5. A school administrator is an educational leader who promotes the success of all students by acting with integrity, fairness, and in an ethical manner.

This program will also strike a balance between the competing tensions of management and leadership. Courses focus on the development of skills that will create savvy principals who understand how one manages schools. At the same time, the program believes in developing leadership that promotes influencing relationships between the leaders and followers. These relationships inspire, challenge, and look at schools as communities of inquiring learners and leaders where individuals become "bonded together by natural will and who are together bound to a set of shared ideas and ideals" (see Sergiovanni, 1996, p. 48).

Within the context of the program goals and the ISLLC Standards for School Leaders, this graduate program values democratic collaboration, diversity, equity, theory, critical inquiry, reflective practice, continuous improvement, pupil success, and ethical practice. These values will build upon the beliefs that emanate from a culture that supports the four I's of transformational leadership – idealised influence, individualised consideration, inspirational motivation, and intellectual stimulation (see Bass & Avolio, 1994). Candidates will embrace a vision that will encourage supportive, collaborative behaviours within their schools' organization and facilitate continuous change. Moreover, it values a deep commitment to leadership that enables candidates to construct meaning and knowledge together (see Lambert et al., 1995).

This program will require candidates to become members of a cohort. Cohorts encourage collegial, collaborative learning environments. The cohort model supports faculty in the program's goal to deliver a coherent, integrated curriculum. In her research on cohort programs, Hersko (1998) found that cohorts are perceived as improving the overall quality of educational administration programs. They provide opportunities to network, continue student relationships, and enhance student-professor relationships. Hersko's study indicated that students and faculty give excellent or good assessments to such programs. Potthoff, Batenhorst, Fredrickson, and Tracy's study (2001) also supports cohort programs at the masters level. In the cohort masters program at WPUNJ, we anticipate that candidates will build supporting networks that will remain strong after they graduate and acquire new positions as school leaders.

Each cohort is required to take the following three-credit courses during the two-year program:

Year One

Summer:

- Leadership in Learning Communities: From Theory to Practice

Fall Semester:

- Contemporary Issues in Schools and Society: Perspectives and Possibilities
- Educational Research: Qualitative and Quantitative Design

Spring Semester:

- Curriculum Design: Theory and Practice
- Understanding Group Processes and the Psychology of Organizations

Year Two

Summer:

- The Principalship: The Person and the Profession
- Current Topics: Comparative Issues in Schools and Societies (elective)

Fall Semester:

- Clinical Projects in Educational Leadership
- Supervision and Evaluation: People, Programs and Performance Appraisal
- School Management Functions: Finance, Structures, Resources

Spring Semester:

- Problems in Practice: Legal and School-Centered Issues
- Power of Policy: Developing Political Acumen
- Field Experiences

Classes meet on Friday evenings and all day Saturday, once a month during the semester for each course. Additionally, there are two one-week summer courses during the two-year program. To develop collaborative, team-centered competences, candidates meet outside of class time in smaller, micro-groups to work on projects, discuss issues, and apply theory to practice in their schools. These micro-groups vary in size from two to six. Courses are taught by university faculty and field-based practitioners.

Courses are a center around which candidates will form smaller groups that vary according to needs and interests. These groups meet throughout each course to focus on specific questions related to the central focus of study. Candidates are given opportunities to become facilitators in each group and lead the other participants in reflective inquiry and problem-based learning activities. This model necessarily restructures course time and develops a value for life-long learning outside the context of classrooms. Candidates meet for longer and less frequent time periods for classroom meetings. Smaller focus groups meet throughout the semester in between scheduled class meetings. Quality and depth become more important than quantity and breadth.

Instructors require a wide variety of readings from texts and current research journals to achieve this objective. Candidates have an opportunity to apply expert theory to practice through assigned group and individual projects in each course.

The program requires the successful completion of 36 graduate credits. Candidates are not able to transfer any credits from other graduate programs into this masters program. Candidates acquire six credits during each of four semesters, and three credits during each of two summer one-week institutes- one prior to the first semester and one between years one and two. Three credits are assigned to either an independent study or a three-credit seminar abroad that examines educational leadership in another country. Three credits are assigned to 150 hours of field-based experiences that occur throughout the program. All candidates participate in the same course experiences with other members of the cohort.

Candidates have to demonstrate their competence with technology. To achieve this objective, there is a technology strand imbedded in each course. Candidates are required to maintain active e-mail accounts and have access to the Internet. There are opportunities for instructors to use the Internet and engage in cyber talk between and among candidates. Chat rooms and threaded discussions, as well as video conferencing and other technologies, become integral teaching strategies. Specific technology competences are listed for each course and taught by technology experts. Candidates have to demonstrate competences in productivity tools (Word, Excel, Access. PowerPoint), the Internet, electronic mail, distance learning, database research, listserves, teaching and curriculum software, web page design, administrative and statistical software, and electronic portfolios[50]. This integrated approach offers candidates opportunities to apply technology skills to authentic, problem-based learning as they complete required program outcomes that include these competences (see Tomei, 1999).

Candidates develop and maintain an Individualized Leadership Plan (ILP) throughout this graduate program. The ILP consists of both long and short range career goals, a personal analysis of one's leadership style, and an analysis of the candidates strengths and areas needing improvement that are necessary for successful organizational management of and leadership in schools. The ILP becomes an ongoing agenda for both individual and collective topics of discussion, action research, and field-based experiences by the members of the cohort throughout this program. Candidates keep a learning journal as a way to promote reflective inquiry and personal growth. Entries focus on field-based experiences and reflections on course work. The purpose of the journal is to give candidates the opportunity to examine their beliefs and to develop understandings about the relationships between their beliefs, actions, and behaviours. The reflective journal becomes a focus of conversation between the candidates and the faculty during each of the courses of study and during the field visits that are completed at least once a semester.

Initially, two full-time faculty will be required for the program. In year two, three full-time faculty will be required. The Department of Educational Leadership will govern collectively rather than hierarchically. All decisions will be made collaboratively and not through individual committees. This will reflect the overall philosophy of leadership that will become an integral part of this graduate program. Clinical faculty who have specific expertise will join the full-time faculty. For example, the course on legal issues will be taught by an attorney, and may include a

---

[50] Technology competencies were developed with the assistance Dr. Mark Rodriguez, a former faculty member in the Department of Curriculum and Instruction at William Paterson University; and adapted from the Recommended Foundations in Technology for All Teachers and the Technology Foundation Standards for All Students (www.iste.org/Standards/NCATE/found.html and http://cnets.iste.org/sfors.htm).

principal or a full time faculty member from the Department of Educational Leadership. Students from other graduate programs at the university will join with the educational leadership candidates. Students may be from the departments of business, psychology, counselling, sociology, and/or special education and counselling.

Table 34: Outline of the Educational Leadership Program of the William Paterson University of New Jersey, USA

Training and Development of School Leaders in New Jersey, USA Example: The Educational Leadership Program of the William Paterson University	
Provider	Department of Educational Leadership, College of Education of the William Paterson University
Target Group	Teachers aspiring to a school leadership position but also teachers who want to improve their leadership competences
Aims	Competence in each of the Standards for School Leaders developed by the Council of Chief State School Officers and endorsed by the Interstate School Leaders Licensure Consortium; and a vision of school leadership that includes beliefs in democratic collaboration, diversity, equity, theory, critical inquiry, reflective practice, continuous improvement, student success, and ethical practice
Contents	Leadership in learning communities; contemporary issues in schools and society; educational research; curriculum design; understanding group processes and the psychology of organisations; the principalship; clinical projects in educational leadership (action research); supervision and evaluation: people, programs and performance appraisal; school management; legal issues; policy; field experiences; technology competences
Methods	Case study; lecture/discussion; group problem solving; micro-conferencing technology; large and small group discussions; reflective inquiry through journal writing; problem-based learning activities; technology communications; action research; and field-based experiences
Pattern	Ca. 90 course days* (36 credit hours = 36 semester hours over a period of 15 weeks = 540 hours) (timetabling: program follows the semester structure of the university; additionally 2 one-week summer courses) plus 25 days of practical experience (150 hours = 30-40 hours per semester) within 2 years (part-time 4 years)
Status	Mandatory; however, candidates may choose to attend approved programs at other universities in New Jersey as well
Costs	Ca. 10 000 euro (9 300 US dollars) per participant; financed by the participants themselves, reimbursement by local school districts is possible depending upon contracted agreements
Certification	Master of Education in Educational Leadership plus New Jersey State Certificate

** If there is no specification by the provider as far as the number of days is concerned, we converted the contact time in hours into the unit course day, taking 6 hours training as one day.*

Learning outcomes that are required for graduation include:

1. The completion of a minimum of 36 credits of graduate study within the Educational Leadership Program course offerings, including three credits assigned to on going field-based experiences, and successful completion of the technology competences listed in the syllabus of each course. All candidates must participate in the required courses as a cohort throughout this program.
2. The completion of an electronic portfolio that will support one's competence both in technology and in each of the courses.
3. An action research final project that will clearly demonstrate one's competence in analysis, synthesis, application, and evaluation of the program's mission, beliefs, values, and goals and that will focus on quantitative decision making as it relates to school settings and school issues.

An exit interview that will demonstrate one's ability to support one's final project and professional portfolio has to be conducted. Faculty will designate one or more course learning outcomes that will be included in the candidate's exit portfolio.

The course fee is 775 US dollars or a total of 9 300 US dollars for the 12 courses or 36 credits (10 000 euro). Some school districts may reimburse candidates for part or all of these expenses. There are no grants or state-supported funds available to the candidates to assist in the tuition costs at this time.

Recent Developments in New Jersey

The M.Ed. in Educational Leadership began in the summer 2000 with an invitation to 23 candidates from over 30 applicants. A second cohort began in the summer, 2001 with 24 candidates selected from over 40 applicants and Cohort II began in May 2002 with 21 candidates. As of Summer 2002, eighteen candidates graduated from Cohort I, 22 candidates from Cohort II are entering their second year, and 21 candidates from Cohort III completed their first course and will enter the fall semester, 2002.

There are two full-time faculty teaching in the program. A third faculty member is needed but budgetary restrictions have prohibited this from happening. However, to accommodate the time required for faculty to visit candidates in their schools as they integrate leadership activities into their work under the mentorship of their principals (or other administrators in some cases such as vice principals, department chairs, and directors), the department will have two adjuncts working with the two, full-time professors beginning in Fall 2002. Courses for Cohorts I and II have included clinical faculty who have expertise in psychology, law, finance, policy, special education, technology, and statistics and measurement. In our summer course on the principalship, new and experienced principals and assistant principals participated with the cohorts.

In all cases, mentor principals (or in some cases supervisors and central office administrators) were cooperative in working with candidates to enhance their field experiences. The relationships between districts and the university is strengthened because of the more than 60 years of school experiences the two full-time faculty in the department bring to the program as teachers, principals and superintendents.

The field experiences by candidates in Cohorts I and II have included a variety of activities: participation as chairpeople of district level committees; presenting and planning professional development activities, as well as presenting to staffs and boards of education on selected topics; budget preparation at department or grade levels; observations and evaluations of teachers and support staff; interviewing prospective teachers and support staff; parent association liaisons and coordinating parent activities at a grade or building levels; teachers in charge assignments including building level and summer schools; visitations to other schools to shadow principals and/or vice principals; members/chairs of planning committees for institutes, district level events, and school-wide events; scheduling activities at a grade or school levels; officially designated representatives to districts' professional development committees; curriculum development at grade, school, or district levels; coordinating grade level or school-wide events such as trips, graduations, and assemblies; initiating new programs for teachers, pupils, and parents; attending leadership conferences on local, state, or national levels; attending boards of education meetings; completing special assignments for principals/district level administrators; management activities related to grade levels and school-wide scheduling; whole school reform facilitators; and members of school-based management teams.

Technology competences have been self-taught, taught to candidates by their own district personnel, or taught by tech experts within the cohort itself. Like field experiences, technology competences are self-initiated and candidates assume responsibility for their own learning. As a result of first year experiences, several technology competences will be reassigned to different courses. Some technology competences are more complex and, to assist our candidates, we have hired a graduate from Cohort I to provide instruction in those skills: developing one's exit electronic portfolio, web page design, and school-based budgeting using spread sheets.

Individualized Leadership Plans and the reflective journals support the program's goal of creating life-long, inquiring educators who have developed a personal vision and capacity to lead as well as to build understandings about initiating and sustaining substantive and lasting change within complex organizational structures. As a result of reading these journal entries, the faculty have concluded that candidates in Cohorts I and II have been able to connect their field experiences to the Standards for School Leaders.

Sixteen candidates in Cohort I have taken and passed the required Praxis exam. The remaining two candidates have chosen not to take the exam at this time. Their success rate is 100%.

Several candidates have participated in organizing and presenting at a WPUNJ Critical Issues Forum on parent and community engagement is schools. One candidate has initiated a new community based, after-school program as part of her action research project. She has been successful in obtaining funding for this urban, church-based initiative for both years one and two.

Candidates from Cohort I were asked to evaluate the program by completing an 82-question survey using a Likert scale. They were also asked to send a summative journal entry to the department chair about their experiences in the program. The mentor principals were asked to respond to open ended questions during the exit interview in the spring 2002. The results of these assessments are being analysed. As the program proceeded through its initial year, two of the six courses offered to Cohort I received less than favourable evaluations from the candidates. As a result, the courses were re-evaluated and new teaching approaches were initiated.

In May 2002, the advisory board of the program met with eight candidates from Cohorts I and II to evaluate the program. As a result of that meeting, the field experiences will be increased to six credits and the 150 hours required will be expanded. An initial meeting with the mentors will take place in the summer before the first full semester to discuss our expectations for both the mentors and candidates.

After three years of this program, there are on-going discussions with several university department chairs and deans to include graduate students from their departments in some of our courses. Graduate students who are preparing for careers as school nurses will join with our candidates in Fall 2003 for one course (Contemporary Issues in Schools and Society). The elective course, Comparative Issues in Schools and Societies, was offered during two summers but, due to an insufficient number of interested candidates, the international trip to Australia was not realised. Finally, the cost of this program is approaching 1 000 US dollars per course.

Statement by Michael Chirichello

The roles and responsibilities of principals in the United States are more complex than ever. Principals today face a daunting task. They are expected to supervise staff, discipline pupils, interact with parents, manage facilities, lead the instructional program, work on special projects assigned by the central office, insure the safety of pupils and teachers, manage budgets, participate in school-wide reform, build partnerships with social agencies in the community, and understand the legal implications of these activities. A 1998 study of elementary school principals (see Doud & Keller, 1998) reported increases over 55 percent in principal responsibilities in each of the following areas: marketing/politics to generate support for schools and education; working with social agencies; planning/implementing site based staff development; development of instructional practice; curriculum development; working with site-based councils and other constituencies; and attention to issues related to potential legal liabilities.

In a recent survey conducted by the NJ Principals and Supervisors Association, 443 teachers with five or more years of experience indicated little desire to become principals (83 percent of the female respondents and 68.7 percent of male respondents). Reasons selected indicated the job is too stressful, parental and board of education demands give the perception that the job is undesirable, and the time required on the job adversely affects their quality of life (see Educational Viewpoints, 2001). As we begin to face a shortage of principals in the next few years (see ERS, 1998; IEL, 2000), who then will aspire to the principalship?

Yet, we know that, behind every exceptional school, there is an outstanding principal; and the first and most important step in turning a troubled school around is to find a terrific principal. Leadership becomes the first- and most essential step to improving the most troubled schools (see Public Agenda, 2001). Nevertheless, how do principals maintain a focus on educating children and staying focused "in an environment where interest group politics, board relations and a regulatory muddle conspire to handcuff their leadership?" (p. 10).

In a survey conducted in both June 2001 and 2002, graduate students in our leadership program distributed surveys and conducted interviews with principals and teachers in K-8, 6-8, and 9-12 schools in New Jersey. The principals were asked to examine a list of activities and choose three in which they spend most of their time and three in which they spend the least time. They were also asked to share their vision- three activities in which they would choose to spend the most and the least time. Teachers were asked to examine the same list of activities and choose three in which they perceive their principal spends the most and least time. They were also asked to share their vision- three activities in which they believed their principal should spend the most and least time. The summary of data for the two years (N=58 principals, 85 teachers) indicated school management activities take up a significant amount of time in the day of the principal[51].

How can principals achieve more of their vision and move away from activities that are more management than leadership? In larger schools, the roar of complexity becomes louder and, even with vice principals or assistant principals, managerial challenges seem to drown out the role of principal as leader. The principalship, as it is currently structured, makes our best educators avoid formal leadership roles (see Donaldson, 2001).

As university programs continue to prepare candidates who aspire to lead as principals in the United States, they must develop a deep understanding of the roles and responsibilities of the principalship with their candidates. They must work more closely with school districts. More importantly, they must nurture and sustain the passion to lead and to follow. Today's schools demand that principals build learning communities and take on the roles of educator, leader, manager, and inner person (see Speck, 1999). Preparation programs must attract leaders who will have the capacity to be effective as principal in each of these roles. Moreover, our programs must focus on a key question- what should principal leaders know, understand, and be able to do?

[51] School management activities was chosen by 60% of principals and 39% of teachers in how principals spend time; this same activity was chosen by 5% of principals and 6% of teachers as the vision in which principals should spend time.

California: University- and State-Supported Professional Development

Stephan Huber and Janet Chrispeels

Example for the State of California: Foundation II of the California School Leadership Academy

In California those aspiring to become administrators (assistant principals, principals, or district administrators) are required to complete a university course of study called the Preliminary Administrative Services Credential. Potential administrative candidates can fulfil the requirements at either public or private universities that offer programs approved by the California Commission on Teacher Credentialing and certified that they meet the state adopted administrative standards. The Preliminary Credential program (Tier I as they are sometimes called) takes about 2 years to complete and consists of traditional university courses that provide a theoretical understanding of leadership, educational policy, politics, organizations and organizational change. Other course requirements address practical issues such as school finance, law, labor relations and human resource management, school-community relations, technology, curriculum, supervision and instruction. In addition, students complete a semester long field experience or with the support of cooperating school districts, a yearlong internship.

Once a Preliminary Administrative Services Credential holder secures an administrative post, she or he has five years to complete the Professional Administrative Services Credential certificate (Tier II). Completing the Tier II credential usually involves both university course work, especially training in how to conduct action research in their schools as well as 120 professional development hours. The requirement for the 120 hours can be completed through state certified programs provided by other non-university agencies such as the Association of California School Administrators or the California School Leadership Academy.

The remainder of this brief report will describe three unique programs offered by the California School Leadership Academy (CSLA): Foundation II, Ventures and School Leadership Teams (SLT). All three programs are designed to enhance the knowledge and skills of aspiring or current educational leaders and to support the development of teacher leaders. CSLA was founded 20 years ago as part of a comprehensive school reform package. Although supported through State Department of Education funds, it operates as a semi-autonomous agency with a core staff housed at WestEd, one of the federal educational labs, and 12 regional centers located at county offices of education throughout the state. The core staff develops training materials and provides ongoing training and support for the regional directors. Through monthly meetings, regional directors have an opportunity to share program ideas, refine materials to meet ever-changing school leader needs, and to hone their own professional development skills. Since its inception, over 10 000 educational leaders (school, district and state administrators as well as teacher leaders) have participated in CSLA programs. The mission of CSLA is:

> To help practising administrators and teachers in leadership positions strengthen their instructional leadership skills and focus their actions on the issues and strategies critical to increasing the achievements of all students in California. (21st Century Leadership Opportunities 1999-2000)

Foundation II

The target group for this CSLA program is newly appointed or aspiring school leaders who have been recommended by their superintendent, or in the case of a teacher leader the principal. The Foundations II program consists of 19 –22 full days of professional development spread over two years. Successful participation in the CSLA programs is mapped against the CSLA Performance Indicators, which include the following:

1. Create an environment in which all individuals are able to articulate a vision of powerful learning, describe its characteristics, and cause the conditions that promote it.
2. Use a standards-based accountability system to drive curriculum and instruction and measure student progress.
3. Shape a culture in which norms, values, and beliefs manifest powerful learning.
4. Build communication structures that support powerful learning.
5. Mobilise relationships within the school to create a dynamic learning environment.
6. Use systems thinking to continuously improve performance within the school.
7. Create a diversity-sensitive environment grounded in multicultural and linguistic multiplicity that is intolerant of racism, sexism, and exclusion.

To achieve these aims, participants complete a three-part curriculum: a core curriculum of ten seminars, elective seminars addressing specialised knowledge in curriculum content areas or special topics and a portfolio day to share with colleagues their development as a school leader. The focus of the entire program is to provide school leaders with the knowledge and skills needed to ensure powerful learning by all students.

The 10-seminar core curriculum includes the following:

- Leading Through Vision
- Building a Vision of Powerful Learning
- Re-culturing to Create Powerful Learning
- Assessment in Service of Powerful Learning
- A Thinking, Meaning–Centered Curriculum in Service of Powerful Learning
- Teaching in Service of Powerful Learning
- Creating a Diversity-Sensitive Environment for Powerful Student Learning

- Systems Thinking in Service of Powerful Learning
- Building Relationships and Communication Structures in Service of Powerful Learning
- Shared Vision and Shared Leadership Service of Powerful Learning

Participants select three electives based on their particular needs or interests in order to enhance their knowledge of the California Curriculum Contents Standards that were adopted beginning in the late 1990s. Electives address sessions in:

- English Language Arts
- History/Social Science
- Science
- Math
- Visual and Performing Arts
- Physical Education

Table 35: Outline of Foundation II of the California School Leadership Academy in California, USA

Training and Development of School Leaders in California, USA Example: Foundation II of the California School Leadership Academy	
Provider	California School Leadership Academy (CSLA)
Target Group	Aspiring and newly appointed school leaders
Aims	Enabling the participants to realise the vision of creating student- respectively learner-centred schools where powerful learning is possible
Contents	Leading through vision; building a vision of powerful learning; re-culturing to create powerful learning; assessment in service of powerful learning; a thinking, meaning–centered curriculum in service of powerful learning; teaching in service of powerful learning; creating a diversity-sensitive environment for powerful student learning; systems thinking in service of powerful learning; building relationships and communication structures in service of powerful learning; shared vision and shared leadership service of powerful learning
Methods	Active learning through conversations; dialogue; problem-solving; short lectures; discussions; videos; readings; reflection techniques and presentations by the participants themselves; writing a personal vision and a professional growth plan; methods of documenting the reality of a school and documenting school development activities in a portfolio
Pattern	19 – 22 course days (divided up into 10 two-days seminars plus 1 portfolio-day) plus literature studies within 2 years
Status	Optional; participants who are doing a university program for aspiring school leaders can hand in the certificate for successful completion of the program to the university in charge as part of their qualification
Costs	Ca. 750 euro (700 US dollars) per participant; financed by the participants themselves, usually funded by the school or the district
Certification	Certificate of successful participation

A range of specialised topics is also available to meet participant's needs such as:

- Creating Safe and Productive Environments for Powerful Student Learning
- Connecting Students to the School Community
- Leading Successful Change Efforts in Your School
- Preparing Students for a Productive Work Life
- Creating a Healthy Learning Community

The Foundations II Program, similar to the other CSLA programs, is designed to help participants experience powerful learning themselves by drawing on important concepts of adult learning, primarily sustained engagement with colleagues in active learning that involves conversations, dialogue, problem-solving, and reflection about complex problems facing today's school leaders in their school contexts. The regional School Leadership Center staffs use a variety of teaching tools including short lectures, discussions, videos, readings, reflection techniques and presentations by the participants themselves. Participants assess their leadership style, develop their own vision statement, collect data and track projects in their schools, develop a professional growth plan, and prepare a portfolio of learning and accomplishments to share with colleagues at the end of the program. The Portfolio Day represents the culminating activity in the Foundations II curriculum. The portfolio serves as a basis of assessing individual achievements. Participants can submit this portfolio to their university advisor if they are also enrolled in a Tier II Professional Administrative Credential Program to help satisfy the 120 professional hours requirement.

The costs of the program per participant are 750 euro (700 US dollars) or 385 euro/350 US dollars per year. After completing the program, participants may permanently use the regional professional network of CSLA and attend seminars free of charge. The school district or school site usually pays the cost of participation.

Ventures

The target group of Ventures is experienced school leaders, who have been committed to the development process in their school for some time. Frequently, participants are graduates of the Foundation II program. The Ventures curriculum is designed to build on the foundational knowledge by providing guidance for participants to apply what they have learned in their own school context. The program lasts three years and again utilises a seminar and discussion format, but the meetings occur less frequently than the Foundation II seminars (that is, there are 10-12 seminars spread over three years).

Ventures, similar to the other CSLA programs, is based on a decidedly system approach and regards school leaders as transformational leaders. This is mirrored in the explicitly formulated aims of the Ventures activities:

- improving the students' learning via the notion of school culture as a system, which can be influenced by symbolic actions and symbolic language,
- improving students' achievements by creating a shared notion of how individuals in the school should act so that the students' performance can be improved.

Participants are taught action research skills and learn how to develop hypotheses about the relationship between their observations at their schools and their knowledge of the influence of organizational culture on the behaviour of those working in the school and on their working and learning conditions, seen against the background of a suitable theoretical framework. The participants are supported in the seminars to work out a school development strategy on the foundation of their analysis of the individual organizational culture of their school, with particular attention to organizational impacts on pupil learning. Thus, Ventures links the development of the school leader closely to the development of her or his school. Fellow participants as well as the CSLA trainers serve as coaches throughout the three-year process.

The program typically unfolds in three stages. In stage or year one, participants determine their emphasis or area of activities and learn how to conduct action research. In the second stage, participants gather and analyze data pertinent to their area of emphasis. Based on the data, an improvement strategy is designed, implemented and evaluated for effectiveness. In stage three, the participants summarise their findings and present their results in a culminating seminar. The presentation of the improvement projects serves as a final assessment along with a written record or school story, which documents the participants' three-year research journey.

Costs for the Ventures Program vary from region to region and range from no cost up to 800 US dollars per participant per year. These costs may be covered by the district or by the individual principal from their professional development budget.

School Leadership Teams

Although the Foundations II and Ventures programs have been well received and continue to draw many participants, the CSLA core staff and regional directors realised that to fully bring about the goal of powerful learning for all pupils required not just individual leadership but team leadership by the principal and teachers. Thus in 1993, the School Leadership Team (SLT) component was added to the list of CSLA offerings. Many of the schools sending a leadership team to the training seminars are led by a principal or assistant principal who has graduated from the Foundations II or Ventures programs. A participating team generally consists of 6 to 10 members including the school leader (required) and teachers or counsellors. A few teams in the state broadened their membership to occasionally include a parent, pupil representatives (grade 7 – 12), a district representative or a board member. Who is on the team is left to the discretion of the school. By the summer of 1996 more than 250 elementary, middle or high school teams with over 2 300 members had participated in the SLT program.

Schools self select to participate or may be encouraged to participate by the superintendent, however, schools are strongly urged to have approval from the local board of education before participating. In a few instances, the superintendent has asked regional CSLA directors to provide training for all schools in the district.

Teams explicitly commit to create a culture of inquiry and learning, a willingness to be open to change, and to give and receive support from other teams and CSLA trainers.

The program's aim is to create a community of leaders and to help the teams develop the competences needed to initiate and carry through a school development process designed to enhance pupils' achievement. Often the leadership teams play a critical role in leading a program quality review or at the high school level, the state accreditation process. Because they have learned how to collect and analyze data, they are in a strategic position to assist in these external review processes. These Communities of Leaders form a statewide network. At their particular schools, they are to serve as catalysts for change, develop the leadership capacity of other teachers and multiply the transformative leadership effect in grade levels or departments as well as school wide. As Thompson (1997) has argued, "It is helpful to think of school improvement not as change but as group learning" (p. 15).

Although each region tailors the SLT team full-day seminars, which typically occur 5 times a year for up to 3 years, to meet specific needs of the region, most regions cover the following topics at some point during this timeframe.

- Understanding the roles and functions of the team and defining the team's place in the school's organization,
- team building, high quality team work, and relating with other school groups,
- data collection and analysis, especially multiple types of measures of student achievement,
- understanding the school development process,
- using data to guide school improvement plans,
- conducting and carrying out action research and engaging in an inquiry process,
- understanding the components of powerful learning and strategies for enhancing student learning,
- understanding the California Curriculum Content Standards,
- learning how to address undiscussables and seeing resisters as friendly critics,
- creating a shared vision of powerful teaching and learning and designing curricula and assessments to foster learning by all,
- changing the school's culture to encourage greater collaboration and continuous school improvement.

At the end of each year, the leadership teams are asked to complete two surveys: one given by the CSLA core staff, the other developed and administered by the University of California, Santa Barbara. In each region at the end of the year, teams often are paired and use a protocol to share with each other their accomplishments. Each team serves as a critical friend to the other providing warm and cool feedback in regard to the school's development process and plan.

The costs amount to about 1 000 US dollars per year (about 1 100 euro) and team consisting of up to six people. For each additional member, 150 US dollars are charged. The fees cover all costs for materials and lunches.

Recent Developments in California

In 2002 there were several dramatic developments in administrative preparation and inservice requirements as well as in the California Leadership Academy. These changes have emerged through state legislated mandates and funding cuts. The legislature passed Senate Bill 328, which allows future administrators who want to earn an administrative credential to do so by passing an examination rather than enrolling in a university certified preparation program. In addition, those wishing to receive the second level of certification and who are already in an administrative position may also complete the requirements through an examination rather than through additional university course work and professional development seminars.

Two forces seemed to be driving these changes. One was the concern that there were too few administrators certified to apply for vacant administrative positions, particularly in the large metropolitan areas like Los Angeles with its many schools facing challenging circumstances and high need students. Rural areas and large secondary schools, especially large urban high schools, also indicated a shortage of job applicants. The second force seemed to be a generalised disdain for the university preparation programs, which proponents of the bill asserted were not effective in preparing future administrators. Opponents of the legislation, especially university faculty who offer the certification programs, countered that it was not the lack of certified personnel in the pipeline, but rather the lack of willingness of those certified to apply for these challenging positions. They argued that the shortage of administrators could best be resolved by increasing compensation, redefining the roles and responsibilities of principals to make the job more manageable, and providing a political climate that was less blaming. Since the examination has not yet been developed, it is uncertain at this time the impact this legislation will have on the number of applicants to university preparation programs. It is also an "unknown" whether districts will want to hire as administrators those who have had even less preparation than demanded of the university-based programs. Previous legislation in the 1970s that allowed certification by examination did not dramatically decrease university enrolments.

A second major shift in state policy was the adoption of Assembly Bill 75, which required districts that had high proportions of under-performing schools receiving state funds to have principals of these schools participate in a state prescribed professional development program. Private or public (e.g.. county offices of education) vendors could offer the training if their curriculum met state specifications and was approved. For some topics, even articles principals were to read were specified by the state. Since the program is in its first year of operation, it is too soon to assess its value or impact. The program may be short-lived, however, because California's budget faces a $34 billion short fall for 2003. A number of districts also chose to take a "wait and see" attitude since the program was untested and many felt did not meet their needs.

A third major change, due in part to funding cuts as well as political turf battles, was the eliminate of the state core staff that had guided and directed the California School Leadership Academy for the past 20 years. Although the 12 regional centers in the state were still funded to offer the School Leadership Foundation and Leadership Team programs it is uncertain how these programs will be sustained in the long-run given more extensive budget cuts expected in 2003. Furthermore, although the regional CSL Centers operationally implemented all the programs, the shape of those programs was very much influenced by the monthly meetings of the regional directors with the state core staff. These meetings provided program coherence and consistency as well as ensured that the program continued to adapt to meet changing state needs and respond to new knowledge.

Statement by Janet Chrispeels

The challenge of developing school leaders is clearly in a state of flux in California. The questions abound and answers are still being sought. What should be the content of administrative preparation programs? What is the right balance between a practical program that enables administrators to enact the state's curriculum standards and prescriptive instructional approaches and a conceptually challenging program designed to prepare principals to guide their schools in the process of continuous improvement? Who should provide it? What should be the role of universities, county offices of education, school districts, and private providers? Will public and private universities, which have traditionally offered administrative preparation programs, still have students now that teachers and others can be certified as school administrators through an exam process offered by the California Commission on Teacher Credentialing? What should ongoing preparation and development of administrators look like and who is best position to provide it? How can be best support teacher leadership development as well as administrative preparation?

There is a need for more administrators in California, especially in Los Angeles, Oakland and other highly impacted urban areas, who are prepared to assume leadership in the neediest urban schools. However, an historical perspective suggests that when administrative certification by exam was instituted in the early 1970s, only some chose this route. Most other future administrators will continue to enrol in university preparation programs. The exam route may provide short-term relief for urban districts, but is unlikely to solve the problem of insufficient numbers of administrators. A quicker route to certification does not address the primary reason that there are too few administrative applicants in the urban areas. The principal's job in urban centers has become almost untenable. Schools are over-crowded and operate on 12 month, multi-track year-round schedules. The demands of the job of being a plant manager for a facility designed for 600 and now serving 1200 students, a human resource administrator overseeing a staff of 100 to 150, and instructional leader are too great for any one or two people. Until fundamental job design and redesign questions are addressed, the "shortage" of highly qualified principals and assistant principals to lead urban schools is likely to persist. Furthermore, another reason for candidates may continue to pursue university training is that suburban school districts, which offer more desirable working environments, have a choice in selecting administrators. They are more likely to choose applicants that have had more training and preparation, not less. (Interestingly, in the testimony before the CCTC, the developers of the administrative certification exam, Educational Testing Services, testified that their exam was never intended as a stand-alone certification process but to be used as an evaluative instrument after a candidate had completed a preparation program.)

As the California Commission on Teacher Credentialing prepares to adopt new standards for administrative preparation programs, which are aligned with the Interstate School Leaders Licensure Consortium (ISLLC) standards, the content of university preparation programs are likely to undergo a shift to bring their programs more in alignment with these standards. The ISLLC standards tend to be more global in scope in contrast with the current standards, which are more aligned to job specific knowledge such as expertise in leadership, educational finance, curriculum and instruction, and home-school-community relations. The challenge and generally unanswerable question is how changing standards significantly and qualitatively alters the preparation of principals and if such alterations actually lead to more qualified candidates. From my own experience of preparing future administrators, there are two changes that could enhance the quality of preparation. One is the requirement that school districts and universities offering the credential must form partnerships that support candidates to be involved in internship during their enrolment in a university program. Although the internship option exists, many districts are reluctant to support teachers in such a position. A second suggestion is for universities and school districts to collaborate much more extensively on recruiting and selecting future administrative candidates to participate in university programs rather than to the general self-selection process that operates currently. Such a collaborative process would promote a more thorough discussion of the qualities, skills and habits of mind that the profession feels will best equip future leaders to fulfil the variety of leadership positions in schools.

In July 2002, the state education budget eliminated funding for the core staff that had coordinated the California School Leadership Academy 12 centers, and this year the budget eliminated funding for the regional centers. These financial and political decisions brought to an end a 20-year history of CSLA support for administrative and teacher leadership development. Who will fill their void? It is ironic that this incredible network of school leadership development has been eliminated just as England's government is investing heavily in establishing a similar Networked Learning Community to support leadership development. A significant loss is the School Leadership Team Training program, one of the few in the country and certainly the one with the longest history. My research of this program, showed that through training, teams could be helped to become focused on improvements in teaching practices and student learning needs, precisely what studies of site-based management has shown is needed if student learning is to be improved.

3. Summary of Washington, New Jersey, and California, USA

School districts in the US are governed by school boards. Members are usually elected officials but sometimes they may be appointed by local government agencies depending upon local laws. This is a decentralised system. Boards are policy making bodies but often times may tend to micro-manage. The work of principals at the local school level is influenced by these governing boards, as well as by the State Departments of Education and local district administrators such as superintendents, assistant superintendents, and directors.

The US has a long and well documented history of decentralised education. Education is largely a state responsibility. In many cases, universities play a significant role in the preparation of school leaders as well as teachers. Currently, there are a number of initiatives that may result in alternative leadership preparation programs that are more school or district centred than university driven.

Currently, most States require prospective principals to have a masters degree and pass a knowledge and/or case study exam. This process can take up to four years for part-time traditional programs although there are new models emerging that take less time and require more field-based experiences. Some programs require full time field experiences and candidates have to leave their current teaching jobs to fulfil that requirement. Some programs integrate the field experiences with current full-time employment. In either case, the school's principal as mentor plays an important role in the development of the knowledge, skills, and dispositions of the candidates.

Some states, like New Jersey, have assessment centers that evaluate the knowledge and skills of new principals and develop a leadership plan with them. In New Jersey, a new principal is also assigned a mentor principal during the first year of employment. Upon successful completion of this process, a permanent certificate is awarded that is good for life. Some states may require additional professional development throughout active employment. In NJ there is no requirement for additional professional development for principals.

The Danforth Educational Leadership Program of the College of Education at the University of Washington, Seattle, was set up in 1987 within the scope of the sponsorship of the Danforth Foundation and since then evaluated and modified continuously. It offers the Washington State initial principal certification and masters degree at the end of an intensive one year course of study and internship. The schedule brings the students together for intensive sessions during the summer and throughout the academic year to participate in activities and discussions linked closely to the practice of leadership in the schools. Field supervisors work with students at their internship schools. At the end of the program, students create a portfolio of their work which includes a leadership platform outlining their beliefs and how they plan to carry out the work of leadership in public schools. The schedule layout changes between several-day block seminars in the summer holidays and half-day or day events throughout the entire school year. The creation of mixed learning groups is encouraged. The methodical repertoire contains mainly participant- and application-oriented procedures. Great store is set by learning in the workplace during the whole-year practice, with the experiences brought into the seminars and thus enables the participant, to participate in concrete leading tasks in the course of the school year.

The Educational Leadership Program at William Paterson University of New Jersey in the College of Education, seeks to balance the competing tensions of management and leadership for aspiring principals. This cohort program requires candidates to work together over a two-year period, develop their leadership skills, and maintain their current job positions, usually as teachers, directors, supervisors, or coordinators. Candidates develop Individualized Leadership Plans that focus on career goals and their strengths as well as areas in need of continuous improvement. The ILP becomes an action plan throughout the field experiences and, along with the ISLLC Standards for School Leaders, shapes the experiences and develops and nurtures the knowledge, skills and dispositions of candidates as they aspire to school leadership positions that require principal licensure. Candidates are visited regularly in the field by faculty advisors. To maintain the quality of the program, no more than 25 candidates will be accepted annually. Since their programs must be individually financed, and because of the complexity of the principalship, a limited number of educators apply for admission. Cohort I began in May 2000 with 23 candidates selected from over 30 applicants and graduated 18 candidates in May 2002. Cohort II began with 24 candidates and is beginning year two with 22 candidates. Our graduate program will support a deep commitment to leadership that will enable candidates to construct meaning and knowledge together. Courses will become a center around which candidates will form smaller, micro-groups that will vary according to needs and interests. These groups will meet throughout each course to focus on specific questions related to the course of study. Candidates will be given opportunities to become facilitators in micro-groups and lead the other participants in reflective inquiry and problem-based learning activities. This model will necessarily restructure course time and develop a value for life-long learning outside the context of classrooms. Candidates will meet with the cohort for longer and less frequent time periods in the macro-group. The smaller micro-groups will meet throughout the semester in between scheduled class meetings. Quality and depth will become more important than quantity and breadth.

The programs of the California School Leadership Academy represent a unique commitment by the California Department of Education to provide ongoing professional development for aspiring and current school leaders. Over its 20-year history the programs have evolved to address changing school needs as well as state policy mandates (e.g., the adoption of Curriculum Content Standards for all grade levels and subjects which schools are expected to implement as well as new testing programs to assess pupil learning and rank schools). These programs compliment the university based pre-service preparation programs and allow school leaders to extend their learning in practical and school-based contexts. Without the demands of a more structured university course system, the CSLA can quickly update practising and aspiring school leaders' knowledge and skills to meet their needs in a rapidly shifting political climate of mandates and calls for accountability.

The three programs have consistently pursued their general aim of promoting instructional leadership and learner-centered schools with slightly different audiences. Foundations II addresses the needs of aspiring and newly appointed school leaders. Once in the post of principal or assistant principal, Ventures pushes these more experienced leaders to a deeper level of analysis, reflection and action through sustained work on a school improvement project. The School Leadership Team program allows principals to develop teacher leaders so that they are not trying to improve their school alone but are able to engage in a vigorous process of school development with the assistance of a strong team of teacher leaders.

Short Summaries of Country Reports: A Juxtaposition

The following table summarises characteristics of school leadership development models in 15 countries. It is meant to provide an easily accessible overview of predominant approaches in use across Europe, Asia, Australia/New Zealand, and North America.

Table 36: Current international approaches of school leadership development

Europe	
Sweden	A national preparatory program offered by universities through an optional basic course plus additional offers by the municipalities.
Denmark	Optional offers made by municipalities, universities and private suppliers without any central framework or delivery system.
England	A centrally organised program delivered by regional training centres; combining assessment and training with a competence-based and standards-driven approach; the program is embedded in a three-phase training model.
The Netherlands	A broad variety of different optional preparatory and continuous development programs by different providers (e.g., universities, advisory boards, school leadership associations) in an education market characterised by diversity and choice.
France	A mandatory, centrally-designed, intensive, full-time, half-year preparation program with internship attachment for candidates who have successfully passed a competitive selection process; completion guarantees a leadership position on probation during which further participation in training is required.
Germany	Courses conducted by the state-run teacher training institute of the respective state, mostly after appointment; differing from state to state in terms of contents, methods, duration, structure, and extent of obligation.
Switzerland	Quasi-mandatory, canton-based, modularised programs offered post-appointment; delivered by the respective provider of the canton, most often the teacher training institute, wherein the aim is nation-wide accreditation (national standards are currently being developed).
Austria	Mandatory centrally-designed, modularised courses post-appointment; delivered by the educational institute of each state; required for continued employment after four years.
South Tyrol, Italy	A mandatory program for serving school leaders to reach another status and salary level as becoming 'dirigente'; delivered by a government-selected, private provider, which combines central, regional, and small group events with coaching attachment.

Region / Country	Program
Asia	
Singapore	A mandatory, centrally-controlled, preparatory, nine-month, full-time program provided through a university; comprised of seminar modules and school attachments.
Hong Kong, China	A centrally-designed, mandatory, nine-day, content-based induction course immediately after taking over the leadership position.
Australia/New Zealand	
New South Wales, Australia	An optional, modularised, three-phase program offered by the Department for Education; centrally-designed, yet conducted decentralised via regional groups; besides there are offers by independent providers.
New Zealand	A variety of programs with variation in contents, methods and quality; conducted by independent providers, but also by institutes linked to universities; no state guidelines, standards or conditions for licensure.
North America	
Ontario, Canada	Mandatory, preparatory, university-based, one-year, part-time programs delivered through several accredited universities following a framework given by the College of Teachers, the self-regulatory body of the profession.
Washington, New Jersey, California, USA	Mandatory, intensive, preparatory, one-to-two-year university programs that include extensive internship attachments; programs use a broad variety of instructional methods.

Summary of the Book

The central importance of school leadership for the quality and the development of the school is increasingly emphasised in a growing number of countries. The relevance, quality, and timelines of school leader preparation programs, carried out by the profession itself, by educationalists, but also by educational policy makers as a critical component of this process can not be underestimated.

According to the results of school effectiveness research, school leadership is a key factor for the school's quality. Current studies on school development also focus on the role of school leadership in the improvement process: school leaders are considered decisive change agents.

Particularly in the last decades, school leaders worldwide have been confronted with new challenges and demands. Internationally, similar social, cultural, political, and economic changes have been emerging. These will eventually affect schools and their stakeholders. Moreover, educational policy measures such as an extended school-based management increase the challenges imposed on school leaders. The role and function of school leadership has undergone swift changes in many countries of the world. New tasks are added to the traditional duties, and the character of the latter ones is changing as well.

Correspondingly, many different skills and competences are required for performing this professional role, which make up the amalgam school leadership competence.

New leadership theories are meant to respond to the manifold demands on school leadership. While, for example, the term 'transactional leadership' has been applied to a concept of 'steady state leadership', 'transformational leadership' is reputed to be particularly successful in school development processes. 'Integral school leadership' aims at an integration of management and leadership tasks. Studies conducted in North America, especially in the field of school effectiveness, have emphasised the relevance of 'instructional leadership'. The new German field of 'organisational education' provides a consistent link of educational actions and organisational actions, based on educational principles. It favours leadership conceptions like 'post-transformational leadership' or, in the light of 'organisational education', the leadership conception of 'organisational-educational management and leadership' in terms of 'cooperative and democratic leadership'.

The great responsibility of school leaders, requires suitable training and development opportunities. A number of countries in the world already offer solid qualification programs, others have just started to do so, or are modifying existing ones.

The aim of this study was to explore the school leadership training and development landscape in a sample of countries. On the basis of this general research interest with its exploratory character, the following central research question was formulated: 'how are (aspiring) school leaders qualified in different countries of the world, what kind of training and development opportunities for school leaders are offered?'

This study should be viewed in the context of the methodological discussion of international comparative educational research. Its methodical approach corresponds to that of the systematic procedures of comparative educational research; the individual steps can be illustrated accordingly: the research question was first of all operationalised. Then, the sample countries and programs were selected. However, the sample was not restricted at the first stage of the study, but was based on the accessibility of information, which depends on language conditions as well as on contacts to experts and persons in charge in the respective countries. Methods of data collection comprised international literature research, expert meetings, systematic collection of documents, a systematic questionnaire developed for this study, open and non-directive questioning, and additional research via telephone and email. The data was screened, sorted, selected according to a classification scheme, systematised in several steps, and was then presented in a matrix, which allowed a systematic comparison.

The analysis resulted in country or program descriptions, which in turn were modified in several steps. An additional validation of the country and program descriptions by renewed feedback from the expert of the respective country was carried out. This educationalist was invited to be the co-author of the country report. Hence, a specific methodical step has been achieved that way, namely the combination of an independent external expert, who assures the consistent approach across the country studies, and an internal expert, who has specific context knowledge.

*

The countries with their individual training and development programs can be compared according to certain classification criteria. The following dimensions are used for comparison: 'provider', 'target group, timing, nature of participation, and professional validity', 'aims', 'contents', 'methods', and 'pattern'. By means of comparative statements, similarities are grouped together and differences are emphasised. The comparative statements are then discussed, central aspects are referred to, investigated in greater detail, and commented on.

The issue of who provides the qualification is solved in different ways in the sample countries. Central provision on the one hand and completely decentralised provision on the other hand, which then features a diversified market of providers, can be explored. Sometimes there are only providers of a certain type (e.g. universities alone) while in other countries various forms of cooperation and interconnected responsibilities exist. The principle of a rather decentralised implementation of the qualification programs concurrently with central institution for quality assurance through nation-wide standards, the accreditation of providers, or a standardised certification procedure for the participants seems particularly interesting. Emphasis should be placed on the staff and the training of the trainers. The combination of the teams of trainers including individuals with different professional backgrounds is supposed to create synergy effects.

In more than half of the countries, the target group of the qualification consists preferably of teachers aspiring to a leadership position or of designated school leaders before taking over a leadership position. A preparatory qualification like this corresponds adequately with the challenges imposed on school leadership as a profession in its own right, which requires an adequate professionalisation. Here, the qualification is given a particularly high professional validity and status. Access to the qualification programs is restricted through a selection or assessment process in a number of countries.

The distinction between programs focusing on the individual school leaders and aiming at providing them with the necessary competences on the one hand, and those closely linking school leadership training and development with school development processes at the individual school on the other hand seems particularly instructive. In the latter programs, the school leader participates together with a team (from staff, sometimes even together with parents or community representatives). It would be worth some consideration whether these teams should not even be arranged beyond the different levels of hierarchy, that is whether they should include representatives from the school authorities as well.

When comparing the mostly explicitly and extensively formulated program aims, a heterogeneous picture emerges. Some have a functional orientation, that is they explicitly take into account the demands of the government. Others have a task orientation and start from quite a pragmatic preparation for the different tasks of school leadership. Partly there is a competence orientation or a school development orientation. Some distinctly aim at the change or development of mental concepts as they have a cognition orientation. Programs with a vision orientation build explicitly on a vision of leadership, or on a vision of school. Partially, new paradigms for leadership like 'transformational leadership' or 'integral leadership' – are enunciated with the formulation of goals. Those having a value orientation include moral and political dimensions of leadership. A clear grouping of the qualification programs to not more than one of the criteria, however, is very rarely possible, for most of the programs take quite different foci into account.

Some programs explicitly use a reflection on the goals and development of school and schooling as a starting point: on the basis of a conception of the educational goals of school and the teaching and learning processes, consequences for the conception of school leader qualification are deduced. School is systematically viewed as a learning community, as a culturally independent organism which is able to develop and has to do so. An organisational-educational adjustment of the aims and objectives of school seems to be implicitly included.

As far as the contents of the training and development programs are concerned, a multiple paradigm shift can be observed: a communicative and cooperative focus replaces the primacy of administrative and legal topics. The concept of an 'educational leader' is at the centre of attention; components of school-based leadership are often represented in the compilations of topics of the programs. Instead of perceiving school as a static institution and understanding school leadership as the administration of the status quo, this concept has been replaced by a new perception of school as a learning organisation and its leader as a driving force and warrantor for effective change processes. Developing school rather than administering school is emphasised. This is mirrored by the selection of topics. Consequently, communication and cooperation are important issues as well.

The methodical implementation of the contents is increasingly based on adult learning principles. Next to rather traditional forms of imparting knowledge and learning (such as lectures and plenary sessions), methods oriented towards processes, participants, and application are often explicitly included, for example, group work, role playing, or simulation exercises. Here, types of collegial learning such as peer-assisted learning, learning communities, and networks seem to be worth mentioning. Methods like cognitive mapping as well as keeping a learning diary or writing a learning journal are extremely process- and reflection-oriented. The problem-based learning approach features a clear problem orientation and strongly relates to applying knowledge. Moreover, the participants can leave the workshop and use the workplace itself as a learning place by means of internships, which are integrated into the program. Next to shadowing, actively trying out leadership behaviour becomes increasingly important, thereby, guidance is provided to the future school leaders by mentors.

When considering the time needed to complete the program, remarkable differences become apparent. In many countries the number of training and development days provided have increased, which takes the identified tasks and duties of school leaders as well as their importance for assuring the quality of school into account quite clearly. Therefore, internationally, the participants need to invest significantly more time and efforts into qualification measures than they used to do.

A number of countries show a tendency towards multi-phase designs and a modularisation of programs. Instead of one single training period, multiple stages of qualification are scheduled, whose individual phases are synchronised and tailored to suit the different stages of the professional career. Then, a certain degree of modularisation is added as well. The various modules can then be chosen according to the individual training and development needs and the needs of the individual school.

*

From the results of this study, certain international trends and development tendencies can be deduced. Some of them are mere shifts of emphasis while others may be even considered changes of paradigm. Such current trends and paradigm shifts in qualifying school leaders include: central quality assurance and decentralised provision; new forms of cooperation and partnership; dovetailing theory and practice; a preparatory qualification; extensive and comprehensive programs; multi-phase designs and modularisation; personal development instead of training for a role; the communicative and cooperative shift; a shift from administration and maintenance to leadership, change and continuous improvement; qualifying teams and developing the leadership capacity of schools; a shift from knowledge acquisition to creation and development of knowledge; experience and application orientation; new ways of learning: workshops and the workplace; adjusting the program to explicit aims and objectives; new paradigms of leadership; an orientation towards the school's core purpose.

Considering the various approaches, concepts, and macro- and micro-didactic realisations of development opportunities for school leaders, research into the effectiveness of the individual programs would be of central importance. This, however, cannot be achieved by this study for various reasons. An exploratory international comparison like this, however, is the conditio sine qua non for extended research into the efficiency of such programs.

What can be accomplished here, however, is to describe best and promising practice against the background of the analysis, the comparison, the discussion of the findings, and the theoretical considerations based on the research paradigms of school effectiveness, school improvement, and on 'organisational education'. The Danforth Educational Leadership Program of the University of Washington, for example, can be considered a highly inspiring example of best practice.

Besides that, one might indicate approaches which explicitly regard the development of school leaders as a continuous process with various phases matching the different phases before and after taking over a school leadership position (e.g. the multiple-phase concept of the School Leadership Strategy in New South Wales featuring modularisation, too). Moreover, the concept's target group does not only include aspiring school leaders, but also teachers who would generally like to expand their leadership competences or strive for or already hold positions with extended responsibility others than principal- or headship. It might even be feasible to include entire school leadership teams or teams within staff in combination with representatives of the parents or the community, as it is the case in the training program School Leadership Teams of the California Leadership Academy. Offerings for leadership teams like this can be considered best practice, since not only increasingly shared leadership competences, but also extended development capacities can be gained from them.

It becomes apparent that a school leadership training and development program needs to use a multi-stage adjustment of aims as a starting point. The first question would be: what are the essential aims of education? From these, the corresponding aims for schools and schooling in general can then be derived: what is the purpose of school and what are the aims of the teaching and learning processes? Considering the perspective of the new field of 'organisational education', one should ask: how does the school organisation need to be designed and developed in order to create the best conditions possible so that the entire school becomes a deliberately designed, educationally meaningful environment? This in turn would enable teaching and learning to take place as well as multi-faceted and holistic educational processes that would lead to achieving the schools' aims.

This, again, leads to the essential concern of school leadership: what are the aims of school leadership regarding the school's purpose and the individual context of each school? To be even more precise: how do school leaders lead if it is the goal of leadership to shape school in a way that makes it possible to reach its aims?

This is explicitly how the aim of a school leader training and development needs to be deduced, which reads as follows: what does a school leader training and development need to be like in order to prepare and support school leaders as well as possible for providing an individual answer to this core issue. This should be the essential or core goal for aligning and evaluating school leadership development programs.

Nineteen fundamental recommendations for designing and conducting school leader training and development programs can be deduced from this international comparative study. These recommendations could even serve as guidelines for the design and conception of future programs, or as standards which have to be considered by providers. Moreover, they could be the criteria for the accreditation of providers and programs or, in case of a certificate for quality assurance offered to the providers, serve as criteria for certification. These recommendations include: centralised guidelines for quality assurance combined with a decentralised implementation; a suitable recruitment of teams of highly qualified trainers with appropriate backgrounds; a selection of participants; a clear and explicitly stated definition of aims, using the core purpose of school as a focus; an alignment according to values and educational beliefs; development as a continuous process; the importance of declarative and procedural knowledge; a suitable balance between theory and practice; the orientation towards the actual needs of the participants; an active involvement of the participants; inspiring collegial learning and intensive collaboration; problem-based training in workshops; learning opportunities at the workplace; a focus on the personal and professional development of the participants as well as on improving their schools; self-organised and reflective learning processes, supported by communication and information technology; academically grounded and authentic training material; the presentations of learning results and self-evaluation of learning processes; the certification of participants; and a conceptually established support for the actual transition.

The involvement of research in the field of school leadership and school leader qualification remains particularly important. A number of research desiderata can be indicated here. There is still some need of further pure research into educational leadership and the training and development of educational leaders, and of research into tasks of and demands on school leaders and the training and development needs of school leaders at different career steps and in different school contexts. Specific research has to be carried out in the ways school leaders develop competences. Further international comparative studies would be helpful. There should be effectiveness analyses as a basis for judgments of the quality, effectiveness, sustainability, and efficiency of development programs. Moreover, it would be interesting to investigate how the development of individual school leaders could be linked effectively to the development of individual schools in terms of qualifying school leadership teams and other change agents in the individual schools. The importance of recruiting or selecting school leaders and its connection with school leader development should not be neglected. Additional research would be desirable here as well.

It would also be useful to link empirical research to concrete conceptions of development programs, in terms of 'research-based training conceptions'. National and international experiences should be combined, and international cooperations should be promoted. As a basis for this, national and international networks should be developed (further) as a forum for exchange and cooperation.

*

A standardised structural pattern is employed for the country reports: they provide information about the educational and school system of the country, recent educational changes, and consequences for the role of school leadership. Then, information about the development provision of the country is given, partially including information about its development and historical course (when relevant), about the range of offerings as well as about the general approach to training and development. After that, the qualification program of the respective country or – if several opportunities exist – exemplary programs are described in detail.

Additionally, a table provides an overview about each program. A summary tries to pull all essential aspects together. Another section provides information about recent developments in the program since the time the description is based on. Finally, the internal expert gives a personal statement.

In Sweden, the qualification is provided by means of shared responsibility, i.e. partly by a national, basic state program conducted at universities and partly by regionally differentiated additional offers made by the municipalities. Participation is optional, but usually expected by the communities employing the school leaders.

In Denmark, school leaders already in this position are qualified through optional offers by the municipalities, universities and private suppliers without any central framework.

In England, qualification takes place in the context of a one-to-three-year, modularised program featuring regular assessments. It is centrally organised, but implemented locally and embedded in a concept comprising three phases, including a second stage for newly appointed school leaders as well as a third for experienced school leaders, who have held that position for some time already.

In the Netherlands, there are a broad variety of different optional preparatory and continuous development programs by different providers (e.g., universities, advisory boards, school leadership associations) in an education market characterised by 'diversity and choice'.

Recruiting and qualifying school leaders in France are the government's responsibility. There is a mandatory, centrally-designed, intensive, full-time, half-year preparation program with internship attachment for candidates who have successfully passed a competitive selection process; completion guarantees a leadership position on probation (during which further participation in training is required).

In Germany, the courses conducted by the state-run teacher training institute of the respective state, mostly after appointment differ from state to state in terms of contents, methods, duration, structure, and extent of obligation.

Switzerland features quasi-mandatory, canton-based, modularised programs offered post-appointment delivered by the respective provider of the canton, most often the teacher training institute, wherein the aim is nation-wide accreditation (national standards are currently being developed).

In Austria, the obligatory qualification is conducted in the individual federal states after appointment to a position as a school leader. The program is designed and implemented in compliance with state guidelines and consists of modularised School Management Courses offered by the respective educational institute of each federal state. It is required for continued employment after four years.

The South Tyrolean mandatory program for serving school leaders to reach another status and salary level as becoming 'dirigente' is delivered by a government-selected provider. It combines central, regional, and small group events with coaching attachment.

In Singapore, the mandatory, centrally-controlled, preparatory, nine-month, full-time program is provided through a university. It comprises seminar modules and school attachments.

School leaders in Hong Kong, China, are qualified through a centrally-designed, mandatory, nine-day, content-based induction course immediately after taking over the leadership position.

The qualification for leadership positions in New South Wales, Australia, is accomplished through an optional, modularised, three-phase program offered by the Department for Education. It is centrally designed, yet conducted decentralised via regional groups. Additionally, there are offers by independent providers.

In New Zealand, a variety of programs with variation in contents, methods and quality are conducted by independent providers, but also by institutes linked to universities. There are no state guidelines, standards or conditions for licensure.

In Ontario, Canada, school leaders are qualified through a mandatory, preparatory, university-based, one-year, part-time program delivered through several accredited universities following a framework given by the College of Teachers (the self-regulatory body of the profession).

In the US States of Washington, New Jersey and California, the obligatory, preparatory qualification is carried out on the basis of standards and can be freely chosen among one-to-two-year full-time courses of study leading to a university degree. They possess a strong link to practice by means of extensive internships and the implementation of participation- and application-oriented methods such as problem-based learning, coaching, shadowing or mentoring.

Information about Co-Authors

Michael Chirichello
is a professor and chair of the Department of Educational Leadership at William Paterson University of New Jersey. He designed the M.Ed. cohort program for aspiring principals. Michael was a County Superintendent of schools, a District Superintendent, Principal, Vice-principal and teacher in urban and suburban districts in New York and New Jersey. He was a Senior Fellow for Citizens for Better Schools in NJ and a Fulbright scholar. He has authored several articles and consults nationally on topics of leadership and curriculum design. He was a collaborating author and WPU's site coordinator for an $8.1 million, five-year, grant awarded to three universities to improve the quality of teacher preparation and the life-cycle of the career teacher.

Janet Chrispeels
is a professor in the Gevirtz Graduate School of Education, University of California Santa Barbara. She directs the certification and leadership development programs for school administrators and teaches, and supervises doctoral students in the Educational Leadership and Organizations emphasis. She is the director of the new Center for Educational Leadership at UCSB and also directs the California Center for Effective Schools, which works with schools and districts to implement systemic school change. Her current research involves a longitudinal study of California School Leadership Teams (SLT) as well as studies of the implementation and impact of the Effective Schools Initiative in one California school district. Janet has served as president of the Board directors of National Center for Effective Schools Research and Development Foundation and of the International Congress for School Effectiveness and School Improvement, is currently on the editorial boards of several international journals, and is series editor of Swets and Zeitlinger's Context in Learning Series.

Peter Cuttance
is director of the Centre of Applied Educational Research and a professor in the University of Melbourne. He operates a diverse portfolio of interests in education and is a director of the Research Australia Development and Innovation Institute and director of Educational Development and Review, both of which operate as teams of private sector educational development and consultancy partners. Peter was the head of the School Review and School Development Programs in South Australia and New South Wales from 1989-96, where he served as Director and Assistant Director-General of Education, respectively. He has been a professor of education and the head of the School of Educational Psychology, Measurement and Technology in the University of Sydney and a professor of education in the School of Educational Psychology, Literacies and Learning in the University of Sydney.

Stephan Huber
is a researcher and consultant at the Research Centre for School Development and Management of the University of Bamberg which he is currently co-directing. He is holding a Master of Philosophy of the University of Cambridge, England, and a Doctorate in Philosophy of the University of Bamberg, Germany. He has been conducting research and consulting in the fields of school development, continuous professional development of teachers, and school leadership and leadership development and assessment nationally and internationally. He is chairing an international network for school leadership development. Among his projects are, for example, reviews of national and international school effectiveness research and school improvement initiatives, a survey of development needs of teachers in Germany, and a feasibility study to establish a National College for School Leadership with a conception including different scenarios.

S. Gopinathan
is the dean of the Foundation Programs at the National Institute of Education, Nanyang Technological University, Singapore. He is responsible for all initial teacher training in Singapore. From 1994 to and citizenship education policy developments in cross-national 2000, he was the Dean of the School of Education, and the head of NIE Centre for Educational Research. His research interests are in education and development, language perspective, school effectiveness and school leadership and teacher education. He has served as a consultant to the World Bank, Asian Development Bank, Asian Institute of Development Studies, and the Commonwealth of Learning.

Jeroen Imants
is an associate professor at the Graduate School of Education / ILS at the University of Nijmegen. In cooperation with GCO fryslân he developed the conceptual framework for the school leader training program Master your school. His recent research interest is professional development of teachers and school leaders. He has initiated varying research projects in this field. He participates in two nationally granted projects on teacher learning and educational steering by school leaders, in which several universities cooperate. Jeroen is also Co-Chair of the International Committee of Division A of American Educational Research Association.

Olof Johansson
is an associate professor of Political Science, chair of the Centre for Principal Development and the director of the National Head Teachers Training Program at Umeå University, Sweden. He is also an international associate of the Centre for the Study of Values and Leadership at Ontario Institute for Studies in Education, University of Toronto, Canada. He also holds an appointment as a visiting fellow of the International Institute for Educational Leadership at the University of Lincoln, United Kingdom. His research interests are political culture, public policy making, administration, and educational leadership. He is at present involved in three different comparative research projects with England, Cyprus, Greece, Hong Kong, Canada, the USA, and Australia.

Kathy Kimball
is an assistant professor and director of the leadership preparation programs for principals and superintendents in the College of Education at the University of Washington. Kathy served five years as the assistant executive director of the Washington State Commission on Student Learning. Her work experience includes K-12 teaching and leadership roles in Maine: school principal, assistant superintendent, and director of special education and curriculum. She has a doctorate in Educational Leadership and Policy Studies from the University of Washington and a master's degree in Special Education from the University of Utah.

Kenneth Leithwood
is professor of Educational Leadership and associate dean for Research at the Ontario Institute for Studies in Education/University of Toronto. He does research and writes about school improvement, educational leadership and large-scale reform. He is the senior author of both the first and second International Handbooks on Educational Leadership and Administration. His recent books include Changing Leadership for Changing Times (1999), Organizational Learning in Schools (with Karen Seashore, 1999) and Making Schools Smarter (2002). Along with Michael Fullan, Lorna Earl and others, he has been responsible for conducting the government-funded external evaluation of England's national literacy and numeracy strategies for the last four years.

Denis Meuret
is professor at the University of Burgundy, Dijon, France. He worked as the head of the department of school evaluation in the Ministry of Education. He has been involved in several international projects on school and systems evaluation. He teaches Evaluation, Comparative Education, and Economy of Education. His main current research topic is how to apprehend and compare the equity of educational systems. He also is engaged as a deputy director in a program for the coordination and improvement of educational research in France.

Lejf Moos
is associate professor and the director of The Research Program on Professional Development and Leadership at the Danish University of Education, Copenhagen. He has been teaching and doing research in school leadership, school development, professional development and school evaluation. He has been created, launched and maintained courses and graduate studies in school leadership and school development. Many research projects have been carried out in trans-national collaboration with colleagues from Scotland, England, Norway, Sweden, Canada, USA, and Germany.

Jan Robertson
is the assistant dean for International Development, and director of the Educational Leadership Centre of the University of Waikato, Hamilton, New Zealand. Her research is in the area of graduate leadership programs, and leadership coaching. She is working nationally and internationally in this field, and was involved in the design of the induction program for New Zealand principals.

Heinz Rosenbusch
is a professor at the University of Bamberg, Germany, where he is the director of the Research Centre for School Development and Management. His areas of professional interest are school inspection, school management, school organisation, the development of school leaders and teachers, non-verbal communication, group processes in the classroom, and organisational education. After his initial teacher training, he served as a classroom teacher and as a school leader for 17 years. Since 1975 he has held various positions at university. He launched the Bamberger School Management Symposia, and worked as a counsellor, evaluator of school development processes in various German states, and as a member of editorial boards of several educational journals.

Michael Schratz
is a professor of Education at the University of Innsbruck, Austria, where he acts as the director of Teacher Education and School Research. His main interests are in educational innovation and change with a particular focus on leadership, self-evaluation and quality development. He taught in Austria and Great Britain, did research at the University of California, San Diego, and worked at Deakin University, Australia. Among his publications are books on teaching and learning, leadership and management, innovation and change, evaluation and quality assurance.

Anton Strittmatter
is the head of the Education Department of the Swiss Teachers' Association. He is also the director of a school leader training program for all types of schools (a joint venture between the Lucerne Academy for Adult Education and the Swiss Teachers' Association) and a consultant for schools for their school development projects, the reorganization of school leadership, and crisis interventions. He taught at universities in Switzerland and Austria, and publishes on school development and school leadership issues.

Mel West
is the dean and director of the Faculty of Education, University of Manchester, and also holds a chair in Educational Leadership at Manchester Business School. Currently he is co-directing the Leadership Development Unit, established by the DfES and the National College for School Leaders, to support leaders in schools facing challenging circumstances. Presently, much of his work is focused around the problems of leading schools in difficult, urban contexts. He has worked on school management and school improvement programs for many years and in many countries, and has particular interests in China, where he is engaged in a number of collaborations and initiatives.

Huen Yu
is a senior lecturer of the Department of Educational Policy and Administration of the Hong Kong Institute of Education. His major research areas include school management, principal leadership, educational change and school-institute collaboration. He has had 16 years' principalship experience. He has been a Central Council member of a large sponsoring body and a school manager of its secondary schools, primary schools and kindergartens for many years. He is currently a supervisor of four primary schools. He was an adviser of the Working Group on the Professional Development of Principals of the Hong Kong Education Department. He has been actively involved in local and overseas professional development projects for newly appointed principals, serving principals, and education administrators.

Appendix: Methodology and Methods

1. Comparison and Comparability

'Comparison' plays an essential role in every process resulting in scientific insights. Fundamentally, insights are gained in comparison. Whenever something is being measured or phenomena are arranged quantitatively or qualitatively, 'comparing' them to a certain standard involved – either explicitly or implicitly – always plays an essential part. In accordance with cognitive science, every perception in daily life is immediately connected to comparisons. An object becomes identifiable because the brain compares the perceived image of this object with learned concepts and previously seen images and thereby realises structural similarities and differences. From there, the categorisation of the object perceived is deduced. Categories and concepts can only be created due to processes of abstraction by recognising similarities and differences. Experiments – as the chief method used for empiric research – depend on comparisons as well. The results of several experiments are compared to the result expected in compliance with the hypothesis. This enables statements concerning the validity of the hypothesis postulated.

A number of scientific disciplines – among them education since the beginning of the 19th century – have systematically used comparisons as a method for gaining scientific knowledge. Hereby, a systematic comparison is supposed to establish a relation between two (or more) phenomena with regard to a 'third' aspect, the 'tertium comparationis'. If the phenomena to be compared differ concerning their structure, the degree of abstraction of this base becomes higher. Newton's example of comparing apples with planets, which indeed did prove to be comparable items as far as gravity is concerned, is often referred to. If complex social systems are included in the comparison, an additional problem arises: phenomena which are similar in their structure serve different functions in different social contexts. Hörner (1993) infers that the actual objective of the comparison cannot be mere data juxtaposition (e.g. the juxtaposition of the products of the education system) but their functional relevance. The function then becomes an essential 'tertium comparationis'. The concept of functional equivalence then facilitates the equalisation of differences as well, as far as their function is concerned. Two things are functionally equivalent when they reach the same goal and solve the same problem even if they are structurally heterogeneous (see Duijker, 1955; Luhmann, 1968).

Busch et al. (1974) pose the problem of the comparability of the sample units on a theoretical as well as on a practical level. The latter faces the difficulty of making the data from different social, cultural and national contexts and the categories they are based on suitable for an international comparison. Additionally, the analytical concepts employed need to be internationally valid and not originate in an ethnocentric perspective.

In this research project the point of reference for the comparison is the function of the different training and development programs for school leaders. They are functionally equivalent if they – no matter how structurally heterogeneous they may be – reach the same aim, which is to appropriately prepare school leaders for their role and tasks in the respective school system.

The functional connections of the different system levels to the outside and to the inside, or those among the levels themselves are included in this work, such as the relations within the education system (see, for example, the information provided about the changes in the various education systems and what impact these had on schools and the role of school leadership in the various countries), as are the relations of the education system to the economic system, the cultural or especially the political and the educational-policy system (only partly, however, see the information in the country reports which refer to the wider background).

The function of different training and development or qualification programs represents a point of reference which can be accessed only limitedly, due to the lack of efficacy research on various levels: on the program level (external evaluations), on the state level (state-specific studies) and on the international level (comparative meta-studies). Therefore, this work is going to primarily recur to the research questions as proposed in I.2 and aims at investigating commonalities and similarities as well as differences in the training and development programs of the chosen countries regarding macro- and micro-didactical aspects. The question will then be how far the overall approach (regarding provider and target group) on the one hand, and the contents, the methods, and the pattern on the other hand comply with the postulated aims. This will be connected with empirical insights from school effectiveness, school improvement, and from the field of organisational education.

2. Approach

In the respective definitions established by various comparative researchers, Schriewer and Holmes (1988) recognise the contrast between 'comparative education' and 'international education' as well as 'comparative education research' and 'international development education' and between 'science of comparative education' and 'educational statesmanship'. Differences supposedly arise from varying methodological approaches and definitions of problems or differing target groups. A major division into two approaches – throughout the history of the development of comparative research – becomes apparent: at the beginning, comparative research was perceived as an acceptable method for the scientification of the theory of education on the one hand. On the other hand, there were expectations that it would provide support and practical guidelines concerning the creation, complementation and consolidation of the respective national education systems, which were in their early stages at that time. Accordingly, Schriewer and Holmes categorise the 'classics' of comparative educational research as belonging to one of these two contrary classes: one part of these studies analysed the differences of the education systems of various countries and deduced explanations based on theoretical aspects. Many studies, however, were devoted to the international struggle for reforms and therefore concentrated on international, cross national descriptions, which aimed at the opportunity to learn from the experiences previously made in other countries.

At the early stages of comparative research as a scientific discipline, Sadler (e.g. 1908) made efforts at combining both, the scientific-analytical and the practical-problem-solving approach. It became apparent, however, how greatly these two methodological paradigms differ, almost to the point of mutual impediment. This 'Sadler-dilemma' has remained an immanent feature of the methodological controversies of comparative research up to this day.

The influence of the modern social sciences resulted in weakening differences: 'academic comparative education' (with the goal of providing explanations on the basis of creating international theories) and 'interventionist comparative education' (aiming at influencing political decision making processes of education) came more and more together. Contrasting notions like 'scientification' versus 'practicality', 'discovery' versus 'assistance', 'academic interests in objective research' versus 'melioristic interests in professional practice and remedy' have continuously disappeared.

According to Reischmann (1997) we could differentiate different phases and types of international comparative works as far as adult education is concerned: firstly, attention is directed to the fact that there were hardly any real comparative studies in the 1970s and 1980s. In contrast, the focus was on the education system of one individual country. Secondly, the same is true for the program studies of the 1980s. These studies included educational organisations and institutions of one or more countries in one volume without actually comparing them. Thirdly, another type presents data from two or more countries through a juxtaposition, for example, in the form of a statistic report, again without any explicit comparison. Following the definition of Charters and Hilton (1989), we classify the first three phases and types of studies as mere pre-forms of 'real' comparisons. Yet another type then features this explicit comparison with the intention of understanding similarities and differences:

> Comparative study is not the mere placing side by side of data [...] such juxtapositions is only the prerequisite for comparison. At the next stage one attempts to identify the similarities and differences between the aspects under study [...] The real value of comparative study emerges only from [...] the attempt to understand why the differences and similarities occur and what their significance is for adult education in the countries under examination. (Charters & Hilton, 1989, p. 3)

Our study also strives for explicit reflection of the individual field context as well as the methods applied. It is of analytical-comprehensive nature and intends to provide approaches of explanation. However, this is not regarded an opposition to the educational-political concerns due to the necessity of reforms in many parts of the education system. A 'problem-oriented approach' with a detailed research question and a specific scope is pursued here. The study focuses on the topic of school leader training and development, which is a 'singular problem' of the education system, deduced from the context as a whole, and cross-refers to it.

Hörner (1993, 1996) indicates a categorisation into four differentiating functions because of the varying intentions underlying comparative studies:

- the 'idiographic function', which originates from an academic interest in the peculiarity of the object studied,
- the 'evolutionist function', according to which trends and developments are to be identified on the basis of educational-policy interests,
- the 'melioristic function', which is derived from the intention of learning from the realities and experiences of other countries and singling out positive aspects which help improve one's own practice,
- the 'experimental function', where the object of the comparison is treated like a study field which is characterised by a specific combination of variables (where the comparison adopts the function of a scientific experiment).

The research interest of our study makes clear that its research approach aims at various intentions: the individual country studies – or program studies – first of all obtain an idiographic function which derives from the academic interest in the peculiarity of the research object. The countries chosen differ clearly from each other. The surveys therefore fulfil the function of providing information. Additionally, our approach follows an evolutionist function as international trends and tendencies are identified. This means that this work furthermore fulfils an inspirational function by providing ideas for an improvement of school leader qualification. The melioristic function of comparative educational research – which lies in applying contexts of effects from other countries to educational policy planning – is seen critically in this study and can only be applied considering the country-specific circumstances. The experimental function of the international comparison is not applied here, as this work does not aim at specifying and examining hypotheses because of its explanatory character.

3. Methods

In pursueing specific comparative educational research goals the following systematic methodical steps have been taken (see for further explanations Hilker, 1962; Bereday, 1969; Mitter, 1979; Wörmann, 1985; Pecherski, 1986; Anweiler, 1974, 1990, 1996; Hörner, 1993):

- developing a clearly defined research question,
- making it operational,
- developing criteria for the choice of the sample (here the sample countries),
- gathering data with different methods and instruments,
- constructing a classification scheme for the processing of information,
- selecting, systematizing and classifying information,
- identifying indicators for interpretation and evaluation,
- utilising country-specific understanding, explanation, evaluation,
- arrangement (juxtaposition),
- comparison (comparing, explaining, and discussing).

The research question with its exploratory interest was clearly defined and made operational. Then, criteria were developed for the choice of the sample (i.e. the countries). The collection of information about the education systems that are of interest and their description differs from a non-scientific approach in so far as it tries to systematise, select and classify the collected data. It also attempts to evaluate and critically examine the credibility of information derived from documents by looking carefully at the sources that were used. The interpretation of information includes finding explanations for why certain phenomena occurred in a certain way, which conditions triggered them and which factors caused them. To gain such information one has to be able to access knowledge of the economic, historical, cultural and general political context of the countries sampled. Moreover, a high degree of independence and critical distance from the official patterns of explanation must be maintained in order to obtain insight into what lies behind the official façade, if necessary.

For this investigation, both understanding the school situation and the role of school leadership, and necessary knowledge about the societal, cultural, historical, political and especially educational-political context have been gained through the following steps:

- interviews with educational scientists and school leaders in several countries about educational-political changes, their effects on school and the role and function of school leadership, and issues concerning qualification for that leadership,
- participant observations during the leadership training and development programs,
- unsystematic and systematic expert talks and symposia during several international conferences[52],
- a survey of similar research (for this purpose about 100 educationalists of the fields of school development, school leadership, and international comparative educational research from about 25 countries were addressed in autumn 1999 to collect information about similar and recent research in this field).

Selection of Countries

A number of problems became evident when developing a comparative research approach. As soon as the subject of study is made up of open, highly complex, learning systems, the situation is fundamentally different than in research constellations of the natural sciences, in which variables can be manipulated experimentally. Moreover it is practically impossible to randomise the choice of countries for the sample survey. The choice is already rather limited because of language, political and administrational givens. International comparative research offers an attempt to solve this problem by conducting the choice of sample surveys very conscientiously in order to avoid, on the one hand, unintentional correlations and to include, on the other hand, as many variables as possible (see Berstecher, 1970; Robinsohn, 1973).

[52] International conferences such as the European Conference on Educational Research 1997, and the International Congress for School Effectiveness and Improvement 1998, 1999, 2000, 2001 and 2002 helped to build up and to deepen contacts with people from different countries.

In this research, it did not seem reasonable to limit the choice of countries (or of training and development programs for school leaders) from the beginning as the research question has an exploratory character. Therefore, from the outset, we have researched extensively in many different countries worldwide. Retrospectively, however, a choice of countries or programs developed due to two prerequisites: on the one hand, due to language constraints, only German, English or French could be analysed (this included information that had been translated into these languages). On the other hand, the availability of liaisons which already existed or which developed during the research project – especially contacts with academics, politicians in the educational arena, representatives of ministries, of the state and non-state institutions of school leader development, and with the representatives from professional associations from the countries represented at international conferences. Other countries could only have been included with the support of additional funds for research travels and translation work. Access of information that developed differently according to the language and the availability of contacts, therefore, can account for the choice of the countries and programs.

If cooperative links did not already exist in the countries selected, we tried to get into contact via Email and letters after the results of the sounding out enquiries and to gain the necessary support.

The decision for the criteria of the final choice of the sample (of the countries and their training and development programs) for the research has been made, neither according to the 'most similar systems design', nor according to the 'most different systems design' (see Przeworski & Tenne, 1970), as the comparison aimed at in this research does not have the function of an 'indirect experiment'. The ultimate choice was made after consultation with four educationalists from different countries, who are conducting research in this field. It is based upon the following criteria. The countries that were chosen appear to:

- represent the developments in Europe and illustrate the differences there,
- be (leading) industrial countries which influence other countries (politically but also in terms of educational-policy, also concerning the former colonies, especially influential in the Asian regions),
- have differentiated school leadership research,
- have been able to gain experience for many years in school leadership training and development,
- represent the new and recent developments in Asia.

Therefore, the following countries have been chosen: Sweden, Denmark, England, the Netherlands, France, Germany, Switzerland, Austria, South Tyrol/Italy, Singapore, Hong Kong/China, New South Wales/Australia, New Zealand, Canada, and some states of the USA.

Methods of Data Collection

International literature research

Library research was conducted for information about Europe in Germany at the University Erlangen-Nuremberg (which houses the special collection 'educational research' for Germany) and in England at Cambridge University at the Copy-Right Library and at Manchester University; for information about America in the US at Trinity University in San Antonio; for information about Asia in Hong Kong at the Hong Kong Institute for Education.

Internet research made it possible to sound out training and development programs in different countries (and search for providers there), to gain information about individual qualification offers (course programs, etc.) and to trace contacts (as preparation for expert inquiries).

Expert meetings

In the past years, different international expert meetings took place: an interactive discussion at ICSEI 1999 in San Antonio, Texas, USA (including the founding of an informal network for school leadership development), three symposia at ICSEI in January 2000 in Hong Kong, China, 2001 in Toronto, Canada, and 2002 in Copenhagen, Denmark, as well as at the annual meeting of AERA 2002 in New Orleans[53].

These discussions were recorded and then transcribed. Moreover, it was possible to establish a network which will officially be embedded within the ICSEI context: the ICSEI-Network School Leadership and Leadership Development.

Systematic collection of documents

Individuals from the countries sampled were asked for cooperation in the research project in a personal letter. The persons chosen included:

- internationally acknowledged educationalists who are experts in the fields of school leadership and school leadership development,
- representatives of Education Ministries who are responsible for the training of school leaders,
- representatives of (state and non-state) training and development education institutions who organise and conduct the offerings,
- representatives of interest groups for school leaders (school leader associations, etc.).

They were asked to provide the following:

- general information about the educational and school system of the respective country,
- documents about the country-specific role of school leadership (also publications),

[53] At ICSEI 2003 in Sydney, Australia, and at AERA 2003 in Chicago, USA, final symposia are already planned, and further research projects will be considered.

- material about development offerings (such as unpublished course conceptions, leaflets, information packages for people interested in participation, curricula and – as far as possible – examples of course and study material, reports of various kinds, etc.),
- written feedback of participants and evaluation reports, in so far as they exist or are available.

Questionnaire

The chosen participants were also asked to respond to a questionnaire, which was central for data collection, the Synopsis of Training and Development Opportunities for School Leaders in Different Countries. The questionnaire was made up from the following questions (which again follow the research questions, see Chapter I.2.)[54]:

1. Provider(s):
- Which institution(s) provide training for school leaders in your country? Is it organised centrally, decentrally, are they government or state-run institutions, local and regional institutions, private sector organisations, universities or colleges or other providers?

2. Target group, timing, nature of participation, professional validity:
- At what point in time are school leaders trained for school leadership, at what point in time (before appointment, between appointment and taking over the post, after taking over the post) do they take part in training and development programs preparing them for their new job?
- What is the nature of participation? Is it mandatory to all school leaders, optional to all or partially mandatory?
- Do deputies, vice principals or other senior staff have the possibility of training and development opportunities as well? If the answer is yes or partly, could you provide some additional information here or elsewhere, or enclose further information?

3. Aims:
- What are the aims and objectives of the program? Is there an explicit vision? Are there clearly defined standards? How are these identified?

4. Contents:
- What content, topics, issues are covered?
- How much time is provided for each unit and for each topic?
- Is every topic mandatory, optional, partially mandatory or partially optional to all participants?
- What were the criteria for selecting the contents?
- Are individual training and development needs of the participants taken into consideration?
- If the answer is yes, how are they determined, and then, how are they taken into account?
- Which training and development opportunities are taken up more frequently, which topics are particularly favoured by school leaders?
- What do you personally think are the reasons for this?

[54] Further questions have been added to the questionnaire which are not relevant for the recent research but of interest for further studies taking place later.

5. Methods:
- What teaching strategies and learning methods are used in each of the trainings units? (Please indicate the range of methods. If some methods are used more often, please indicate by ranking.)
- What training material is used? (If possible, please give examples, or enclose printed materials if available.)
- Who is responsible for the delivery of the training and development program, who are the trainers? What is their professional background?
- How many participants are on one single course, how large is each training group (e.g. 20 participants per group)?
- Is training differentiated according to type or size of school or according to region, individual needs, aspiring school leaders and those already in post or other?

6. Pattern:
- What is the pattern of the training for these aspiring or newly appointed school leaders like? How is it structured with regard to time (e.g. six weekends or three one-week trainings) before appointment, between appointment and taking over, after taking over the post?
- How many days are provided in total for the program offered? Please give the answer in terms of full days (e.g. 3 one-week periods = 15 full days).
- Over what period of time does the training and development opportunity take place for the individual participant (e.g. the 3 one-week periods are spread over 18 months)?

7. Costs (additional questions):
- What financial resources can the participant rely on (public funding, privately paid by participant or other)?
- Is there any statistical information available of the costs of the program?
- If the answer is yes, what is the approximate cost per school leader for the days of training provided?

8. Strengths and weaknesses (additional questions):
- What, in your opinion, are the strengths of this kind of school leadership training in your country? Where do you see weaknesses?
- What do you know about the quality and effectiveness of this program? What kind of experiences do you have?
- Is there provision for feedback from participants? If yes, in what form?
- Has the program been evaluated? If yes, in what form, and by whom?

9. Interest and motivation of participants (additional questions):
- How high do you rate the interest of school leaders in the training and development opportunity which is described above (high, medium, low or no comment)?
- How high do you rate the interest of school leaders in general to take part in training and development opportunities before appointment, between appointment and taking over the post, after taking over the post, after 5 years in post?
- What could be the reasons for these levels of interest?

10. Availability (additional questions):
- How many training places are provided annually for participants in total, and how many participants take part?
- Does the number of places in the programs cover the current demands by school leaders?
- If the answer is no, should the offer be extended?
- Do you consider an enlargement of the currently available training as necessary, unnecessary, not yet investigated?

11. Looking ahead (additional questions):
- What kind of tasks and responsibilities do you think will be most challenging for school leaders in years to come?
- What tasks are most demanding for newly appointed school leaders?
- What competences are most asked for by new school leaders in your country?
- What kind of training and development opportunities could, should be offered to supplement current provision?
- Do you know of any planned changes in the situation of training and development opportunities in your country in the near future?
- If the answer is yes, what are these planned changes?

12. Selection procedure (additional questions):
- How are school leaders selected in your country?
- Are there more applicants than vacancies, and why?
- Is a preparatory qualification mandatory, helpful, or not relevant for the application?
- Is a mandatory qualification before application under consideration?

13. Training and development opportunities for school leaders in post for more than six years (additional questions):
- What institutions offer programs for school leaders in your country in general?
- What themes, contents, topics are offered?
- What teaching and learning methods are used?
- Is the number of places available sufficient for the current demand?
- If the answer is no, why do you think there is such a deficiency?
- How often do school leaders who have been in post for more than three years take part in training and development programs?

14. Personal view (additional question):
- What – in your view – is the most interesting approach to school leadership training and development?

These questions were formulated in a way that should be easy to understand and intended to avoid ambiguity, bias and constriction, but facilitate clear communication. With regard to the content of the questions, as well as the terminology used, expressions were chosen which were not country-specific or jargon. One had to be conscious of the differences of the school systems and the terminology for school leadership, school types, etc.

For the questionnaire the order of the questions was revised in order to allow the interviewees to develop their thoughts more clearly and to facilitate the completion of the questionnaire: easily accessible questions were posed first. The first page provided the interviewees with important information about the research topic (background, etc.) (this was also outlined in the accompanying letter), and explanations for answering the questions. A well-developed and clear layout seemed to be important as well. Extra space was provided in order to enable the interviewees to give additional information and provide feedback. Moreover, additional information material could be enclosed, which would allow the interviewee to complete the questionnaire more quickly by referring to the respective information. They could also express their interest in the research results.

After carefully evaluating clarity, exactness and linguistic appropriateness of the questionnaire by an international group of teachers and (some future) school leaders, comprising of 10 people, especially from the USA and England, during a seminar of the National Louis University's Heidelberg Campus, it was evaluated for its usefulness in one country. Modifications did not seem necessary according to the positive feedback, so ICSEI 2000 in Hong Kong was used to distribute the questionnaire. It was passed on to the participating representatives from the chosen countries where possible and posted or sent as an email attachment. Also included was the handout of the symposium session in Hong Kong.

Nearly all questionnaires were answered and returned. There was a varying amount of additional information enclosed. The high reply quota may be due to the personal delivery of many questionnaires, as well as the correspondence via email and telephone, which may have had a cooperation-supporting effect.

Open and non-directive questioning

From the sample survey mentioned above, some people – with whom a closer mutual cooperation relationship had developed – were asked to compose a one- to three-page (or longer) 'individually formulated summary' of the school leader development situation in their country. This additional (and very open) method made it possible to gain country-specific information, which might not have been recorded by very structured questionnaire responses alone. That is to say, this form of open questioning made subjective choice and emphasis possible, and also allows an integration of certain information into a larger context. This might not have been achieved by the answers from the systematic questionnaire alone.

Additional research via telephone and email

If – after a first analysis of the questionnaires, the individually formulated summaries and the material received – it was seen that single aspects were still left blank or not sufficiently covered by the responses, additional research via email or telephone was conducted in order to supplement the data.

Analysis

The analysis is based on well planned steps and provides a systematic comparison in addition to a juxtaposition. Bereday (1969) defines juxtaposition as "preliminary matching of data from different countries to prepare them for comparison, including the systematizing of these data, so that they may be grouped under identical or comparable categories for each country under study" (p. 5).

> Juxtaposition [...] is preliminary comparison or a confrontation of partners to comparison. Its purpose is not to draw comparative conclusions, but to determine whether comparison is possible at all. Juxtaposition is really intended to provide an answer to the question, 'compare in terms of what?'. The search here is for what Robert Ulich has called the tertium comparationis. Juxtaposition is designed to establish comparability or basic consistency of data. (Bereday, 1969, p. 9)

According to Pecherski (1986), the aim is to reach – beyond the juxtaposition – the comparison of the found facts and figures. The purpose of the comparison, then, is the systematic study of similarities and differences. Bereday identifies a difference between 'balanced comparison' and 'illustrative comparison'. The former achieves a comparison in small steps by extreme precision. Step by step, countries are compared to each other concerning the subject of investigation. "The essence of this method is that every type of information from one country must be matched, balanced, by comparable information from other countries." (Bereday, 1969, p. 9) It seems reasonable to assume that this procedure is not suited for all subjects of investigation. The second variant, 'illustrative comparison', is less precise, but it offers – if conducted thoroughly – many supportive insights: "In this method, educational practices in different countries are drawn at random as illustrations of comparative points suggested by the data." (ibid.)

A very rigorous comparison, which relates the studied countries to each other by very small steps, does not seem to be possible regarding the explorative character of the research and the complexity of the research subject. Having received the information we organised the material according to countries and programs and then studied it ideographically in order to assure a country-specific understanding. Afterwards we composed a classification scheme in order to select and systematise the information gained from the different methods of data collection.

The scheme used for classification is divided into the following fields according to the questions relevant in the context of the research question: 1. Provider(s); 2. Target group, timing, nature of participation, professional validity; 3. Aims; 4. Contents; 5. Methods; 6. Pattern.

As the investigated data was intended to be evaluated specifically for each country and be displayed clearly in order to facilitate comparisons between the countries, we designed a matrix that organises the data. In this kind of complex table each column contains a country's program and the rows are organised according to the dimensions for comparison. The countries are in horizontal, the differentiated classification scheme is along the vertical axis.

Following this, a second country-specific analysis of the material regarding the countries and the qualification programs was conducted. The material was structured according to the classification scheme, analysed in steps and the results were summarised into the matrix. Afterwards country and program descriptions were formulated individually. Then they were revised in order to enable a similarly structured text as far as possible. This was important for conducting juxtaposition. To check the intracontextual dimension and the general coherence, the country and program descriptions as well as the facts from the matrix were read from the perspective of the original material. This involved reading those once more as well.

In order to be able to compare single aspects in a more systematic manner, we returned again to our matrix for the dimension-related country-comparative analysis. Also, we made partial use again of the original material.

The collected data about the individual countries and their qualification programs was analysed independently by another person in order to increase the value of the content analyses with a rather high analyser overlap. It could be seen that both analysts were in agreement.

A possible additional validation of the country and program descriptions by renewed feedback from the experts from the different countries was regarded as very important. Hence, all country and program descriptions originally written in German were translated into English, either here in Bamberg or by the validating expert. This validating expert was invited to be the co-author for the publication of the country report. Additionally, she or he was asked to add a section about recent developments and a personal statement, which comments on the general situation or on the program(s).

In addition to the validation, a second methodological step has been achieved that way, namely the combination of an independent external expert, who assures the consistent approach in the country studies, and an internal expert, who has specific context knowledge.

Concerning the quality of the material collected, it should be kept in mind that in parts the descriptions made by training and development institutions themselves sometimes displayed a too optimistic view. This bias had to be considered by critical judging and by additionally using information from other sources, as, for example, from other contacts from the same country.

References

Abächerli, A. & Kopp, S. (1997). *Führen: eine Schule leiten – AEB-LCH: Evaluationsbericht*. Diplomarbeit LSEB, eingereicht bei Prof. Dr. J. Oelkers am Institut für Pädagogik an der Universität Bern.

AEB/PA-LCH. (1999). *Aus- und Weiterbildung für Schulleiter und Schulleiterinnen: Projektarbeit*. Unveröffentlichtes Kursmaterial, Luzern.

Akademie für Lehrerfortbildung und Personalführung (1998). *Schulleitung und Schulaufsicht in Bayern: Vorbereitung, Begleitung und Fortbildung*. Dillingen: ALP.

Akademie für Lehrerfortbildung und Personalführung (1999). *Lehrerfortbildung in Bayern*. Dillingen: ALP.

Altrichter, H. & Posch, P. (1999). *Wege zur Schulqualität: Studien über den Aufbau von qualitätssichernden und qualitätsentwickelnden Systemen in berufsbildenden Schulen*. Innsbruck: StudienVerlag.

Altrichter, H., Schley, W. & Schratz, M. (1998). *Handbuch zur Schulentwicklung*. Innsbruck: StudienVerlag.

Amplatz, L. (1999). Die Direktorenausbildung aus der Sicht eines Direktors. *Forum Schule heute*, *13*(5), 19-20.

Anderson, J.R. (1983). *The architecture of cognition*. Cambridge, MA: Harvard University Press.

Anderson, J.R. (1987). Skill acquisition. Compilation of weak-method problem solutions. *Psychological Review*, *94*, 192-210.

Anderson, M.E. (1989). Training and selecting school leaders. In S.C. Smith & P.K. Piele (Eds.), *School leadership* (pp. 53-84). Eugene, OR: ERIC Clearinghouse on Educational Management.

Antal, A. (1997). *The live case: A method for stimulating individual, group and organizational Learning*. FS II 97-112. Wissenschaftszentrum Berlin für Sozialforschung GmbH.

Anweiler, O. (1974). Konzeptionen der Vergleichenden Pädagogik. In A. Busch, F. Busch, B. Krüger & M. Krüger-Potratz (Eds.), *Vergleichende Erziehungswissenschaft: Texte zur Methodologie-Diskussion* (pp. 19-26). Pullach: UTB Verlag Dokumentation.

Anweiler, O. (1986). Systemvergleich und Bildungsforschung. In G. Gutmann & pp. Mampel (Eds.), *Probleme systemvergleichender Betrachtung* (pp. 229-234). Berlin: Duncker & Humblot.

Anweiler, O. (1990). *Wissenschaftliches Interesse und politische Verantwortung: Dimensionen vergleichender Bildungsforschung. Ausgewählte Schriften 1967-1989*. Opladen: Leske + Budrich.

Anweiler, O., Boos-Nünning, U., Brinkmann, G., Glowka, D., Goetze, D., Hörner, W., Kuebart, F. & Schäfer, H.-P. (1996). *Bildungssysteme in Europa*. Weinheim: Beltz.

Apel, H.-J. (1995). *Theorie der Schule*. Donauwörth: Auer.

Arbeitsgemeinschaft der Schulleiterverbände Deutschlands (ASD). (1989). Berufsbild Schulleiter. Berliner Erklärung der ASD. *Schul-Management*, *4*, 5-6.

Arbeitsgemeinschaft der Schulleiterverbände Deutschlands (ASD). (1994). *Schulleitung in Deutschland. Profil eines Berufes*. Bonn: ASD.

Arbeitsgemeinschaft der Schulleiterverbände Deutschlands (ASD). (1999). *Schulleitung in Deutschland. Ein Berufsbild in der Entwicklung*. Stuttgart: Raabe.

Avolio, B.J. (1999). Full leadership development. In B.M. Bass & B.J. Avolio (Eds.), *Improving organizational effectiveness through transformational leadership*. Thousand Oaks, CA: Sage.

Bachmann, H., Iby, M., Kern, A., Osinger, D., Radnitzky, E. & Specht, W. (1996). *Auf dem Weg zu einer besseren Schule. Evaluation der Schulautonomie in Österreich – Auswirkungen der 14. SCHOG-Novelle*. Innsbruck: StudienVerlag.

Barth, R. (2001). *Learning by heart*. San Francisco, CA: Jossey-Bass.

Barth, R.S. (1986). The principalship. *Educational Leadership*, *42*(6), 92-104.

Barth, R.S. (1990). *Improving schools from within: Teachers, parents and principals can make a difference*. San Francisco: Jossey-Bass Publishers.

Bass, B.M. & Avolio, B.J. (Eds.). (1994). *Improving organizational effectiveness through transformational leadership*. Thousand Oaks, CA: Sage.

Bateson, G. (1954/1982). *Geist und Natur*. Frankfurt a.M.: Suhrkamp.

Baumann, U. (1981). Historik und Komparistik – Parallelen zwischen einer historischen und einer komparativen Sozialwissenschaft. In U. Baumann, V. Leonhart & J. Zimmermann (1981), *Vergleichende Erziehungswissenschaft* (pp. 3-11). Wiesbaden: Akademische Verlagsgesellschaft.

Baumert, J. & Leschinsky, A. (1986). Zur Rolle des Schulleiters. *Schul-Management*, *6*, 18-24.

Baumert, J. (1984). Schulleiter-Karriere. *Schul-Management*, *6*.

Beare, H., Caldwell, B.J. & Millikan, R.H. (1989). *Creating an Excellent School*. London: Routledge.

Beck, U. (1986). *Risikogesellschaft – Auf dem Weg in eine andere Moderne*. Frankfurt a. M.: Klett.

Becker, H. (1962). Die verwaltete Schule. In H. Becker (Ed.), *Quantität und Qualität* (pp. 147-174). Freiburg: Rombach.

Begley, P.T. & Johansson, O. (2000, November). *Using what we know about values: Promoting authentic leadership and democracy in schools*. Paper presented at the Annual UCEA Conference, Albuquerque, New Mexico.

Begley, P.T. (2003, in print). In pursuit of authentic school leadership practices. In P.T. Begley & O. Johansson (Eds.), *The Ethical Dimensions of School Leadership* (Chapter 1). Deventer: Kluwer.

Bennis, W. & Nanus, B. (1985). *Leaders: The strategies for taking charge*. New York: Harper & Row.

Bereday, G. (1969). Reflections in comparative methodology, 1964-1966. In M. Eckstein & H.J. Noah (Eds.), *Scientific investigations in comparative education* (pp. 3-24). Toronto: Macmillan.

Beredey, G. (1964). *Comparative method in education*. New York: Holt, Rinehart & Winston.

Bereiter, C. (1990). Aspects of an educational learning theory. *Review of Educational Research*, *60*, 603-624.

Bernfeld, S. (1925/1967). *Sisyphos oder die Grenzen der Erziehung*. Frankfurt: Suhrkamp.

Berstecher, D. (1970). *Zur Theorie und Technik des internationalen Vergleichs. Das Beispiel der Bildungsforschung*. Stuttgart: Klett.

Berstecher, D. (1972). Bemerkungen zur Logik vergleichender Forschung. *International Review of Education*, *18*(3), 286-296.

Berstecher, D. (1974). Bemerkungen zur Logik Vergleichender Forschung. In A. Busch, F.W. Busch, B. Krüger & M. Krüger-Potratz (Eds.), *Vergleichende Erziehungswissenschaft. Texte zur Methodologie-Diskussion* (pp. 36-43). Pullach bei München: Verlag Dokumentation.

Bessoth, R. (1982). Berufliche Interessen von Schulleitern. *Schul-Management, 1*, 43-29.

Bevoise, W. de. (1984). Synthesis of research on the principal as instructional leader. *Educational Journal, 41*(5), 14-20.

Bildungsplanung Zentralschweiz. (2000). *Ausbildung von Schulleiterinnen und Schulleitern: Ausschreibung*. Ebikon: Bildungsplanung Zentralschweiz.

Blum, R.E. & Butler, J.A. (Eds.). (1989). *School leader development for school improvement*. Leuven: ACCO.

Boak, G. (1991). *Developing managerial competences: The management learning contract approach*. London: Pitman.

Boam, R. & Sparrow, P. (1992). D*esigning and achieving competency*. Maidenhead: McGraw Hill.

Bolam, R. (1993). School-based management, school improvement and school effectiveness: Overview and implications. In C. Dimmock (Ed.), *School-based management and school effectiveness* (pp. 219-234). London: Routledge.

Bolam, R., McMahon, A., Pocklington, K. & Weindling, D. (1993). *Effective management in schools: A report for the Department for Education via the School Management Task Force Professional Working Party*. London: HMSO.

Bonsen, M. & Pfeiffer, H. (1998). Schulleitungsforschung in Deutschland: Forschung und Ergebnisse im Überblick. In H. Bucher, L. Horster & H.-G. Rolff (Eds.), *Schulleitung und Schulentwicklung* (Kapitel 16, 2-18). Stuttgart: Raabe.

Bordieu, P. & Passeron, J.-Cl. (1970). *La reproduction. Élements pour une théorie du système d'enseigments*. Paris: Ed. de Minuit.

Boyatzis, R.E. (1982). *The competent manager*. New York: Wiley.

Bransford, J.D., Goldman, S.R. & Vye, N.J. (1991). Making a difference in people's ability to think: Reflections on a decade of work and some hopes for the future. In R.J. Sternberg & L. Okagaki (Eds.), *Influences on children* (pp. 147-180). Hillsdale, NJ: Erlbaum.

Bredeson, P.V. (1992, April). *Responses to restructuring and empowerment initiatives: A study of teachers' and principals' perceptions of organizational leadership, decision making and climate*. Paper presented at the Annual Meeting of the American Educational Research Association, San Francisco, CA. (ERIC Document Reproduction Service No. ED 346 569)

Bridges, E. & Hallinger, P. (1995). *Implementing problem-based learning in leadership development*. Eugene, OR: ERIC Clearinghouse on Educational Management.

Bridges, E. & Hallinger, P. (1997). Using problem-based learning to prepare educational leaders. *Peabody Journal of Education, 72*(2), 131-146.

Bridges, E. (1992). *Problem-based learning for administrators*. Eugene, OR: ERIC Clearinghouse on Educational Management.

Broadbent, D. E., Fitzgerald, P. & Broadbent, M. H. P. (1986). Implicit explicit knowledge in the control of complex systems. *British Journal of Psychology, 77*, 33-50.

Brookover, W., Beady, C., Flood, P., Schweitzer, J. & Wisenbaker, J. (1979). *School social systems and student achievement: Schools can make a difference*. New York: Praeger.

Brown, G. & Irby, B.J. (1997). *The principal portfolio*. Thousand Oaks, CA: Sage.

Buchen, H. (2000). Leitbild- und strategieorientierte Ausbildung von Führungskräftenachwuchs: NRW setzt bei Qualitätsentwicklung und Qualitätsicherung von Schule auf die Professionalität von Schulleitung (Kapitel C 2.7). In H. Buchen, L. Horster & H.G. Rolff. (Eds.), *Schulleitung und Schulentwicklung*. Stuttgart: Raabe.

Bullock, A. & Thomas, H. (1997). *Schools at the centre? A study of decentralisation*. London: Routledge.

Burns, J.M. (1978). *Leadership*. New York: Harper and Row.

Busch, A., Busch, F., Krüger, B. & Krüger-Potratz, M. (Eds.). (1974). *Vergleichende Erziehungswissenschaft: Texte zur Methodologie-Diskussion*. Pullach: UTB Verlag Dokumentation.

Button, H.W. (1966). Doctrine of administration: A brief history. *Educational Administration Quarterly*, *2*(3), 378-391.

Caldwell, B.J. & Spinks, J.M. (1988). *The self-managing school*. London: Falmer Press.

Caldwell, B.J. & Spinks, J.M. (1992). *Leading the self-managing school*. London: Falmer Press.

Charters, A.N. & Hilton, R.J. (Eds.). (1989). *Landmarks in international adult education. A comparative analysis*. London: Routledge.

Chirichello, M. (2001). Collective leadership: Sharing the principalship. *Principal, 81*(1), 46-51.

Codd, J. & Gordon, L. (1991). School charters: The contractualist state and education policy. *New Zealand Journal of Educational Studies*, *26*(1), 21-34.

Coleman, J.S. (1986). *Die asymmetrische Gesellschaft. Vom Aufwachsen mit unpersönlichen Systemen*. Weinheim: Beltz.

Collins, A., Brown, J. & Newman, S.E. (1989). Cognitive apprenticeship: teaching the crafts of reading, writing, and mathematics. In L.B. Resnick (Ed.), *Knowing, learning, and instruction. Essays in honor of Robert Glaser* (pp. 453-494). Hillsdale: Erlbaum.

Conger, J.A. (1999). Charismatic and transformational leadership in organizations: An insider's perspective on these developing streams of research. *Leadership Quarterly*, *10*(2), 145-179.

Corder, C. (1990). *Teaching hard, teaching soft*. Aldershot: Gower Press.

Corrigan, D. (1980). *In-service education and training of teacher. Towards new policies*. Paris: Center for Educational Research and Innovation, OECD.

Council of Chief State School Officers (1996). *Interstate school leaders licensure consortium: Standards for school leaders*. Washington, DC: Author.

Creemers, B. (1994). The history, value and purpose of school effectiveness studies. In D. Reynolds, B. Creemers, P. Nesselrodt, E. Schaffer, S. Stringfield & C. Teddlie (Eds.), *Advances in school effectiveness research and practice* (pp. 9-23). Oxford: Pergamon.

CSLA (California School Leadership Academy). (Ed.). (1999). *21st century leadership opportunities 1999-2002*. Santa Barbara, CA: CSLA.

Dalin, P. & Rolff, H.G. (1990). *Das Institutionelle Schulentwicklungsprogramm*. Soest: Soester Verlag-Kontor.

Daresh, J.C. & Playko, M.A. (1989). *Administrative mentoring: A training manual.* Columbus, OH: The Ohio LEAD Center.
Daresh, J.C. & Playko, M.A. (1992). Mentoring of headtechers: A review of major issues. *School Organisation, 12*(2), 145-152.
Daresh, J.C. (1998). Professional development for school leadership: The impact of U.S. educational reform. *International Journal of Educational Research, 29*(4), 323-333.
Dawson, G. (1999, Januar). *School leaders for the 21st century.* Paper presented at the Twelfth International Congress on School Effectiveness and Improvement, San Antonio, Texas, USA.
Delors, J., Al Mufti, I., Amagi, A., Carneiro, R., Chung, F., Geremek, B., Gorham, W., Kornhauser, A., Manley, M., Padron Quero, M., Savané, M.-A., Singh, K., Stavenhagen, R., Suhr, M.W. & Nanzhao, Z. (1996). *Learning: The treasure within.* Report to UNESCO of the International Commission on Education for the Twenty-first Century. Paris: UNESCO.
DES (Department for Education and Science). (1990). *Developing school management: The way forward. A report by the School Management Task Force.* London: HMSO.
Deutscher Bildungsrat. (1970). *Empfehlungen der Bildungskommission: Strukturplan für das Bildungswesen.* Stuttgart: Klett.
DfEE (Department for Education and Employment). (1997). *National standards for headteachers.* London: DfEE.
Diederich, J. & Tenorth, H.E. (1997). *Theorie der Schule.* Berlin: Cornelsen Scriptor.
Dilger, B., Kuebart, F. & Schäfer, H.-P. (Eds.). (1986). *Vergleichende Bildungsforschung. DDR, Osteuropa und interkulturelle Perspektiven.* Berlin: Berlin Verlag Arno Spitz.
Dimmock, C. & Walker, A. (1998). Transforming Hong Kong schools: Trends and emerging issues. *Journal of Educational Administration, 36*(5), 476-491.
Doerry, G., Kallmeyer, G., Isslinger, H., Riedel, J., Westermann, R., Schmidt, B., Gunster, R. & Lautz, G. (1981). *Bewegliche Arbeitsformen der Erwachsenenbildung.* Braunschweig: Westermann.
Donaldson, Jr. G.A. (2001, October 3). *The lose-lose leadership hunt.* Education Week. Retrieved November 24, 2002, from www.edweek.org/ew/ew_printstory.cfm?slug=05donaldson.h21.
Doud, J.L. & Keller, E.P. (1998). *A ten-year study: The K-8 principal in 1998.* Alexandria, VA: NAESP.
Duijker, H.C.J. (1955). Comparative research in social science with special reference to attitude research. *International Social Science Bulletin, 4*, 555-556.
Durkheim, E. (1885). *Les règles de la méthode sociologique.* Paris: Pr.Univ.Paris.
Durkheim, Emile. (1961). *Die Regeln der soziologischen Methode.* In neuer Übersetzung und eingeleitet von René König. Neuwied: Luchterhand.
Dweck, C.S. (1986). Motivational processes affecting learning. *American Psychologist, 41*, 1040-1048.
Dweck, C.S. (1996). Implicit theories as organizers of goals and behavior. In P.M. Gollwitzer & J.A. Bargh (Eds.), *The psychology of action* (pp. 69-90). New York: The Guilford Press.
Earley, P., Weindling, D. & Baker, L. (1990). *Keeping the raft afloat: Secondary headship five years on.* Slough: NFER.

Ebner, H. (2000). Vom Übermittlungs- zum Initiierungskonzept: Lehr-Lernprozesse in konstruktivistischer Perspektive. In C. Harteis, H. Heid & S. Kraft (Eds.), *Kompendium Weiterbildung* (pp. 111-120). Opladen: Leske+Budrich.

ED (Education Department). (1999). *Education in Hong Kong. A brief account of the educational system with statistical summary*. Hong Kong.

Edding, F. (1970). Vorwort. In D. Berstecher (Ed.), *Zur Theorie und Technik des internationalen Vergleichs. Das Beispiel der Bildungsforschung* (pp. 7-8). Stuttgart: Klett.

Edmonds, R. (1979). Effective schools for the urban poor. *Educational Leadership*, *37*(1), 15-27.

Educational Research Service, NAESP, NASSP (2000). *The principal, keystone of a high-achieving school: Attracting and keeping the leaders we need*. Arlington, VA: ERS.

Eikenbusch, G. (1995). *Schulentwicklung und Qualitätssicherung in Schweden*. Bönen: Verlag für Schule und Weiterbildung.

Eisner, E.W. (1991). What really counts in schools. *Educational Leadership*, *48*(5), 10-17.

Elmore, R.F. (1996). Getting to scale with good educational practice. *Harvard Educational Review*, *66*(1), 1-26.

Elmore, R.F. (Ed.). (1991). *Restructuring schools: The next generation of educational reform*. San Francisco: Jossey-Bass.

Esp, D. (1993). *Competences for school managers*. London: Kogan Page.

EURAC (Europäische Akademie Bozen). (1999). *Führungskräfteschulung für SchulleiterInnen 1999-2000. Konzept und Angebot*. Bozen: EURAC.

Eurydice. (2000). http://www.eurydice.org/Eurybase/Application/eurybase.htm

Fend, H. (1980). *Theorie der Schule*. München: Urban & Schwarzenberg.

Fend, H. (1987). "Gute Schulen – schlechte Schulen" – Die einzelne Schule als pädagogische Handlungseinheit. In U. Steffens & T. Bargel (Eds.), *Beiträge aus dem Arbeitskreis Qualität von Schule* (Heft 1, pp. 55-79). Wiesbaden: Hessisches Institut für Bildungsplanung und Schulentwicklung (HIBS).

Fend, H. (1998). *Qualität im Bildungswesen. Schulforschung zu Systembedingungen, Schulprofilen und Lehrerleistung*. Weinheim: Juventa.

Fischer, F. (1998). *Mappingverfahren als kognitive Werkzeuge*. Frankfurt a.M.: Peter Lang.

Fischer, W. & Schratz, M. (1993/1999). *Schule leiten und gestalten. Mit einer neuen Führungskultur in die Zukunft*. Innsbruck: StudienVerlag.

Fischer, W.A. & Schratz, M. (1993). Innovationen von Schulleitern zur Entwicklung ihrer Schulen. *Pädagogische Forschung*, *4*(5), 217-219.

Froese, L. (1965). Bildungsstrukturen in Ost und West. *Paedagogica Europea*, *1*, 209-219.

Froese, L. (1967). Paradigmata des Selbstverständnisses der Vergleichenden Erziehungswissenschaft – kritisch-vergleichende Bestandsaufnahme. *Zeitschrift für Pädagogik*, *13*(4), 315-324.

Fuchs, J. (1998). Das schwedische Bildungswesen. *Schul-Management*, *4*, 20-29.

Fullan, M. (1988). *What's worth fighting for in the principalship*. Toronto: Ontario Public School Teachers' Federation.

Fullan, M. (1991). *The new meaning of educational change*. London: Cassell.

Fullan, M. (1992a). *Successful school improvement*. Buckingham, Philadelphia: Open University Press.

Fullan, M. (1992b). *What's worth fighting for in headship*. Buckingham, Philadelphia: Open University Press.
Fullan, M. (1993). *Change forces, The school as a learning organisation*. London: Falmer Press.
Fullan, M. (1995). Schools as learning organizations: Distant dreams. *Theory into practice*, *34*(4), 230-235.
Fullan, M. (1998). Leadership for the 21st century. Breaking the bonds of dependency. *Educational Leadership*, *55*(7), 6-10.
Fürstenau, P. (1969). Neuere Entwicklungen der Bürokratieforschung und das Schulwesen. Ein organisationssoziologischer Beitrag. In P. Fürstenau, C.-L. Furck, C.W. Müller, W. Schulz & F. Wellendorf (Eds.), *Zur Theorie der Schule* (pp. 47-66). Weinheim: Beltz.
Gaudig, H. (1917). *Schule im Dienste der werdenden Persönlichkeit*. Leipzig: Quelle & Meyer.
Giroux, H.A. (1988). *Teachers as intellectuals: Toward a critical pedagogy of learning*. Granby, MA: Bergin & Garvey.
Glatter, R. (1987). Tasks and capabilities. In N.E. Stegö, K. Gielen, R. Glatter & S.M. Hord (Eds.), *The role of school leaders in school improvement* (pp. 113-121). Leuven: ACCO.
Glenn, C.L. (1989). Putting school choice in place. *Phi Delta Kappan, 71*(4), 295-300.
Glowka, D. (1972). Vergleichende Erziehungswissenschaft und methodologische Reflexion. *International Review of Education*, *18*(3), 305-315.
Glowka, D. et al. (1995). *Schulen und Unterricht im Vergleich: Rußland/ Deutschland*. Münster: Waxmann.
Glumpler, E. & Luchtenberg, S. (Eds.). (1997). *Jahrbuch der Grundschulforschung*, Band 1. Weinheim: Beltz.
Glumpler, E. (1997a). Vergleichende Grundschulforschung: Von der Reflexion über Schulreform zum interkulturellen Vergleich. In E. Glumpler & S. Luchtenberg (Eds.), *Jahrbuch der Grundschulforschung* (Band 1, pp. 89-102). Weinheim: Beltz.
Glumpler, E. (1997b). *Vergleichende Bildungsforschung im Dienst interkultureller Bildungsplanung*. Dortmund: Institut für Allgemeine Didaktik und Schulpädagogik.
Goens, G.A. & Clover, S.I.R. (1991). *Mastering school reform*. MA: Allyn and Bacon
Grace, G. (1995). *School leadership: Beyond education management*. London: Falmer Press.
Gräsel, C. & Mandl, H. (1999). Problemorientiertes Lernen. Anwendbares Wissen fördern. *Personalführung*, *32*(6), 54-62.
Gräsel, C. (1997). *Problemorientiertes Lernen. Strategieanwendung und Gestaltungsmöglichkeiten*. Göttingen: Hogrefe.
Gray, J. (1990). The quality of schooling: Frameworks for judgements. *British Journal of Educational Studies*, *38*(3), 204-233.
Greenfield, W. (1985). The moral socialization of school administrators: Informal role learning outcomes. *Educational Administration Quarterly*, *12*(4), 99-120.
Greeno, J. G. (1989). Situations, mental models, and generative knowledge. In D. Klahr & K. Kotovsky (Eds.), *Complex information processing. The impact of Herbert A. Simon* (pp. 285-318). Hillsdale, NJ: Erlbaum.

Greeno, J. G., Smith, D. R. & Moore, J. L. (1993). Transfer of situated learning. In D.K. Detterman (Ed.), *Transfer on trial – Intelligence, cognition and instruction*. Norwood, NJ: Ablex Publication.
Gregg, R.T. (1969). Preparation of Administrators. In R.L. Ebel (Ed.), *Encyclopedia of Educational Research* (pp. 993-1004). London: MacMillan.
Griffith, D.E., Stout, R.T. & Forsyth, P.B. (1988). *Leaders for America's schools: The report and papers of the National Commission on Excellence in Educational Administration*. Berkely, CA: McCutchan.
Gruber. (2000). Erfahrung erwerben. In C. Harteis, H. Heid & S. Kraft, *Kompendium Weiterbildung* (pp. 121-130). Opladen: Leske+Budrich.
Haase, S.I. & Rolff, H.G. (1980). Schulleitungstätigkeiten und Organisationsklima. In H.G. Rolff (Ed.), *Soziologie der Schulreform* (pp. 157-170). Weinheim: Beltz.
Habeck, H. (2001). Aus- und Weiterbildung von Schulleitungen. In H. Buchen, L. Hoster & H.G. Rolff (Eds.), *Schulleitung und Schulentwicklung* (Kapitel C4.5). Stuttgart: Raabe.
Hall, G.E. & Hord, S. (1987). *Change in schools: Facilitating the process*. Albany: State University of New York Press.
Hall, V., Mackay, H. & Morgan, C. (1986). *Headteachers at work*. Milton Keynes: Open University Press.
Hallinger P. & Bridges, E. (1997). Problem-based leadership development: Preparing educational leaders for changing times. *Journal of School Leadership, 7*, 592-608.
Hallinger, P. & Murphy, J. (1985). Assessing the instructional management behaviour of principals. *Elementary School Journal*, *86*(2), 217-247.
Hallinger, P. (1992) The evolving role of American principals: From managerial to instructional to transformational leaders. *Journal of Educational Administration*, *20*(3), 35-48.
Hallinger, P., Leithwood, K. & Murphy, J. (Eds.). (1993). *Cognitive perspectives on educational leadership*. New York: Teachers' College Press.
Hameyer, U. (2001). Schulentwicklung fördern und begleiten – Aufgabenfelder der Schulleitung. *Bayerische Schule*, *54*(2001), 8-10.
Hans, N. (1959). The historical approach to comparative education. *Thoughts on Comparative Education*, 43-53.
Hargreaves, A. (1994). *Changing teachers, changing time: Teachers' work and culture in the post modern age*. London: Cassell.
Hargreaves, D.H. (1994). The new professionalism: The synthesis of professional and institutional development. *Teaching and Teacher Education*, *10*(4), 423-438.
Häring, L. (1997). Akademie Dillingen: Personalführung als zentraler Aufgabenbereich. *Schulreport*, *4/5*, 10-11.
Harold, B. (1998). 'Head'-ing into the future: The changing role of New Zealand principals. *International Journal of Educational Research*, *29*(4), 347-357.
Harteis, C., Heid, H. & Kraft, S. (2000). *Kompendium Weiterbildung*. Opladen: Leske + Budrich.
Hatano, G. & Inagaki, K. (1992). Desituating cognition through the construction of conceptual knowledge. In P. Light & G. Butterworth (Eds.), *Context and cognition: Ways of learning and knowing* (pp. 115-133). Hillsdale, NJ: Erlbaum
Heller, R., Conway, J. & Jacobson, S. (1988). Executive Educator Survey. *The Executive Educator*, 18-22.
Henninger, M. & Mandl, H. (2000). Vom Wissen zum Handeln – ein Ansatz zur Förderung kommunikativen Handelns. In H. Mandl & J. Gerstenmaier (Eds.),

Die Kluft zwischen Wissen und Handeln. Empirische und theoretische Lösungsansätze (pp. 198-219). Göttingen: Hogrefe.
Hentig, H. von. (1993). *Schule neu Denken*. München: Hanser.
Hersko, A.P. (1998). *A descriptive study of cohort programs in educational administrative preparation programs*. Unpublished doctoral dissertation, Seton Hall University, NJ.
Heyman, R. (1979). Comparative Education from an Ethnomethodological Perspective. *Comparative Education, 15*(1), 241-249.
Hilker, F. (1962). *Vergleichende Pädagogik. Eine Einführung in ihre Geschichte. Theorie und Praxis*. München: Hueber.
Hilker, F. (1963). Was kann die vergleichende Methode in der Pädagogik leisten? *Bildung und Erziehung, 16*, 511-526.
Höher, P. & Rolff, H.-G. (1996). Neue Herausforderung an Schulleitungsrollen: Management – Führung – Moderation. In H.-G. Rolff, K.-O. Bauer, K. Klemm & H. Pfeifer (Eds.), *Jahrbuch der Schulentwicklung* (Band 9, pp. 187-220). Weinheim: Juventa.
Holmes, B. (1965). *Problems in education: A comparative approach*. London: Routledge & Kegan.
Holmes, B. (1977). The positivist debate. Comparative education – an Anglo-Saxon perspective. *Comparative Education, 13*(2), pp. 115-132.
Holmes, B. (1981). *Comparative education: Some considerations of method*. London: Allen & Unwin.
Hopes, C.W. (1983). *Kriterien, Verfahren und Methoden der Auswahl von Schulleitern am Beispiel des Landes Hessen – ein Beitrag zur Begründung der Relevanz von Schulleiterausbildung*. Inaugural-Dissertation zur Erlangung des Grades eines Doktors der Philosophie im Fachbereich Erziehungswissenschaft der Johann Wolfgang Goethe-Universität zu Frankfurt am Main.
Hopes. C.W. (1982). *Professional development of school management*. Unpublished paper, Frankfurt a.M.
Hopkins, D. (1996). Towards a theory for school improvement. In J. Gray, D. Reynolds, C. Fitz-Gibbon & D. Jesson (Eds.), *Merging traditions: The future of research on school effectiveness and school improvement* (pp. 30-50). London: Cassell.
Hopkins, D., Ainscow, M. & West, M. (1994). *School improvement in an era of change*. London: Cassell.
Hopkins, D., West, M. & Ainscow, M. (1996). *Improving the quality of education for all: Progress and challenge*. London: David Fulton Publishers.
Hörner, W. (1993). *Technische Bildung und Schule. Eine Problemanalyse im internationalen Vergleich*. Köln: Böhlau.
Hörner, W. (1996). Einführung: Bildungssysteme in Europa – Überlegungen zu einer vergleichenden Betrachtung. In O. Anweiler, L.I. Boos-Nünning, G. Brinkmann, D. Glowka, D. Goetzl, D. Wittörner, F. Kliebert & H.P. Schäfer. (Eds.), *Bildungssysteme in Europa* (pp. 13-29). Weinheim: Beltz.
Huber, S.G. & Hameyer, U. (2000). Schulentwicklung in deutschsprachigen Ländern – Stand des Forschungswissens. In H. Altrichter & H.-G. Rolff (Eds.), *Theorie und Forschung in der Schulentwicklung* (pp. 78-96). Innsbruck: StudienVerlag.
Huber, S.G. & West, M. (2002). Developing school leaders – A critical review of current practices, approaches and issues, and some directions for the future. In P. Hallinger & K. Leithwood (Eds.), *International Handbook of Educational*

Leadership and Administration (pp. 1071-1101). Dordrecht: Kluwer Academic Press.

Huber, S.G. (1997a). *Initial teacher training and teaching competence: Some lessons from England for Bavaria?* Cambridge: School of Education, University of Cambridge.

Huber, S.G. (1997b). *Headteachers' views on headship and training: A comparison with the NPQH*. Cambridge: School of Education, University of Cambridge.

Huber, S.G. (1998). *Dovetailing school effectiveness and school improvement. Towards a model of effectiveness and improvement features: A panorama of the educational effectiveness and improvement landscape*. In P. Clarke (Ed.), Compilation of Papers at the 11th International Congress for School Effectiveness and Improvement (ICSEI) 1998. Manchester, England: University of Manchester.

Huber, S.G. (1999a). School Effectiveness: Was macht Schule wirksam? Internationale Schulentwicklungsforschung (I). *Schul-Management*, *2*, 10-17.

Huber, S.G. (1999b). School Improvement: Wie kann Schule verbessert werden? Internationale Schulentwicklungsforschung (II). *Schul-Management*, *3*, 7-18.

Huber, S.G. (1999c). Effectiveness & Improvement: Wirksamkeit und Verbesserung von Schule – eine Zusammenschau. Internationale Schulentwicklungsforschung (III). *Schul-Management*, *5*, 8-18.

Huber, S.G. (1999d). *Qualifizierung von Schulleiterinnen und Schulleitern in den deutschen Bundesländern – Eine Synopse*. Unpublished report, Bamberg.

Huber, S.G. (1999e). *Schulleitung international*. Studienbrief im Studium "Vorbereitung auf Leitungsaufgaben in Schulen". Hagen: Fernuniversität Hagen.

Huber, S.G. (2001a). Vom Wissen zum Handeln – Problemorientiertes Lernen in der Qualifizierung von Schulleiterinnen und Schulleitern. In H. Altrichter und D. Fischer (Eds.), *Journal für LehrerInnenbildung – Praxis in der LehrerInnenbildung* (pp. 49-55). Innsbruck: Studienverlag.

Huber, S.G. (2001b, January). Introduction to the symposium. In S.G. Huber, M. Chirichello, J. Cox-Millet, R. Ikin, J. Imants, O. Johansson, M. Memon, L. Moos, T. Townsend, M. West (2001, January), *Preparing school leaders – training and development opportunities in different countries*. Paper presented at the International Congress for School Effectiveness and Improvement 2001, Toronto, Canada.

Huber, S.G. (2002a). Qualifizierung von Schulleiterinnen und Schulleitern: Wie verfahren die deutschen Bundesländer? In H.-G. Rolff & H.-J. Schmidt (Eds.), *Schulaufsicht und Schulleitung in Deutschland* (pp. 251-269). Neuwied: Luchterhand.

Huber, S.G. (2002b). Schulentwicklung in England und Wales. *Pädagogik*, *2*, 43-47.

Huber, S.G. (2002c). Schulleitung im internationalen Trend – erweiterte und neue Aufgaben. In P. Daschner (Ed.), *Journal für Schulentwicklung – Schulleitung und Schulaufsicht* (pp. 7-19). Innsbruck: Studienverlag.

Huber, S.G. (2002d). *Preparing school leaders for the 21st century: An international comparison of needs, development programs and difficulties*. Paper prepared for the American Educational Research Association Annual Meeting, New Orleans, USA.

Huber, S.G. (2002e). *Machbarkeits- und Konzeptionsstudie zur Gründung einer Länderakademie für pädagogische Führungskräfte: Bericht für die Cornelsen Stiftung Lehren und Lernen*. In the series Länderakademie für pädagogische

Führungskräfte (Band 2) edited by H.-G. Rolff, H.S. Rosenbusch & S.G. Huber. Bamberg: Länderakademie für pädagogische Führungskräfte.

Huber, S.G. (2002f). Trends in der Qualifizierung von Schulleiterinnen und Schulleitern – Ausgewählte Ergebnisse einer international-vergleichenden Studie. In J. Wissinger & S.G. Huber (Eds.), *Schulleitung – Forschung und Qualifizierung* (pp. 215-233). Opladen: Leske & Budrich.

Huber, S.G. (2002g). Schulleitungsqualifizierung: wann und wie? Praxis in Deutschland und ein Argumentarium zum Zeitpunkt. In T. Strittmatter (Ed.), *Journal für Schulentwicklung – Mythen und Wirklichkeiten* (pp. 31-41). Innsbruck: Studienverlag.

Huber, S.G. (2003a). School Leadership Development – Current trends from a global perspective. In P. Hallinger (Ed.), *Reshaping the Landscape of Educational Leadership Development: A Global Perspective* (pp. 273-288). Lisse: Swets & Zeitlinger.

Huber, S.G. (2003b). International trend of school management. The complex range of tasks and responsibilities: An international perspective. *Journal of the Japanese Association for the Study of Educational Administration (JASEA), 45*, 212-230.

Huber, S.G. (2003c). *Qualifizierung von Schulleiterinnen und Schulleitern im internationalen Vergleich: Eine Untersuchung in 15 Ländern zur Professionalisierung von pädagogischen Führungskräften für Schulen.* In the series Wissen & Praxis Bildungsmanagement. Kronach: Wolters Kluwer.

Huber, S.G., Chirichello, M., Cox-Millet, J., Cuttance, P., Leithwood, K., West, M., Yu, H. (2002, April). *Preparing School Leaders for the 21st Century: An International Comparison of Needs, Development Programs and Difficulties.* Paper presented at the American Educational Research Association Annual Meeting, New Orleans, USA.

Huber, S.G., Chirichello, M., Cox-Millet, J., Ikin, R., Imants, J., Johansson, O., Memon, M., Moos, L., Townsend, T., West, M. (2001, January). *Preparing school leaders – Training and development opportunities in different countries.* Paper presented at the International Congress for School Effectiveness and Improvement 2001, Toronto, Kanada.

Huber, S.G., Cox-Millet, J., Cuttance, P., Imants, J., Johansson, O., Schratz, M., West, M. (2002, January). *Preparing school leaders for the 21st century: Country reports and vision for the future.* Paper presented at the International Congress for School Effectiveness and Improvement 2002, Copenhagen, Denmark.

Huber, S.G., Cox-Millet, J., Dessieux, G., Griffith, A., Imants, J., Johansson, O., Memon, M., Moos, L., Townsend, T., West, M. (2000, January). *Preparing school leaders for the 21st century: Training and development Opportunities in different countries.* Paper presented at the International Congress for School Effectiveness and Improvement 2000, Hong Kong, China.

Huberman, M. (1992). Critical introduction. In M. Fullan (Ed.), *Successful school improvement* (pp. 1-20). Milton Keynes: Open University Press.

Hughes, M.G., Carter, J. & Fidler, B. (1981). *Professional development provision for senior staff in schools and colleges: Final report.* Birmingham: University of Birmingham.

Hurrelmann, K. (1975). *Erziehungssystem und Gesellschaft.* Reinbek: Rowohlt.

Imants, J. & de Jong, L. (1999, January). *Master your school: the development of integral leadership.* Paper presented at the International Congress for School Effectiveness and Improvement, San Antonio, Texas.

Imants, J. (1999). *De School Meester*. Leeuwarden: Eduforce.

INES-data base/DIPF. (2000a). *Denmark*. http://www.dipf.de/datenbanken/ines/ines_v_daen.htm.

INES-data base/DIPF. (2000b). *France*. http://www.dipf.de/datenbanken/ines/ines_v_fr.htm.

INES-data base/DIPF. (2000c). *USA*. http://www.dipf.de/datenbanken/ines/ines_v_usa.htm.

INES-data base/DIPF. (2000d). *Sweden*. http://www.dipf.de/datenbanken/ines/ines_v_schwe.htm.

Institut für schulische Fortbildung und schulpsychologische Beratung. (2000). *Lehrerfort- und -weiterbildung*. Programm des IFBS Rheinland Pfalz, I, 2001.

Institute for Educational Leadership (2000, October). *Leadership for student learning: Reinventing the principalship*. Washington, DC: Author.

Jackson, D. & West, M. (1999, January). *Learning through leading: Leading through learning. Leadership for sustained school improvement*. Paper presented at the International Congress for School Effectiveness and Improvement 1999, San Antonio, Texas, USA.

Jirasinghe, D. & Lyons, G. (1996). *The competent head: A job analysis of heads' tasks and personality factors*. London: The Falmer Press.

Johansson, O. (2000) 'Om rektors demokratiskt reflekterande ledarskap', I: Görandets lov Lova att göra. In L. Lundberg (Ed.), *Skrifter från Centrum för skolledarutveckling,* Umeå: Umeå universitet.

Johansson, O. (2001). Swedish School Leadership in transition: In search of a democratic, learning and communicative leadership? In *Pedagogy, Culture & Society*, *9*(3).

Johansson, O. (2003, in print). School leadership as a democratic arena, in P.T. Begley & O. Johansson (Eds.), *The ethical dimensions of school leadership* (Chapter 12). Deventer: Kluwer.

Johansson, O., Moos, L. & Møller, J. (2000). Visjon om en demokratisk reflekterende ledelse. Noen sentrale perspektiver i en forståelse av nordisk skoleledelse. In L. Moos, S. Carney, O. Johansson, and J. Mehlbye (Eds.), *Skoleledelse i Norden – en kortlægning af skoleledernes arbejdsvilkår, rammebetingelser og opgaver. En rapport til Nordisk Ministerråd. In print.*.

Jones, A. (1987). *Leadership for tomorrow's schools*. Oxford: Basil Blackwell.

Joyce, B. & Showers, B. (1988/1995). *Student achievement through staff development*. New York: Longman.

Joyce, B. & Weil, M. (1986/1996). *Models of teaching*. Englewood Cliffs, NJ: Prentice-Hall.

Joyce, B. (1991). The doors to school improvement. *Educational Leadership*, *May*, 59-62.

Kandel, I.L. (1959). The methodology of comparative education. *International Review of Education*, *5*, 270-280.

Katz, R.L. (1974). The skills of an effective administrator. *Harvard Business Review*, *52*, 90-102.

Keck, R. & Sandfuchs, U. (Eds.). (1994). *Wörterbuch Schulpädagogik*. Bad Heilbrunn: Klinkhardt.

Kidd, J.R. (1970). Developing a methodology for comparative studies in adult education. Paper presented at the Pugwash Conference on Education for International Understanding. *Convergence*, *3*, 12-27.

Kidd, J.R. (1975). *How adults learn*. New York: Falmer Press.

Klafki, W. (1989). Gesellschaftliche Funktionen und pädagogischer Auftrag der Schule in einer demokratischen Gesellschaft. In K.-H. Braun, K. Müller & R. Odey (Eds.), *Subjekt – Vernunft – Demokratie* (pp. 4-33). Weinheim: Beltz.

Knoll, J.H. (1978). Bedeutung und Nutzen der Vergleichenden Erwachsenenbildungsforschung im Rahmen akademischer Ausbildungsvorgänge. *Theorie und Praxis der Erwachsenenbildung*, *2*, 37-47.

Knowles, M. (1980). *The modern practice of adult education.* From Pedagogy to Andragogy. New York: The Adult Education Company.

Koehler, J.W. & Pankowski, J.M. (1997) *Transformational leadership in government.* Delray Beach, FL: St Lucie Press.

Kolb, D.A. (1984). *Experiential learning*. Englewood Cliffs, NJ: Prentice Hall.

Kolb, D.A., Rubin, I.M. & McIntyre, J.M. (1971). *Organizational psychology: An experiential approach.* Hemel Hempstead: Prentice Hall International.

Kolodner, J.L. (1983). Towards an understanding of the role of experience in the evolution from novice to expert. *International Journal of Man-Machine Studies*, *19*, 497-518.

Kolodner, J.L. (1997). Educational implications of analogy. A view from case-based reasoning. *American Psychologist*, *52*, 57-66.

Kommission "Zukunft der Bildung – Schule der Zukunft" beim Ministerpräsidenten des Landes Nordrhein-Westfalen. (Ed.). (1995). *Zukunft der Bildung – Schule der Zukunft.* Neuwied: Luchterhand.

König, E. & Vollmer, G. (2000). *Systemische Organisationsberatung. Grundlagen und Methoden*. Weinheim: Beltz.

König, E. & Vollmer, G. (2002). *Systemisches Coaching*. Weinheim: Beltz.

König, E. & Zedler, P. (1998). *Theorien der Erziehungswissenschaft.* Weinheim: Beltz.

Kowalski, T.J. & Reitzug, U.C. (1993). *Contemporary school administration.* White Plains, NY: Longman.

Krainz-Dürr, M. (1999). *Wie kommt Lernen in die Schule? Zur Lernfähigkeit der Schule als Organisation*. Innsbruck: StudienVerlag.

Krainz-Dürr, M. (2000). Wie entstehen Netzwerke? Fortbildung als Netzwerkarbeit. *Journal für Schulentwicklung*, *3*, 20-25.

Krainz-Dürr, M., Krall, H., Schratz, M. & Steiner-Löffler, U. (Eds.). (1997). *"Was Schulen bewegt". 7 Blicke ins Innere der Schulentwicklung*. Weinheim: Beltz.

Krüger, H.-H. (1996). Strukturwandel des Aufwachsens – Neue Anforderungen für die Schule der Zukunft. In W. Hesper, M. DuBois-Reymand & W. Bathke (Eds.), *Schule und Gesellschaft im Umbruch* (pp. 253-276). Weinheim: Beltz.

Krüger, R. (1983). Was tut der Rektor? Zum Berufsbild und Selbstverständnis des Schulleiters. *Schul-Management*, *4*, 32-36.

Kunert, U. & Kunert, K. (1998). Schulleitung in Frankreich. *Blickpunkt Schulleitung*, *62*(3), 3-7.

Lafond, A. (1993). Die Ausbildung von Schulleitern in Frankreich. *Pädagogische Führung*, *4*, 222-223.

Lambert, L., Walker, D., Zimmerman, D., Cooper, J., Lambert, M., Gardner, M. & Slack, P.J. (1995). *The constructivist leaders.* New York: Teachers College Press.

Landesinstitut für Schule und Weiterbildung (1998). *Fortbildung für Leitungsmitglieder für Schule und Studienseminar in Nordrhein-Westfalen: Das Fortbildungskonzept (Entwurf).* Soest: Landesinstitut für Schule und Weiterbildung.

Landesinstitut für Schule und Weiterbildung (1999). *Aufgaben und Struktur des Landesinstituts Soest*. Soest: Landesinstitut für Schule und Weiterbildung.

Leithwood et al. (1987). *Improving classroom practice using innovation profiles*. Toronto: OISE Press.

Leithwood, K.A. & Montgomery, D.J. (1986). *Improving principal effectiveness: The principal profile*. Toronto: OISE Press.

Leithwood, K.A. (1992a). The principal's role in teacher development. In M. Fullan & A. Hargreaves (Eds.), *Teacher development and educational change* (pp. 86-103). London: The Falmer Press.

Leithwood, K.A. (1992b). The move toward transformational leadership. *Educational Leadership*, *49*(5), 8-12.

Leithwood, K.A. (2000). *Understanding Schools as Intelligent Systems*. Stanford: Connecticut JAI Press.

Leithwood, K.A., Jantzi, D., Silins, H. & Dart, B. (1992, January). *Transformational leadership and school restructuring*. Paper presented at the Annual Meeting of the International Congress for School Effectiveness and Improvement, Victoria, British Columbia, Canada. (ERIC Document Reproduction Service No. ED 342 126)

Levine, D.U. & Lezotte, L.W. (1990). *Unusually effective schools: A review and analysis of research and practice*. Madison: National Centre for Effective School Research.

Liebel, H. (1992). Psychologie der Mitarbeiterführung. In E. Gabele, W. Oechsler & H. Liebel (Eds.), *Führungsgrundsätze und Mitarbeiterführung: Probleme erkennen und lösen* (pp. 109-161). Wiesbaden: Gabler.

Liebermann, A. & Mc Laughlin, M.W. (1992). Network for educational change: Powerful and problematic. *Phi Delta Kappan*, *73*(9), 673-677.

Liket, T.M.E. (1992). *Vrijheid & Rekenschap – Zellfevaluatie en externe evaluatie in het voortgezet onderwijs*. Amsterdam: Meulenhoff Educatief bv.

Liket, T.M.E. (1993). *Freiheit und Verantwortung: Das niederländische Modell des Bildungswesens*. Gütersloh: Bertelsmann Stiftung.

Lortie, D. (1975). *Schoolteacher. A sociological study*. Chicago: The University Press.

Louis, K.S. & Miles, M.B. (1990). *Improving the urban high school: What works and why*. New York: Teachers' College Press.

Luhmann, N. & Schorr, K.-E. (1979). *Reflexionsprobleme im Erziehungssystem*. Stuttgart: Klett-Cotta.

Luhmann, N. (1968). *Zweckbegriff und Systemrationalität*. Tübingen: Mohr.

Lutzau, M. von & Metz-Göckel, S. (1996). Wie ein Fisch im Wasser. Zum Selbstverständnis von Schulleiterinnen und Hochschullehrerinnen. In S. Metz-Göckel (Ed.), *Vorausdenken – Querdenken – Nachdenken* (pp. 211-236). Frankfurt: Campus.

Mandl, H & Gerstenmaier, J. (Eds.). (2000). *Die Kluft zwischen Wissen und Handeln. Empirische und theoretische Lösungsansätze*. Göttingen: Hogrefe.

Mandl, H. & Fischer, F. (Eds.). (2000). *Wissen sichtbar machen*. Göttingen: Hogrefe.

Mandl, H. & Reinmann-Rothmeier, G. (1999). *Individuelles Wissensmanagement. Strategien für den persönlichen Umgang am Arbeitsplatz*. Bern: Hans Huber.

Mandl, H., Gruber, H. & Renkl, A. (1993). Misconceptions and knowledge compartmentalization. In G. Strube & F. Wender (Eds.), *The cognitive psychology of knowledge*. (pp. 161-176). Amsterdam: Elsevier.

Masemann, V. (1976). Anthropological approaches to comparative education. *Comparative Education Review, 20*, 368-380.

Meuret, D., Broccholichi, S. & Duru-Bellat, M. (2001). Autonomie et choix des établissments scolaires: finalités, modalités, effets. *Cahiers de l'IREDU, 62*, 304ff.

Meyer, H. (1996). *Was ist eine Lernende Schule?* Oldenburg: Carl von Ossietzky-Universität Oldenburg.

Meyer, H. (1997a). *Schulpädagogik. Band I: Für Anfänger*. Berlin: Cornelsen Scriptor.

Meyer, H. (1997b). *Schulpädagogik. Band II: Für Fortgeschrittene*. Berlin: Cornelsen Scriptor.

Milstein, M. (1990). Rethinking the clinical aspects in administrative preparation: From theory to practice. In S.L. Jacobson & J. Conway (Eds.), *Educational leadership in an age of reform*. New York: Longman.

Milstein, M. (Ed.) (1993). *Changing the Way We Prepare Educational Leaders: The Danforth Experience*. Newbury Park, CA: Corwin Press.

Milstein, M.M., Bobroff, B.M. & Restine, L.N. (Eds.). (1991). *Internship programs in educational administration: A guide to preparing educational leaders*. New York: Teachers' College Press.

Ministerium für Schule und Weiterbildung des Landes NRW, Landesinstitut für Schule und Weiterbildung. (1999). *Fortbildung für Leitungsmitglieder in Schule und Studienseminar in NRW: Ziele, Inhalte, Arbeitsformen*. Unveröffentlichte Konzeption.

Ministerium für Schule und Weiterbildung des Landes NRW. (1997). *Professionalität stärken: Rahmenkonzept "Staatliche Lehrerfortbildung in Nordrhein-Westfalen"*. Düsseldorf: MSW.

Mitter, W. (1979). Überlegungen zur Theorie und Praxis der vergleichenden Bildungsforschung. In T. Hanf & W. Mitter (Eds.), *International Vergleichende Bildungsforschung. Zur Theorie und Forschungspraxis erziehungswissenschaftlicher Komparatistik* (pp. 23-42). Frankfurt/Main: GEPF-Materialien 10.

Morgan, C., Hall, V. & Mackay, H. (1983). *The selection of secondary school headteachers*. Open University Press: Milton Keynes.

Mortimore, P., Sammons, P., Stoll, L., Lewis, D. & Ecob, R. (1988). *School matters: The junior years*. Wells: Open Books.

Mulford, B. (1984). On teaching educational administration. *The journal of educational administration, 22*(2), 223-246.

Müllner, M. (2000). *Schulmanagement in Österreich. Berufsbegleitende Weiterbildung für SchulleiterInnen*. Paper presented at the SICI-Workshop, Salzburg.

Murphy, J. (1991). *Restructuring schools: Capturing the phenomenon*. Beverley Hills, CA: Sage.

Murphy, J. (1992). *The landscape of leadership preparation: Reframing the education of school administrators*. Newbury Park, CA: Corwin Press.

Naisbitt, J. & Aburdene, P. (1990). *Megatrends 2000*. New York: William Morrow.

Naisbitt, J. (1982). *Megatrends*. London: Futura Press.

National Commission on Excellence in Education. (1983). *A nation at risk*. Washington, DC: Government Printing Office.

NCEEA/UCEA (National Commission on Excellence in Educational Administration/University Council for Educational Administration).

(1987/1988). *Leaders for America's schools.* Tempe, AZ: University Council for Educational Administration.
NEAC (National Educational Assessment Centre). (1995). *The competencies.* Oxford Brooks University: NEAC.
Negt, O. (1968). *Soziologische Phantasie und exemplarisches Lernen.* Frankfurt: Europa Verlags-Anstalt.
Neuberger, O. (1990). *Führen und geführt werden.* Stuttgart: Enke.
Neulinger, K.U. (1990). *Schulleiter – Lehrerelite zwischen Job und Profession: Herkunft, Motive und Einstellungen einer Berufsgruppe.* Frankfurt: Haag und Herchen.
Nevermann, K. (1982). *Der Schulleiter. Juristische und historische Aspekte zum Verhältnis von Bürokratie und Pädagogik.* Stuttgart: Klett.
New Jersey Principals and Supervisors Association. (2001). Principal shortage survey: Analysis of data. *Educational Viewpoints, 1*, 28.
NIE (National Institute of Education), Nanyang Technological University. (2000). *Diploma in Educational Administration.* Singapore: NIE, Nanyang Technological University.
Noah H. & Eckstein, M.A. (1969). *Toward a science of comparative education.* New York: MacMillan.
NPBEA (National Policy Board for Educational Administration). (1989). *Improving the preparation of school administrators: The reform agenda.* Charlottesville,VA: NPBEA.
NSO (Nederlandse School voor Onderwijsmanagement). (Ed.). (1998/1999). *Information on the two-year program in educational management.* Amsterdam: NSO
NSW Department of Education and Training. (Ed.). (1998a). *School leadership strategy.* Sydney: Training and Development Directorate.
NSW Department of Education and Training. (Ed.). (1998b). *Interdistrict school leadership group handbook.* Unveröffentlichtes Handbuch. Sydney.
Nygren, A.M. and Johansson, O. (2000). Den svenske rektorn efter 1945 – Kvalifikationer, arbetsuppgifter och utmaningar. In L. Moos, S. Carney, O. Johansson and J. Mehlbye (Eds.), *Skoleledelse i Norden – en kortlægning af skoleledernes arbejdsvilkår, rammebetingelser og opgaver. En rapport til Nordisk Ministerråd, Köpenhamn.*
OFSTED (Office for Standards in Education). (1995). *Headteacher training in France.* London: OFSTED Publications.
O'Neill, J. (1992). Putting performance assessment to the test. *Educational Leadership, 49*(8), 14-19.
Palmer, D., LeMahieu, P.G. & Eresh, J. (1992). Good measure: Assessment as a tool for educational reform. *Educational Leadership, 49*(8), 8-13.
Paulston, R.G. (1993). Comparative education as an intellectual field: Mapping the theoretical landscape. *Comparative Education, 23*, 101-114.
Pecherski, M. (1986). Einige Bemerkungen zur Methodologie der vergleichenden Bildungsforschung. In B. Dilger, F. Kuebart & H.P. Schäfer (Eds.), *Vergleichende Bildungsforschung. DDR, Osteuropa und interkulturelle Perspektiven.* Festschrift für Oskar Anweiler zum 60. Geburtstag. Berlin: Arno Spitz.
Pedersen, U.G. (1995). Qualitätssicherung des Schulsystems in Dänemark. In Landesinstitut für Schule und Weiterbildung (Ed.), *Schulentwicklung und*

Qualitätssicherung in Dänemark (pp. 7-11). Bönen: Verlag für Schule und Weiterbildung.

Peterson, K. (2001). The roar of complexity. *Journal of Staff Development, 22*(1), 18-21.

PI Salzburg. (2000). *Schulmanagement-Lehrgang (berufsbegleitender Weiterbildungslehrgang gemäß § 26a (3) LDG und § 207 (4) BDG)*. Salzburg: PI Salzburg.

Portin, B.S. (1998). From change and challenge to new directions for school leadership. *International Journal of Educational Rsearch, 29*(4), 381-391.

Potthoff, D.E., Batenhorst, E.V., Frederickson, S.A. & Tracy, G.E. (2001). Learning about cohorts – A masters degree program for teachers. *Graduate Teacher Education: The Journal of the Association of Teacher Educators, 23*(2), 36-42.

Przeworski, A. & Teune, H. (1970). *The Logic of Comparative Social Inquiry*. New York: Krieger.

Public Agenda. (2001). *Trying to stay ahead of the game: Superintendents and principals talk about school leadership*. New York: Author.

Reischmann, J. (1995). Internationale Gemeinsamkeiten der Weiterbildung und die Methodologie vergleichender Erwachsenenbildungsforschung. In J.H. Knoll (Ed.), *Internationales Jahrbuch der Erwachsenenbildung* (Band 18, pp. 107-129). Köln: Böhlau.

Reischmann, J. (1997). Von Anfängen: ISCAE – International Society for Comparative Adult Education. *Bildung und Erziehung, 50* (3), 273-280.

Reischmann, J. (2001). *Lernen hoch zehn – wer bietet mehr? Von "Lernen en passant" zu "kompositionellem Lernen" und "lebensbreiter Bildung*. http://www.treffpunktlernen.de/render.asp?menu=ddb&subject=ddb_ftn_vorrueber

Renkl, A. (1996). Träges Wissen: Wenn Erlerntes nicht genutzt wird. *Psychologische Rundschau, 47*, 78-92.

Renkl, A. (2001). Träges Wissen. In D. Rost (Ed.), *Handwörterbuch Pädagogische Psychologie*. Weinheim: BeltzPVU.

Reynolds, D. (1976). The delinquent school. In P. Woods (Ed.), *The process of schooling*. London: Routledge & Kegan.

Reynolds, D., Bollen, R., Creemers, B., Hopkins, D., Stoll, L. & Lagerweij, N. (Eds.). (1996). *Making good schools: Linking school effectiveness and school improvement*. London: Routledge.

Rhyn, H. (1998). Länderbericht Schweiz. In Bundesministerium für Unterricht und kulturelle Angelegenheiten. (Ed.), *Schulleitung und Schulaufsicht. Neue Rollen und Aufgaben im Schulwesen einer dynamischen und offenen Gesellschaft* (pp. 163-187). Innsbruck: StudienVerlag.

Riedel, K. (1998). *Schulleiter urteilen über Schule in erweiterter Verantwortung*. Weinheim: Beltz.

Robertson, J.M. (1991). *Developing educational leadership: An action research study into the professional development of the New Zealand school principal*. Unpublished Master's Thesis, University of Waikato, Hamilton, New Zealand.

Robertson, J.M. (1995). *Principal partnerships: An action research study on the professional development of New Zealand school leaders*. Unpublished Doctoral Dissertation, University of Waikato, Hamilton, New Zealand.

Robertson, J.M. (1998). From managing impression to leadership perspectives. *International Journal of Educational Research, 29*(4), 359-370.

Robinsohn, S.B. (1973). Vergleichende Erziehungswissenschaft. In F. Braun, D. Glowka & H. Thomas (Eds.), *Erziehung als Wissenschaft* (pp. 313-355). Stuttgart: Klett.

Robinsohn, S.B. (Ed.). (1970/1975). *Schulreform im gesellschaftlichen Prozess. Ein interkultureller Vergleich.* 2 Bde. Stuttgart: Klett.

Rolff, H.-G. (1995). Ein pädagogisches Management in der Schule – Selbststeuerung in bestehenden Strukturen. In H. Buchen, L. Horster & H.-G. Rolff (Eds.), *Schulleitung und Schulentwicklung*. Stuttgart: Raabe.

Roschelle, J. (1992). Learning by collaboration. Convergent conceptual change. *Journal of the Learning Society*, *2*, 235-276.

Rosenbusch, H.S. & Huber, S.G. (2001). Qualifizierungsmaßnahmen von Schulleiterinnen und Schulleitern in den Ländern der Bundesrepublik Deutschland. *Schul-Management*, *4*, 8-16.

Rosenbusch, H.S. & Huber, S.G. (2002). *Schulmanagment im Wandel – Herausforderungen für die Zukunft.* Speech and paper presented at the oldenburgisch-ostfriesischen Schulmanagement-Conference, Publication in Oldenburger Vor-Drucke. Oldenburg: Carl von Ossietzky Universität Oldenburg.

Rosenbusch, H.S. & Schlemmer, E. (1997). Die Rolle der Schulaufsicht bei der pädagogischen Entwicklung von Einzelschulen. *Schul-Management*, *6*, 9-17.

Rosenbusch, H.S. (1992). Schulqualität und Schulleiterausbildung. *Schul-Management*, *4*, 9-16.

Rosenbusch, H.S. (1994a). Zur Herausbildung der Schulleitung in Deutschland. In H. Buchen, L. Horster & H.G. Rolff (Eds.), *Schulleitung und Schulentwicklung* (Kapitel A.2.1). Stuttgart: Raabe.

Rosenbusch, H.S. (1994b). *Lehrer und Schulräte. Ein strukturell gestörtes Verhältnis*. Bad Heilbrunn/Obb.: Klinkhardt.

Rosenbusch, H.S. (1995). Reform der Schulverwaltung aus organisations-pädagogischer Sicht. Schulleitung und Schulaufsicht als erzieherisch bedeutsame Wirklichkeit. *Schul-Management*, *4*, 36-42.

Rosenbusch, H.S. (1997a). Die Qualifikation pädagogischen Führungspersonals. In E. Glumpler & H.S. Rosenbusch (Eds.), *Perspektiven der universitären Lehrerausbildung* (pp. 147-165). Bad Heilbrunn/Obb.: Klinkhardt.

Rosenbusch, H.S. (1997b). Organisationspädagogische Perspektiven einer Reform der Schulorganisation. *SchulVerwaltung*, *10*, 329-334.

Rosenbusch, H.S. (1999). Schulleitung und Schulaufsicht. In E. Rösner (Ed.), *Schulentwicklung und Schulqualität. Kongressdokumentation 1. und 2. Oktober 1998* (pp. 243-258). Dortmund: IFS-Verlag.

Rosenbusch, H.S. (2002). Schulmanagement im Wandel. *Schul-Management*, *1*, 20-22.

Ruesch, J. & Bateson, G. (1951/1995). *Kommunikation.* Heidelberg: Carl-Auer-Systeme.

Rumpf, H. (1966). *Die administrative Verstörung der Schule. Neue Deutsche Schule.* Essen: Neue deutsche Verlags-Gesellschaft.

Rust, V.D., Soumare, A., Pescador, O. & Shibuya, M. (1999). Research Strategies in Comparative Education. *Comparative Education Review*, *43*(1), 86-109.

Rutter, M., Maughan, B., Mortimore, P. & Ouston, J. (1979). *Fifteen thousand hours*. London: Open Books.

Rutter, M., Maughan, B., Mortimore, P. & Ouston, J. (1980). *Fünfzehntausend Stunden: Schulen und ihre Wirkung auf ihre Kinder.* Weinheim: Beltz.

Sadler, M.E. (1908). *Moral instructions and training in schools. Report of an international inquiry*. London: Longmans, Green & Co.
Sammons, P., Hillman, J. & Mortimore, P. (1995). *Key characteristics of effective schools: A review of school effectiveness research*. London: OFSTED.
Scheerens, J. & Bosker, R. (1997). *The foundations of educational effectiveness*. Oxford: Pergamon.
Schmitz, K. (1980). Gegenwärtige Schulprobleme – dargestellt am Wochenlauf eines Schulleiters. *Bildung und Erziehung, 33*(6), 536-549.
Schneider, F. (1958). Prospektive Pädagogik. *International Review of Education, 4*(1), pp. 36-50.
Schneider, F. (1961). *Vergleichende Erziehungswissenschaft. Geschichte, Forschung, Lehre*. Heidelberg: Quelle & Meyer.
Schön, D. (1983). *The reflective practitioner*. New York: Basic Books.
Schön, D. (1984). Leadership as reflection-in-action. In T. Sergiovanni & J. Corbally (Eds.), *Leadership and organizational culture* (pp. 36-63). Chicago: University of Illinois Press.
Schratz, M. & Steiner-Löffler, U. (1998). *Die Lernende Schule*. Weinheim: Beltz.
Schratz, M. (1998a). Länderbericht Österreich. In Bundesministerium für Unterricht und kulturelle Angelegenheiten. (Ed.), *Schulleitung und Schulaufsicht. Neue Rollen und Aufgaben im Schulwesen einer dynamischen und offenen Gesellschaft* (pp. 189-235). Innsbruck: StudienVerlag.
Schratz, M. (1998b). Schulleitung als change agent: Vom Verwalten zum Gestalten von Schule. In H. Altrichter, W. Schley & M. Schratz (Eds.), *Handbuch zur Schulentwicklung* (pp. 160-189). Innsbruck: StudienVerlag.
Schriewer, J. & Holmes, B. (1988). *Theories and Methods in Comparative Education*. Frankfurt: Peter Lang.
Schriewer, J. (1982). "Erziehung" und "Kultur" – Zur Theorie und Methodik vergleichender Erziehungsissenschaft. In W. Brinkmann & K. Renner (Eds.), *Die Pädagogik und ihre Bereiche* (pp. 185-236). Paderborn: Schöningh.
Seidel, G. (1997). Herausforderung für die Fortbildung von Schulleitern. *Schul-Management, 2*, 41-42.
Seljelid, T. (1982). *Handbook for the working environment and the running of school (AMS)*. Oslo: Ministry of Culture and Scientific Affairs.
Senge, P. (1990). *Die fünfte Disziplin*. Stuttgart: Klett-Cotta.
Senge, P. (1990). *The fifth discipline*. New York: Doubleday.
Sergiovanni T. J. (1994). *Building community in schools*. San Francisco Ca: Jossey-Bass.
Sergiovanni, T.J. (1996). *Leadership for the schoolhouse: How is it different? Why is it important?* San Francisco, CA: Jossey-Bass.
Sharpe, L. & Gopinathan, S. (1997). Effective island, effective schools: Repair and restructuring the Singapore school system. In J. Tan, S. Gopinathan & H.W. Kam (Eds.), *Education in Singapore: A book of readings* (pp. 369-383). Singapore: Prentice Hall.
Sharpe, L. & Gopinathan, S. (2000). Leadership in high achieving schools in Singapore: The influence of societal culture. *Asia Pacific Journal of Eduaction, 20*(2), 87-98.
Sheppard, B. (1996). Exploring the transformational nature of instructional leadership. *Alberta Journal of Educational Research XLII, 94*, 325-44.
Siebert, H. (1993). *Theorien für die Bildungspraxis*. Klinkhardt: Heilbrunn.

Siebert, H. (1996). *Didaktisches Handeln in der Erwachsenenbildung: Didaktik aus konstruktivistischer Sicht*. Neuwied: Luchterhand.

Silver, P.F. (1978). Trends in program development. In P.F. Silver & D.W. Spuck (Eds.), *Preparatory programs for educational administrators in the United States* (pp. 178-201). Columbus University Council for Educational Administration.

Sirotnik, K.A. & Kimball, K. (1996). Preparing educators for leadership: In praise of experience. *Journal of School Leadership*, *6*(2), 180-201.

Skolverket (The Swedish National Agency for Education). (1999). *Head Teachers for Tomorrow*. Stockholm: Skolverket.

Southworth, G. (1998). *Leading improving primary schools: The work of head teachers and deputy heads*. London: The Falmer Press.

Speck, M. (1999). *The principalship: Building a learning community*. Upper Saddle River, NJ: Merrill, an imprint of Prentice Hall.

Staatliches Institut für Lehrerfort- und -weiterbildung, Rheinland-Pfalz. (1999). *Führungskolleg: Programm 1999/ 2000*. Staatliches Institut für Lehrerfort- und -weiterbildung.

Steen, L.A. (1991). Mathematics in the U.S. and the Soviet Union. *Educational Leadership*, *48*(5), 26-27.

Stegö, N.E., Gielen, K., Glatter, R. & Hord, S.M. (Eds.). (1987). *The role of school leaders in school improvement*. Leuven: ACCO.

Stenhouse, L. (1979). Case study in comparative education: Particularity and generalisation. *Comparative Education*, *15*(3), 5-10.

Stoll, L & Fink, D. (1996). *Changing our schools: Linking school effectiveness and school improvement*. Buckingham: Open University Press.

Storath, R. (1994). *"Praxisschock" bei Schulleitern? Eine qualitativ ausgerichtete Befragung zur Rollenfindung neu ernannter Schulleiter an Volksschulen in Bayern*. Inaugural-Dissertation zum Erlangen des Grades eines Doktors der Philosophie in der Fakultät Pädagogik, Psychologie, Philosophie der Otto-Friedrich-Universität Bamberg.

Strittmatter, A. (2002). Bitte meiden: Neun Stolpersteine auf dem Weg zur guten Schulleitung. *Bildung Schweiz*, *12-13*, 28-33.

Suchodolski, B. (1961). Education for the future and traditional pedagogy. *International Review of Education*, *7*(4), 420-431.

Szaday, C., Büeler, X. & Favre, B. (1996). *Schulqualitäts- und Schulentwicklungsforschung: Trends, Synthese und Zukunftsperspektiven*. Bericht im Rahmen des NFP 33. Aarau: Schweizerische Koordinationsstelle für Bildungsforschung.

Teddlie, C. & Stringfield, S.C. (1993). *Schools make a difference: Lessons learned from a 10-year study of school effects*. New York: Teachers' College Press.

Terhart, E. (1986). Organisation und Erziehung. Neue Zugangsweisen zu einem alten Dilemma. *Zeitschrift für Pädagogik*, *32*(2), 205-223.

Terhart, E. (1994). *Berufsbiographien von Lehrern und Lehrerinnen*. Frankfurt am Main: Lang.

Terhart, E. (1997). Schulleitungshandeln zwischen Organisation und Erziehung. In J. Wissinger (Ed.), *Schulleitung als pädagogisches Handeln* (pp. 7-20). München: Oldenbourg.

Thanh, K. L. (1981). *L'education comparée*. Paris: Colin.

Thody, A. (1993). Mentoring for school principals. In B. Caldwell & E. Carter (Eds.), *The return of the mentor. Strategies for workplace learning* (pp. 59-76). London: The Falmer Press.

Thompson, S.T. (1997). Site-based development. In J. Caldwell (Ed.), *Professional development in learning – Centered schools*. Oxford, OH: National Staff Development Council.

Tomei, L.A. (1999). Concentration and infusion – Two approaches for teaching technology for lifelong learners. *The Journal*, *26* (9), 72-76.

Trider, D. & Leithwood, K.A. (1988). Influences on principal's practices. *Curriculum Inquiry*, *18*(3), 289-311.

Trotter, A. (1999). Demand for principals growing, but candidates aren't applying. *Education Week*, *1*, 20-22.

Tschenett, S. (2001). *Die Schulverwaltung vor Ort. Auf dem Weg zur lernenden Verwaltung unter dem Gesichtspunkt der Entwicklung der Schulautonomie in Südtirol*. Unpubl. Master Thesis at Innsbruck University.

TTA. (1995a). *Headteachers' leadership and management programme (HEADLAMP): Procedures*. London: TTA.

TTA. (1995b). *HEADLAMP: An initiative to support newly appointed headteachers*. London: TTA.

TTA. (1997a). *How to apply: Information about the National Professional Qualification for Headship*. London: TTA.

TTA. (1997b). *National standards for headteachers*. London: TTA.

Van Agten, P.H.W., Dresselaars, C.F.M. & Hamann, G. (1997). *Schoolleider: een vak apart*. Deventer: Kluwer.

Van de Grift, W. (1990). Educational leadership and academic achievement in elementary education. *School Effectiveness and School Improvement*, *1*(3), 26-40.

Van Kessel, N., van Kuijk, J. & Mensink, J. (1999). *Nog meer scholing of toch maar niet. Scholingsbehoeften van schoolleiders en besturen in het Primair Onderwijs*. Nijmegen: Instituut voor Toegepaste Sociale Wetenschappen, Katholieke Universiteit Nijmegen.

Van Velzen, W.G. (1979). *Autonomy of the school*. S'Hertogenkosch: PKC.

Van Velzen, W.G., Miles, M.B., Ekholm, M., Hameyer, U. & Robin, D. (Eds.). (1985). *Making school improvement work: A conceptual guide to practice*. Leuven: ACCO.

Verbiest, E. (1998). *De schoolleider in beweging, veranderingen in visie en praktijk van het primair onderwijs*. Alphen aan de Rijn: Samson.

Vogel, P. (1977). *Die bürokratische Schule*. Kastellaun: Henn.

Vogelsang, H. (1992). Ausbildung für Schulleitung und Schulaufsicht. In Verband Bildung und Erziehung (Ed.), *Schulaufsicht für die Schule von morgen* (pp. 113-118). Bonn: Teutsch.

Walker, A. & Stott, K. (1993). Preparing for leadership in schools: The mentoring contribution. In B. Caldwell & E. Carter (Eds.), *The return of the mentor. Strategies for workplace learning* (77-90). London: Falmer Press.

Wasden, D. & Muse, I. (1987). *The mentoring handbook*. Provo, UT: Brigham Young University.

Wee Heng Tin & Chong Keng Choy. (1990). 25 Years of school management. In J. Yip Soon Kwong & Sim Wong Kooi (Eds.), *Evolution of educational excellence. 25 years of education in the republic of Singapore* (pp. 31-58). Singapore: Longman Singapore.

Weick, K.E. (1976). Educational Organizations as loosely coupled systems. *Administrative Science Quarterly*, *21*, 1-159.
Weick, K.E. (1982). Administering education in loosely coupled systems. *Phi Delta Kappan*, *63*, 673-676.
Weindling, D. & Earley, P. (1987). *Secondary headship: The first years*. Berkshire: NFER-Nelson.
Weindling, D. (1998). Reform, restructuring, role and other 'R' words: The effects on headteachers in England and Wales. *International Journal of Educational Research*, *29*(4), 299-310.
Weindling, D., Earley, P. & Baker, L. (1994-95). Secondary headship: Ten Years on. A series of ten articles in *Managing Schools Today, 4*(1)-*5*(1).
Weinert, F.E. (1998). Neue Unterrichtskonzepte zwischen gesellschaftlichen Notwendigkeiten, pädagogischen Visionen und psychologischen Möglichkeiten. In Bayerisches Staatsministerium für Unterricht, Kultus, Wissenschaft und Kunst (Ed.), *Wissen und Werte für die Welt von morgen. Dokumentation zum Bildungskongress des Bayerischen Staatsministeriums für Unterricht, Kultus, Wissenschaft und Kunst am 29./30. April 1998* (pp. 101-125). München: Bayerisches Staatsministerium für Unterricht, Kultus, Wissenschaft und Kunst.
West, M & Ainscow, M. (1997). *Tracking the moving school: challenging assumptions, increasing understanding*. Paper presented at the European Conference on Educational Research, Frankfurt, September 1997.
West, M. & Ainscow, M. (1994). *Managing school development*. London: David Fulton.
West, M., Jackson, D., Harris, A. & Hopkins, D. (2000). Learning through Leadership, Leadership through learning. In K.A. Riley & D. Seashore-Louis (Eds.), *Leadership for Change and School Reform* (pp. 30-49). London: Routledge Falmer.
White, D. (1978). Comparisons as cognitive process and the conceptual framework of the comparativist. *Comparative Education*, *14*(2), 93-108.
Whitehead, A.N. (1929). *The aims of education*. New York, NY: Macmillan.
Whitty, G. & Willmott, E. (1991). Competence-based teacher education: Approaches and issues. *Cambridge Journal of Education*, *21*(3), 309-318.
Wilson, D. (1996). Precis of the spirit of the reform: Managing the New Zealand state sector in a time of change (The Schick Report). *Public Sector*, *19*(4), 2-6.
Wissinger, J. (1996). *Perspektiven schulischen Führungshandelns. Eine Untersuchung über das Selbstverständnis von SchulleiterInnen*. Weinheim: Juventa.
Wissinger, J. & Huber, S.G. (Eds.). (2002). *Schulleitung – Forschung und Qualifizierung*. Opladen: Leske & Budrich.
Wolfmeyer, P. (1981). *Die schulinterne Verwaltungstätigkeit der Lehrer*. Kastellaun/Hunsrück: Henn.
Wolgast, H. (1887). Der Bureaukratismus in der Schule. *Preußische Reform*, *46/47*, ohne Seitenzahlen.
Wong, Ping-Man. (2000, January). *The evaluation of a principals' training program in Hong Kong*. Paper presented at the 13th International Congress for School Effectiveness and Improvement, Hong Kong.
Wörmann, H.-W. (1985). *Zwischen Arbeiterbildung und Wissenschaftstransfer. Universitäre Erwachsenenbildung in England und Deutschland im Vergleich*. Berlin: Argument-Verlag.

Wylie, C. (1990). *Developing school management: The way forward. A report by the School Management Task Force*. London: HMSO.
Wylie, C. (1994). *Self-managing schools in New Zealand: The fifth year*. Wellington, New Zealand: New Zealand Council for Educational Research.
Wylie, C. (1997). *Self-managing schools seven years on: What have we learnt?* Wellington, New Zealand: New Zealand Council for Educational Research.
Yu, H. (2000). *Transformational leadership and Hong Kong teachers' commitment to change*. Unpublished Ph.D. Dissertation. Toronto: OISE/UT.

Index

CONTEXTS OF LEARNING
Classrooms, Schools and Society
ISSN 1384-1181

1. Education for All
   Robert E. Slavin
   1996 ISBN 90 265 1472 7 (hardback)
   ISBN 90 265 1473 5 (paperback)

2. The Road to Improvement - Reflections on School Effectiveness
   Peter Mortimore
   1998 ISBN 90 265 1525 1 (hardback)
   ISBN 90 265 1526 X (paperback)

3. Organizational Learning in Schools
   Edited by Kenneth Leithwood and Karen Seashore Louis
   1998 ISBN 90 265 1539 1 (hardback)
   ISBN 90 265 1540 5 (paperback)

4. Teaching and Learning Thinking Skills
   Edited by J.H.M. Hamers, J.E.H. Van Luit and B. Csapó
   1999 ISBN 90 265 1545 6 (hardback)

5. Managing Schools towards High Performance: Linking School Management Theory to the School Effectiveness Knowledge Base
   Edited by Adrie J. Visscher
   1999 ISBN 90 265 1546 4 (hardback)

6. School Effectiveness: Coming of Age into the Twenty-First Century
   Edited by Pam Sammons
   1999 ISBN 90 265 1549 9 (hardback)
   ISBN 90 265 1550 2 (paperback)

7. Educational Change and Development in the Asia-Pacific Region: Challenges for the Future
   Edited by Tony Townsend and Yin Cheong Cheng
   2000 ISBN 90 265 1558 8 (hardback)
   2000 ISBN 90 265 1627 4 (paperback)

8. Making Sense of Word Problems
   Lieven Verschaffel, Brian Greer and Erik De Corte
   2000 ISBN 90 265 1628 2 (hardback)

9. Profound Improvement: Building Capacity for a Learning Community
   C. Mitchell and L. Sackney
   2000 ISBN 90 265 1634 7 (hardback)

10. School Improvement Through Performance Feedback
    Edited by A.J. Visscher and R. Coe
    2002 ISBN 90 265 1933 8 (hardback)

11. Improving Schools Through Teacher Development. Case Studies of the Aga Khan Foundation Projects in East Africa
    Edited by Stephen Anderson
    2002 ISBN 90 265 1936 2 (hardback)

12. Reshaping the Landscape of School Leadership Development - A Global Perspective
    Edited by Philip Hallinger
    2003 ISBN 90 265 1937 0 (hardback)

13. Educational Evaluation, Assessment and Monitoring: A Systemic Approach
    Jaap Scheerens, Cees Glas and Sally M. Thomas
    2003 ISBN 90 265 1959 1 (hardback)

14. Preparing School Leaders for the 21st Century: An International Comparison of Development Programs in 15 Countries
    Stephan Gerhard Huber
    2004 ISBN 90 265 1968 0 (hardback)